Photo by] [*The Author*

A LIKELY SPOT.

Frontispiece.

MICROSCOPIC FRESH WATER LIFE

In a simple and untechnical language, the book embodies the experience and observations of the author spanning over a period of more than 3 decades, in the field of microscopy. All those pursuing studies in freshwater microorganisms need to go through the volume for necessary and fuller description and the intricacies of what they could encounter during the course of their study.

After an introduction to the apparatus used in collecting freshwater microorganisms and the different species of aquatic plants, separate chapters have been devoted to Lovely Organisms, Algae, Diatoms, Desmids, Infusoria, Rotifers, Bryozoa, Entomostraca, Water Mites and Water Bear.

Numerous illustration and a glossary of technical terms, enhance the utility of the volume.

MICROSCOPIC FRESH WATER LIFE

F. J. W. PLASKITT

2023
BIOTECH BOOKS

First Indian Edition 2023

ISBN 81 - 7622 – 006 – x

Published by **BIOTECH BOOKS**
4762-63/23, Ansari Road, Darya Ganj,
New Delhi - 110002
Phones : 011 4300 3222
Email : biotechbooks@yahoo.co.in

PRINTED IN INDIA

AUTHOR'S PREFACE

In the following pages the author has endeavoured to present the tyro in microscopy with a reliable volume to have beside him while pursuing his studies in fresh water micro-organisms.

Many times while busy at his microscope the writer has felt a need for a rather fuller description and a more up-to-date account of these which would give, in a simple way, the chief points to look for and any peculiarity specially to be noted. The usual academic papers have so cut-and-dried a shorthand for delineating objects that it becomes quite startling and almost military in the brevity of its nomenclature.

There will be a host of objects not included here, the ground being so wide that the specialist must be referred to larger works dealing with such. For the beginner, however, and the layman in microscopy there is included most of the commoner objects likely to be encountered.

The descriptions embody the experience and observations of the author during a period of over thirty years' studies, which have always been enjoyable, even if intermittent, coupled with notes made from time to time from various sources.

For the specific names and their accurate classifications the author is much indebted to Messrs. Ward and Whipple's great work on Fresh Water Biology; and for the general arrangement of the work to Dr. A. C. Stokes' *Aquatic Microscopy*, now unfortunately out of print.

For the lucid chapter and illustrations of Fresh Water Mites a deep gratitude is due to Mr. C. D. Soar, F.L.S., F.R.M.S. An acknowledged specialist in Great Britain upon the subject and joint author of the standard work on the "Hydracarina," first part published 1925 by the Ray Society, as well as a master in draughtsmanship, no further introduction is necessary. He has entered into the homely spirit of the volume, and all microscopists will value his addition.

Lastly, and not least is the author's obligation for the kind assistance received in many ways from his sister, Madame Eva Algernon, who has in no small measure helped him in his task, in a manner always encouraging and with material from various sources that have been prolific with many fine specimens, thus reducing his labours considerably.

With regard to the illustrations, in no case are they intended to be to scale, although each object in the text has had its measurements given as near as possible.

Few microscopists rely upon the book size to correspond, even relatively, with the original; both may vary considerably. The blockmaker would be limited and the growth of individual organisms restricted, if such a method had been observed. Each one is drawn simply to give the general outline and some of the principal details of the object, as seen under favourable conditions.

Many new points of structure from the writer's personal observations have been included, notably in Epistylis, Halteria, Actinophrys, Stylonychia, Euplotes and others.

Finally, if the book is found helpful and the work in its simple untechnical language worthy an extended compendium may be added of additional organisms at some future date.

In all his studies the microscopist should endeavour to inculcate into them something more than mere names and parts, for with all our wisdom there must be understanding. "Books must follow science, not science books," says Bacon.

THE AUTHOR.

July, 1926.

CONTENTS

GLOSSARY

AKINETES. A resting cell. A single cell whose walls become thickened and separates from a young shoot or branch (thallus) of algæ, fungi, lichens, and yeasts, for the purpose of propagation.

ALVEOLAR. Air cells; cellular, hollow.

ANDRO-GAMETOPHORE. A plant bearing male sexual organs. A male plant.

ANDRO-GONIDIA. A gonidium or sexual cell bearing androspores.

ANDRO SPORES. A peculiar Zoospore produced asexually which gives rise to a small male plant, known as a "Dwarf Malc," this latter developing true Spermatozoids.

APLANOSPORES. A nonmotile asexual Spore formed by a rejuvenescence in green algæ or the Chlorophyceæ.

ASEXUAL. Without sex. Parthenogenesis. Without two sexes operating. An unfertilized cell.

CHLOROPHYLL. The green colouring of plants, and only occurs in the chloroplasts of cells exposed to light. It is fluorescent.

CHLOROPLAST. A colourless ground substance, a base, transparent, saturated with chlorophyll granular pigments. Varying intensities of light causes them to exhibit photo-tactic movements. They are minute flattened granules, occurring in great numbers in the cytoplasm or cell protoplasm, apart from the nucleus.

CHROMATOPHORES. A coloured body, or plastid, common in Desmids and plant cells generally, varying greatly in shape and size and include green, red, and yellow chloroplasts. They may be Star-shaped, spiral, axial plates, scolloped, or longitudinal bands, etc., usually parietal, i.e. attached to the inner coats of the cells and afford sustenance to the organism, which uses and reduces them eventually to colourless bodies.

CŒNOBIUM. A colony of unicellular organisms surrounded by a common membrane or investment. A spherical colony such as Volvox and Pandorina.

CORTICAL. External layer, resembling the bark.

CRENATE. Notched, indented, scolloped.

DIŒCIOUS. Having the male reproductive organs in one individual and the female in another. Sex organs on separate gametophytes.

GAMETE. A sexual cell, or germ cell. A conjugating cell, which unites with another of, either like or unlike character, to form a new individual.

Gametophyte. The conspicuous part of the plant body in algæ.

Generic. Characters in which there is agreement.

Gonidia. (Gonidium = singular.) An asexual reproductive spore or cell in algæ.

Gynogonidia. An asexual cell forming oospores or egg spores (female cells).

Intercalary. A growth which takes place elsewhere than at the growing point or apex, separately introduced or inserted.

Isogamete. A type of gamete, or sexual cell that does not exhibit sexual or other differentiation. A neutral body.

Metabolism. Sum total of the vital processes of life.

Moniliform. Bead-like, like a string of beads.

Monœcious. Having both male and female reproductive organs or cells, in one species or individual plant, as the Mosses and Ferns.

Nematocysts. Stinging cells of Hydra and the Jelly Fishes. A lasso or nettling cell developed in "Cnidoblasts," which are special cells embedded in the outer coat or ectoderm of the tentacles and exposed parts of the body. They consist of an ovoid, fluid-filled capsule having a long spirally coiled, hollow thread within and barbed near the base like an arrow. See Trichocysts.

Oospore. Sexual spore resulting from the fertilization of an egg cell or oosphere (female) by a sperm cell (male).

Paramylum. A carbo hydrate allied to starch.

Parietal. Growing from the inner walls or cavity side.

Partheno gonidia. (Partheno = maiden, virgin.) A gonidium reproduced without fertilization. Eggs which develop without spermatozoids or antherozoids (male cells). Partheno spores, or Embryos from unfertilized Eggs.

Peristome. The fringe or region around the mouth in Infusoria. The segment next behind the prostomium usually bearing the mouth, in Worms.

Phylum. One of the primary divisions of the animal or vegetable kingdom. Having a common descent. The Phyla of plants are the Schizophyta, Thallophyta, Bryophyta, Pteridophyta, and Myxophyta.

Pyrenoid. A colourless, refractive, proteid body within the Chromatophores, serving as a centre, or factory, for the synthesis and deposition of starch, as food for the plant.

Punctate. Having the surface dotted with minute spots or depressions.

Rejuvenescence. Cell formation in which the entire protoplasm of an old cell escapes by rupture of the cell wall and then develops a new cell wall. In Algæ by Zoospores frequently.

Reticulate. Netted, hexagonal, lozenge-like or geometrically shaped.

Retuse. Blunt, terminating in a round end.

SARCODE. Protoplasmic contents of a unicellular animal organism, including the nucleus.

SCROBICULATE. Pitted, depressions, full of small hollows or holes.

SPOROPHYTE. The ultimate product of the Oospore after sexual fertilization and undergoing a resting period.

THALLOPHYTA. A Phylum of plants including the Algæ, Fungi, Lichens.

THALLUS. The plant body of stems, leaves, and roots united and undifferentiated.

TRICHOCYSTS. Minute stinging organs, needle-like, often acutely bent at the apex occurring on the body of many Infusorians Paramecium, Bursaria, etc., and not occupying a separate cell, as with the Nematocysts.

URCEOLATE. Pitcher shaped.

VERRUCOSE. Knobbed, warty, bossed, or pimpled.

VESTIBULUM. A hall or lobby, a porch entrance.

ZOOSPORE. An asexual spore with one or more cilia or flagella.

ZYGOSPORE. A spore formed by conjugation of two similar gametes (sexual cells) serving commonly as resting spores.

MICROSCOPIC FRESH WATER LIFE

CHAPTER I

INTRODUCTION

A HOBBY! How natural for we humans to want some kind of hobby, one that will be at once a relaxation and delight and to which we can turn in leisure moments. Something in which we can be absorbed and which may be taken up or set down as opportunity occurs. To such among the dwellers of our great cities and towns I would confidently recommend the use and study of the microscope, with all its possibilities, as affording just such a scope. It will become a never-failing source of entertainment and an intellectual enjoyment ready to hand, and a fast friend as time rolls on. But how are we to begin? I hear some one query. Where there's a will, etc. If you already possess a dusty apology lying idle on the shelf, or are deciding to possess a simple but more modern one, you are within easy distance of the goal. As a little boy once said after watching his father play the violin, "Daddy, let me play; I can do it if you'll just start me on." Yes, that's all; we just want starting on, and it is there where the object and utility of this volume is intended to come in and help you. To present before you one, at least, of the alluring paths upon which you may pursue your pleasant hobby and so to interest you that, like Oliver Twist, you will be eager for more, and possibly find yourself eventually making a more particular effort, and specializing in some other of the many sides of microscopy.

The meadows and uplands resplendent with wild flowers, their structures and pollens, the brooks and streams teeming with active and incessant miniature life, open at once an illimitable field of supply for research, becoming a charming offset to the otherwise humdrum of village life.

To the city workers it provides a compensation to their daily routine, sweeping from the minds the musty cobwebs of commercialism, beguiling them to many healthy rambles in country lanes and fields that otherwise would have been missed—there to refresh their bodies and recuperate their minds.

To commune with Nature, absorbing into oneself some of her tranquil moods, her sober teachings and artistic temperaments occasionally, will react wholesomely on the human system, imbue it with a wider outlook and a better understanding; and in the study of her simplest creatures the microscope brings one within the closest touch.

In the collecting of micro-organisms much enjoyment may be found in its pursuit. Apart from the rarer specimens captured, or the prolific gatherings of particular species, there is the sportsman's side: those too venturesome dampings, the many draghooks "without return halves" left at the bottom, the handle impromptu fitted to the dipper the better to reach out farther, which kindly or unkindly takes hold of the weed and as calmly stops with it, leaving you the half as memento. These and many other incidents add piquancy to the pastime, even if of some little temporary inconvenience. By and by, as we pass beneath the prying eye of our Compound Microscope some of the material, what a wonder view delights our sight when all preliminaries have been successful.

Many such pictures will form indelible imprints on the mind and become a treasure store for future reference, from which we shall obtain a lasting satisfaction in having made the effort to accumulate and revive in later years the memories of those happy times gone by.

So much is there to witness in the world of minute life, ever busy "carrying on" beneath one's normal sight to view, and to most people unknown, but yet which lies within easy reach. It is necessary, therefore, to work step by step, observing, noting and sketching, if you will, the many forms presented.

As our journey here together takes us among the realms of fresh water organisms we shall learn the names of many, their appearances and something of their habits, and so be enabled to recognize them again on future occasions, with the same confidence and familiarity as we have of a garden flower.

In passing let us not forget the debt we owe for the opportunities and comparative ease and simplicity by which our researches to-day are made possible, due to the able and expert opticians whose exacting work and labours in computing those valuable aids to sight, objectives, condensers, eyepieces, and the various instruments of precision for their use. Lenses which have emerged from simple curves of transparent mineral with all its faults of curvature, colour, spherical aberration, and other technical imperfections to a now almost perfect combination, so much so indeed as to render the human eye chromatic in comparison.

To be capable of observing objects to a thousand times their actual size with as much clearness as our normal eye is wont to do in faltering light to one hundred is an achievement only to be realized as one undertakes to probe and see these things for oneself with the appliances now obtainable.

During many years of microscopy often has the writer focused down upon some fresh water material, no larger than 1/100th inch across, to be breathlessly overawed at the sudden beauty and wealth of brilliant detail so unexpectedly revealed and unfolded before him. These are the psychic moments of rapture and pure delight that the earnest lover of nature gives his time, his trouble and his money freely to have beheld and received.

The microscope, like all good things, needs study. It is an instrument of precision, only as it is used aright, requiring careful manipulation, careful practice, experiment and perseverance to make it speak its best, and to obtain from its delicate parts the aggregate of its finest performance.

Do not be disheartened; makers are ever willing to give sound advice in the use of any instrument of their own construction; there are also numerous books dealing with the microscope and its parts.

I will add one personal note in connexion with the study of aquatic organisms : become thoroughly conversant and adept in the use and setting up of the microscope for dark ground illumination. Any time utilized to this end is never wasted. It cannot be too strongly urged upon the microscopist to join some club or society devoted to microscopy. There he will hear and see matters relating to all that pertains to the instrument and its capabilities, where too, possibly, will be a library of books connected with the subject, where kindred spirits are met, acquaintances struck up " tuned in " to the same station of thought, objects of interest displayed and sometimes exchanged, one between the other, with much more of kindly fellowship that makes all for the success and happiness of each individual in the study and pursuit of his hobby. Altogether an intelligent, uplifting and pleasant way of acquiring the knowledge he seeks. And it is in the acquiring and pursuit of knowledge strangely enough that man finds his greatest delight. Once a thing is known, much of its interest is lost.

Nature, however, is very chary of yielding up her secrets, keeping one climbing on and on before the goal may be in sight. Microscopy is no exception, and many fruitless side tracks will be made before real success is reached.

Most of the microscopical clubs partake in occasional excursions together for specimen gathering, and at its meetings may be heard an announcement something similar to the following: First train after two o'clock Saturday next, meet at the Eastern Station, number one platform, for Lingford and the Forest Pools! What hosts of happy memories this recalls. What a reverberant ring it brings to mind.

A number of jovial comrades fully equipped thus meet accordingly and in unison for a ramble together. There to see again the old familiar faces among the new, to chat once more amidst nature's charming surroundings as the party wends its way along the countryside.

The sun smiles beamingly, and vigilant eyes are ready to catch any tit-bits of observation and to essay some appropriate thoughts. But here we have arrived beside our first halting place. It is a sedge-edged pool midst bracken and mosses. Out comes the net and rod, the vasculum with its bottles and tubes, the walking-stick and its fitments, and soon one and all are disposed around the sides, taking a dip here, sweeping amongst the Myriophyllum there, or the Water Crowfoot, hooking along the bottom for dead branches on which to search for the beautiful Bryozoa with their graceful tentacles.

One dip over there and a friend's tube comes up with water of a lovely blue. What is it? Some gather near, and with pocket lens in hand await a peep and to hear the verdict. It is a mass of Volvox, and in such crowded numbers as to give a decided hue to the tube of water taken. These tiny revolving globes, as you will read further on, are covered with very fine lashes constantly whipping to and fro, and the very fineness and number of these give an aggregate of vibrations to the light received from the water as to leave the impression of blue on the eye, the colour of its body, a greeny shade, also assisting.

So the time flies; presently a change of venue is heralded and we are away to another pool farther along. Here stentors, or the Nitella plant with its wondrous Cyclosis, may be the "take" amongst many another specimen. No one without some object or other, that is the surety of microscopic angling, as distinct from the larger fishing.

Presently the smiling chief from out his inner recesses draws forth a silvered trumpet and with a Robin Hood's true and lusty blow calls up his straggling huntsmen near, there to inform them time has arrived to follow with him for refreshments, discreetly arranged beforehand at the village inn or restaurant. Still the chatter does not flag; tubes and bottles get overhauled by one another, and thus, loath to leave "Tom Tiddler's" ground, we make our way homewards, there at leisure to place beneath the microscope some of the spoils of the afternoon, or to add as stock to the aquarium. The keeping of small aquaria is almost forced upon the "micro-fisher"; it is impossible to see carefully through all one can gather at a time. It need not be of an elaborate description, for specimens may be replenished in most cases. It is necessary at all times to remember not to overcrowd your vessels. The tubes and bottles you have filled at the pond side contain concentrated solutions of the organisms generally speaking, and need thinning, in wider receptacles, with a portion of aquatic plants to each, and an extra bottle of the water from where they were obtained is a useful precaution. Keep the vessels preferably out of doors in a shady place free from dust and soot as much as possible, but allowing a good surface for the air to rest upon. Glass dishes or clean white porcelain, photographic dishes (see N, Fig. 3) are quite admirable and serviceable. The latter are especially handy when Water Lily leaves or other large leaves are to be looked over. Upon these many of those beautiful Rotifers, Stephanoceros and the Floscules may be found, which cannot be examined in glass jars or tanks with such ease and comfort. With these simple arrangements a constant source of supply is kept up ready to be examined as time permits, and in the quietude of the home, surrounded maybe by the children's happy faces and inquiring interests, you sit beside the evening fire, passing one dip after another through the live box on the stage of your microscope, seeing some fresh object or new structures revealed to the intellectual enjoyment of all around.

As the poet aptly attunes:

> Sweet is the lore which Nature brings;
> Our meddling intellect
> Mis-shapes the beauteous forms of things;
> We murder to dissect.
>
> * *Enough of sections; stainèd art,*
> *Full nature ne'er deceives.*
> Come forth, and bring with you a heart,
> That watches and receives.
>
> WORDSWORTH.

These living slides, self obtained, simply arranged and observed in surroundings approximating their homes in nature, can never

* The italics are mine.

be surpassed in beauty or truthful reality by intricate mounting or tedious manipulations of their dead selves, foremost because they are alive, and thus beside their forms seen disporting themselves, you have their ways, eccentricities and little foibles, dispositions which are expressive of innate satisfaction in their transient lives as they are so fearlessly and guilelessly displayed. Here we may see a mere speck laughing along with its tiny whip, playing hide-and-seek amid the very jaws of death caught at last in the maelstrom of some unknown giant, gambolling to the last, to its inevitable doom. We cannot measure their understanding one with the other, nor their instincts, but we can at least judge that many movements they do are not without significance, and even of evident enjoyment to themselves. Thus they fulfil their particular sphere faithfully and without demur, in which it has pleased nature to place them. Of some, for instance, the Diatoms with their exquisite delicate traceries, or the Desmids with their decorative garnitures, one can but wonder at their hidden purpose, beauty and design.

"Full many a flower is born to blush unseen," says Gray ; happily the world mundane has not absorbed the ideal yet, and the instrument which aids so much to elevate the mind at the same time helps to enlarge the heart and scope of man. The microscope thus bows to a purpose much nobler and worthy above and beyond its material fashionings when coupled with a personality. In the micro world there lies the same field, the same breadth for study open, as in the larger macro world around. The artistic colourings and vanishing shades of all degrees. From water white and creams, through yellows to reds and browns, or the delicate tints of green in harmony. The glistening spheres and tiny pearls flashing with star-like brilliancy ; the delft shades of blues to purples.

Art never had a more entrancing realm in which to watch in miniature the play of light upon intensely fine divided lines of structure and see their harmony ecstatic so purely emitted. Those living, streaming threads of crystal clarity less than 1/20,000th of an inch in breadth waving in rapid unison and rhythm, encircling with mystic halos the crowns of some minute speck of life, are not, and cannot be adequately described in so many words to lovers of the real, the natural and the least trammelled by man. It must be seen.

In the study of propulsion and its manifold methods exhibited by these tiny creatures there is still much the mechanic may find worthy of investigation. Every kind we use in our arts and crafts to-day will find its replica amongst them. The screw, the turbine, the direct and indirect pull and thrust, the rapid reversion. The siphon and compressed air. Pulsating vacuoles forestalling pumps and cylinders. The caterpillar wheels and hosts of others, some beyond the capacity of man, as yet, to accomplish.

Those endless cyclosic, amœbic and protoplasmic potencies, which carry on without apparent source of energy, and yet within propelling bodies through substances many times their own weight where the individual cell has beaten the aggregate and ego of man in the run for perpetual motion, to duplicate. Power applied that has reduced friction to a minimum, so much that a thin cellulose envelope has not been abraided by the smallest part of a hair's breadth, after prolonged usage, nay in all probability has actually strengthened in its substance. There are movements due simply to the fineness of the particles engaged and may be seen dancing merrily in one continued restless motion night and day, and have been doing so for countless ages. These are "Brownian movements," which are known to occur in bubbles of quartz (sand), sealed up definitely æons ago, that have tiny matters suspended in the vacuum, or in some cases water, never wholly evaporated, busy when we were born and will doubtless be so when we have passed the veil.

In the decaying vegetation similar tiny bodies work incessantly night and day, jigging their existence away, which last until the final trace of substance is left. What becomes of themselves is unknown, but their movements can be readily observed in the partially empty cells of Algæ, Desmids and other plants. This stage we understand and speak of as the decomposition of matter, and the minuteness of the busy workers is sufficient to keep them agitated while suspended and with little or no friction in the water.

In the arts and crafts of warfare there are not wanting examples. That slow-moving creature, for instance, the Hydra, might fare very ill if it had nothing more than its lazy tentacular arms alone to protect it. Soft gelatinous substance that could readily be nibbled at and even lopped off by several colleagues in the water surrounding, but for an arrangement of darts provided by special cells that can be exerted into long finely pointed weapons at a moment's notice, barbed like arrows to hold fast once they have caught in. As if this was not security enough they are charged with a poison deftly supplied and secreted within by the cells discharging the missiles. Occasionally they may be used as mines thrown out around the suspended body of the animal, and while it lives in safe immunity from its own weapons it is woe betide the unlucky ones that near them while active. Cladocera, which constitute the water flea group, are large objects comparatively, yet they will succumb frequently.

There are innumerable instances in the economizing of power and energy, in the adaptability of organisms to utilize many varied devices and stratagems for the more profitable and efficient means of procuring food, and for the propagation of its species. Slow-going creatures will attach themselves to the swifter-moving ones and thus be transported into fresh fields and pastures new. The swift-moving ones, on the other hand, will ofttimes anchor themselves by threads and so steady their whirlwind careers. Where a spot offers a favourable ground for obtaining agreeable foods they show desire to stay, and not be propelled or compelled to hasten by.

In the Rotifera the same wheels used for propulsion are also the instruments for procuring and securing their sustenance. It is necessary therefore these must not be stopped, so the resourceful organism counters the idea by stopping himself awhile in such cases.

Among aquatic plant life special organs are found elaborated to capture animal foods, as in Utricularia, which may be used to continue life when normal means are out of action or unavailable.

In cases of drought arrangements established whereby an organism will encase itself more thickly and securely and rest dormant until suitable season arrives, then to revive and literally burst out anew and resume once more its active state.

Examples of parasitism, as in Trichodina, commonly observed gliding up and down along the tentacles of Hydra, performing the part of groom to its host, uninjured among the many stinging threads and cells, both living comfortably and amicably together.

Of Symbiosis in Euglenidæ, etc., where two dissimilar species exist agreeably in a mutual social pact, each thriving better for the company of the other. Of the power of selection, so pronounced in Vampyrella, penetrating into Algal cells, extracting their substance with discrimination and Machiavellian adroitness, there is further evidence of the fitness to survive.

In the infusorians the use and function of contractile vacuoles, spherical vesicles or cavities set apart, pulsating methodically, extracting and eliminating waste fluids and gases, inducing and relieving pressure upon food substances generally, and instrumentally aiding in digestion.

These constitute another ingenious expedient in the existence and persistence of those minute living organisms. They form an analogy to our lungs, simple arrangements wherewith to breathe positioned in curious and divers places about the body at times.

In the Entomostraca we find the lungs situated, in part, within the legs of the animals. How quaint we should think man if his respiratory apparatus were so placed.

Many more such interesting points will crop up as we pursue our investigations, and in our studies let the various objects bring to us as much instruction and even amusement as they will ; we must regard them all essentially as but a transitory phase of life, active but unrealized, a stage simply, along a great cycle of events they are consummating among myriads of others in the grand design of things with an evident trend upward, a struggling onward toward some further and more perfect plane than the present in their evolution, some Ultima Thule no one has searched or even contemplated.

Glimpses will occur to the earnest, thoughtful student, compelling him at times to reflect upon the whys and wherefores, the laws, ideas and scheme of things which co-ordinate the universe, and though the answer must necessarily be negative, the effort philosophically is good, broadening the horizon of one's mind to fuller, vaster proportions, expanding one's mental realm, and lifting one temporarily, at least, away from the whirl of life and affairs terrene. Drawing together the infinite with the deeply finite gives a mental unity, a roundness that makes one feel whole.

CHAPTER II

APPARATUS USED IN COLLECTING FRESH WATER MICRO-ORGANISMS

It may not be generally known to the tyro in microscopy what a vast amount of material teeming with the most wonderful active life is to be obtained both for his hobby and for study from all kinds of fresh waters. Such waters include the canals, rivers, lochs, reservoirs, lakes, ponds, etc., to mountain streams and their moss-laden tarns and pools, hidden away in silent nooks perchance, moorland marshes and trickling rivulets in glens, all fertile and abounding with fresh forms that will open out a new vista of life for him to revel in and only waiting for the mere collecting.

As a commencement it will not be time wasted if upon the next constitutional stroll he sets out " pond hunting " in literal earnest.

Having found the localities of the most likely ponds with plenty of robust aquatic plants growing within or about its margin, maybe a cover of brilliant emerald-green of Lemna whereon to feast the eye, he must make a mental note of them, and when next he calls again with a simple arrangement of collecting apparata he will be able, doubtless, to find more than can be looked through at one sitting of his microscope, and so his home aquarium will be automatically instituted.

The word pond hunting has been superseded here by the compound one " micro-fisher," as applied to the person seeking microscopic organisms by fishing, and may be used in its proper sense whether the water be fresh or salt.

The first simple outfit may be the walking-stick or the umbrella, whichever may be with you, and to either is attached a wide-mouthed bottle that can be effectively stoppered by a cork for safe carriage of the contents afterwards (Fig. 1). If the bottle has a lip to it, this is preferable, and a circular and fairly strong rubber band is then looped round the neck with the remaining portion given a couple of turns round the stick. This is quite sufficient to hold it and to forage among Algæ, Riccia, Lemna, and such plants that may be near the surface. The first dip will probably give you a number of Entomostraca with other life such as Rotifers, etc., and also a few fronds of the tiny plants.

This will be quite sufficient for you to take home and inspect for a first excursion. Do not keep the bottle corked for any undue length of time in your pocket. Not only does this increase the temperature suitable to the habitat of the organisms, but, like all other living creatures, they need air, either from the atmosphere direct or in solution. If this becomes used up the delicate objects decay and die, and will be unable to show you their pristine beauties to advantage upon arrival home.

Another simple arrangement by which to attach the bottle to the stick may be made after the manner of the cornet forceps by

Fig. 1.

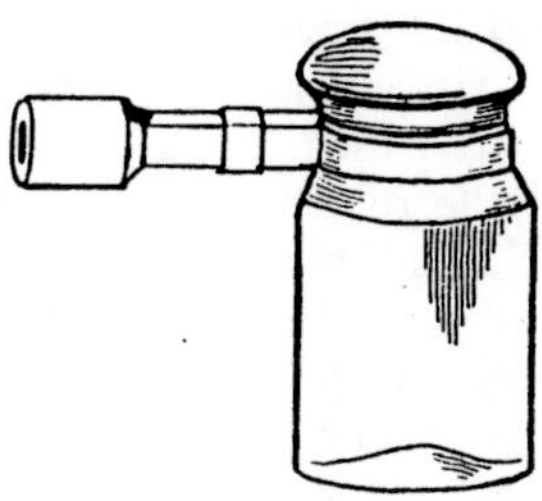

Fig. 2.

two cross scissor-like strips of flat springy metal, or as in the illustration, Fig. 2.

Still another used many years ago, perhaps fifty, was a strip of whalebone bent round the neck of the bottle and fastened to the end of the stick by the ends being held in place with stout rubber umbrella rings, three, a short distance apart, being sufficient. This was a handy contrivance, as the whalebone could be conveniently stowed away in the pocket along with the bottle and rings. It might have metal clips in lieu of cane or whalebone to hold it.

Having now made a beginning, our next visit to the pond or other waters may have a more elaborate equipment, which had better be purchased as an outfit complete from one of the numerous optician houses that deal in the accessories for " micro-fishing." It is not expensive when its usefulness and lasting nature is considered. The writer generally carries one of these, and has had it in constant use nearly thirty years now. A few home-made additions have been added from time to time, but the principle of it remains the same as purchased. It is a strong bamboo walking-stick with a hook handle, useful in suspending on the arm when examining a gathering or for carrying more easily (Fig. 3B). The centre through its length is hollow, and within this another rod is made to slide the full distance, which gives a double extension for reaching out farther from the side. Both the outer and the inner lengths have a thread upon the ends; the outer takes a screw ferrule (Fig. 3F) to be used when closed, as is usual for walking with, and the inner a female screw in which fits the various attachments that go with it.

The first and most generally used of these is a stout iron ring of round metal upon which a conical bag of strong bolting cloth is sewn, which may be still further strengthened by a broad strip of strong canvas or braid (Fig. 3A). To the smaller end of the cone a 4-inch tube having a lip is firmly sewn in (Figs. 3C and K), and the whole is then complete. This is made to screw to the end of the stick, and the apparatus is ready, after being extended to the full length. There is a stop which prevents the inner stick being completely withdrawn, and a good length of brass tube about the outer case at its ferrule end, so that the whole is very strong and compact.

A further adaptable piece is a screw ring (Fig. 3E) which will take one of the screw-top wide-mouth bottles (Fig. 4H) in which chemists frequently put up vaseline, pomades, etc., and which has the further advantage of having a wooden screw or stopper, which may be used in place of a cork quite securely. Another piece is a sharp flat hook (Fig. 3D) for cutting water plants low down and bringing them to the side with their adherent organisms such as Vorticella, Carchesium, Melicerta, etc., as well as many Desmids occasionally which adhere closely to their sides, as Micrasterias and others.

The cornet clip mentioned for holding the bottle is frequently supplied in lieu of the screw ring to obviate any difficulty in obtaining the screw-stoppered bottles.

In using a wide cork to allow of a free access of air without spilling the contents, two short pieces of hollow quill, as used for pen handles, may be inserted in the cork and allowed to project an inch or so above and well into the water inside. The bottles may then be carried safely upright with little or no leakage, unless upset.

Some " micro-fishers " use a draghook, a simple contrivance used at the end of a length of strong fishing line, to draw in branches from the bottom that may be prolific with Polyzoa or Bryozoa. The writer has some misgivings in suggesting the use of this, having left several at the bottom owing to hooking many objects that persisted in retaining the implement, but it is a very useful article when and while it works all right.

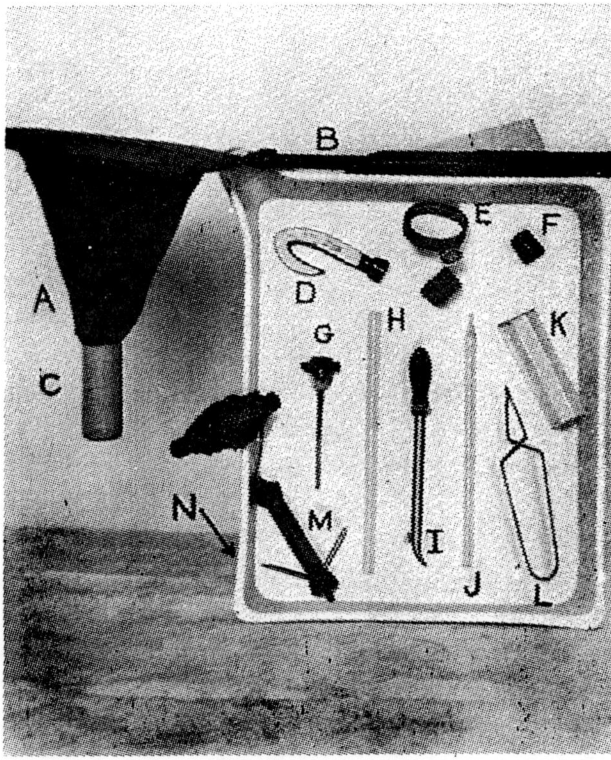

Fig. 3.

[*To face page* 12

A very neat and readily made specimen is a short length of ½-inch tubing, say 3 inches, which has a screw cap at one end, and around that end close to this is drilled three holes, somewhat slanting towards you, so that three strong nails with heads may be pushed through from the inside (Fig. 3M). The cap is then screwed on; this holds them from slipping out, and with a stout piece of string or cord attached through two holes at the other end (which piece may be of any convenient length to suit the depth required to reach the bottom and a little over), and the whole is then complete. It is simply thrown out a distance in a likely spot and then withdrawn towards you, and may you have the best of luck.

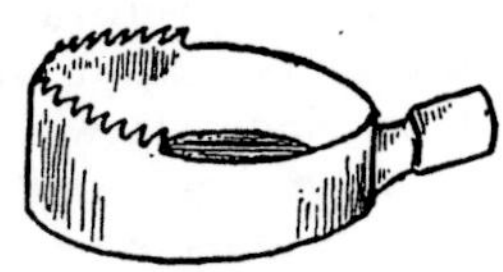

Fig. 5.

A very useful adjunct, rarely made now, but which may usually be obtained from the opticians or fishing tackle manufacturers upon request, is a small circular ladle (Fig. 5) either of iron or stout tin 2 inches in diameter, having a flat bottom and a slanting top about 1 inch deep on its higher side and ¾ of an inch on its lower, fitted with a similar arm and screw to affix it to the stick as the net, etc. Upon the deeper side of the ladle and for a third the circumference a number of teeth ⅛ inch deep is made and all bent slightly inwards, like a saw edge in appearance. This is used in scraping the sides of lock gates, reservoir and sluice sides, and also upon rush and sedge stems and many other prolific places which cannot be so successfully reached by other apparata. The writer can vouch for its general utility in many situations. The dimensions may, of course, be varied to suit one's own particular tastes, or it may be fitted to a separate rod. The average bamboo cane made by the trade is not sufficiently strong to carry a very heavy, bulky ladle.

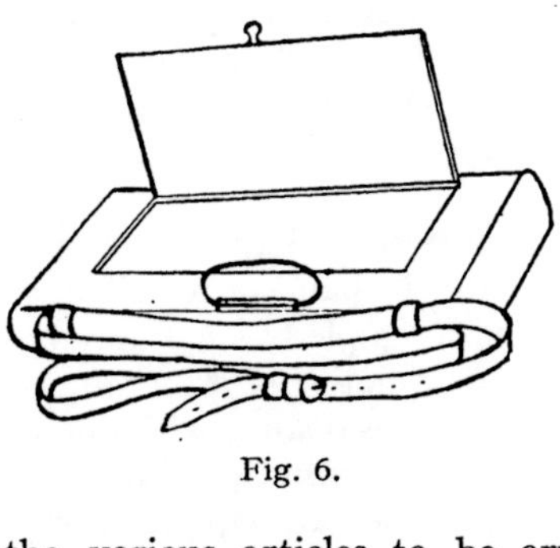

Fig. 6.

The next necessary equipment is some kind of case to carry the tubes, bottles, etc., and the collecting case used by botanists, called a "vasculum" (Fig. 6), is quite suitable with a certain amount of fitment, special for the various articles to be arranged inside. This is an oblong japanned iron or tin receptacle from about 12 inches by 6 to 21 inches by 10 in size, having a lid at the side, a handle to carry it by and also a strap suitable to sling it from the shoulder with, for ease in allowing both hands freedom as well as for travelling. Another useful article answering the same purpose can be made from a photographic camera bag, fitted up at home with wooden recesses for the various implements and impedimenta, net, cloth for drying the hands, and the separate parts for collecting stick, a shelf or shelves with circular holes to allow the bottles and tubes to slip in, etc.

In rambles with a "micro-fishing" party of some "society" many simple expedients will be learned and utilized, and in this manner simplicity will be attained to lessen anything of the nature of a heavy bulky outfit which may hamper your enjoyment and give you no more real facility for the collection of a sufficiency of material upon any one expedition. Herein lies the desideratum of company to the beginner attuned upon the same congenial errand and in unison of thought in the quest of some of nature's minute wonders in her fresh water microscopical world of life.

As an instance, in making a draghook, the writer has seen quite an efficient appliance, consisting of a piece of ½-inch lead piping about 2 to 3 inches long through which several pieces, perhaps half a dozen, of strong stiff wire 12 inches long bent all together in one loop and passed through the inside, having the cord passed through the loop at one end and the other ends bent at an angle towards the pipe, giving an effective grappling arrangement. The loop end (Fig. 7) was hammered so that the cord would not slip through, and was made almost on the spur of the moment. It further had the advantage that when unfortunately entangled with some heavy obstruction at the bottom the strain allowed the wires to pull through and the cord at least was saved while the lead only was left below.

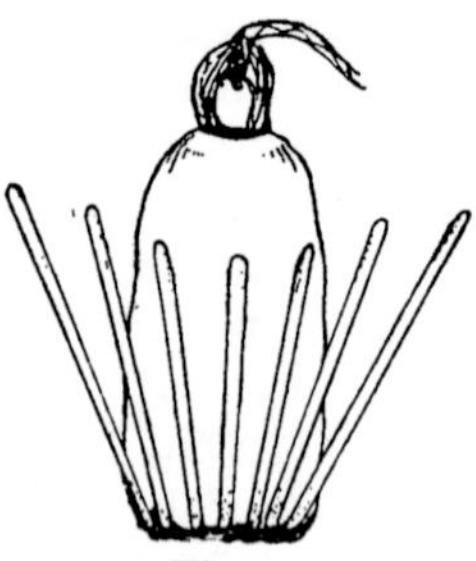

Fig. 7.

Sometimes for skimming the surface waters in large lakes and reservoirs, etc., a drag net (Fig. 8), either for the hand or for affixing to the rowing boat, is often required, and a simple tow net may be devised by making a conical bag of bolting cloth (used by millers in sifting flours and obtainable from the opticians dealing with micro apparatus which may be had of varying fineness) sewn to an iron ring from which three strong cords or hempen ropes are equidistantly fixed. These may come to a single knot about equal to the length of the bag and attached there to a single line. It may have a lipped-glass tube at its apex tied in, or the cone may be intact, in which case you turn the whole out into a jar now and again and apportion to smaller tubes as desired.

It is fortunate for the "micro-fisher" that the bulk of organisms habit the edges of the ponds and streams where generally the aquatic plants and vegetation is most abundant, and so the hand rod and net is most frequently in use.

When the tube has been swept near the surface back and forth a few times it is lifted out and held for examination, or it may be emptied into a flat-sided glass trough, often very convenient, and the pocket lens used to search the gathering roughly for suitable objects to retain or otherwise, and here only experience and the practised eye can help to form the judgment. If it appears likely, then it should be emptied into one of the bottles to carry home. In this way working round in various parts of the water an abundance of material may be gathered and all the containers filled before the return journey begins.

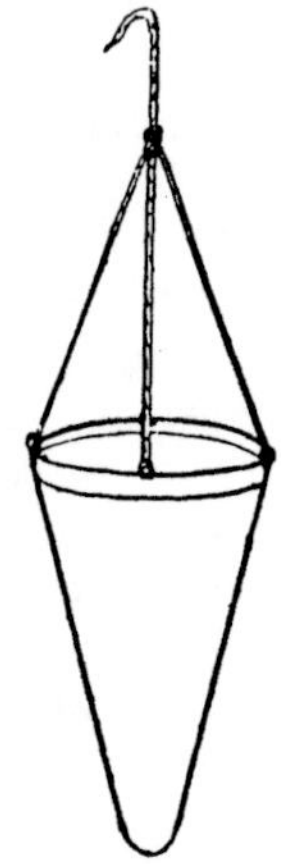

Fig. 8.

If a superfluity of some particular specimens be come across, do not take more than is really needed, there may be others, and a sufficiency is as good as a feast, and when stored in smaller compass than is their habit will often end in the demise or deterioration of the whole. The writer has known of this occurring and a whole stock of Volvox spoilt, and mentions this as a kindly warning to the over enthusiastic collector.

Having now arrived at the stage of preparing a "slide" for inspection under the microscope, a few glass tubes (Figs. 3G, H, I and J) of various shapes are needed and a "live box," as it is termed. The tubes may be had very cheaply from the optician, and may consist of a plain one, one with the end drawn out into a fine bore, a shorter and similar one having a rubber teat at the wider end, and finally a long fine bore one, with a wide bulb at one end covered across with a stretch of taut india-rubber. These are for picking out any particular portion or specimen in the "catch." If the tyro has not seen a glass tube open at both ends and of the same diameter used he will perhaps wonder how it can remove anything from his bottle or container. The method is to place the first finger over one end, holding the tube by the thumb and second finger, keeping it there until it nears the object sought and as close over this object as possible. Then remove the finger and the water and objects near will rush in. The finger is then quickly replaced before withdrawing the tube from the water, and the contents will come out with the tube ready to be placed in your "live box." If your object, say a Water Mite, is at the top of the water in the tube you may let some of it run back into the bottle gradually by slightly relieving the pressure of the forefinger and permitting a small quantity of air in ; this will press the water lower until your specimen is near the end. This permits of a smaller volume of unnecessary liquid filling your "live box," your endeavour being to retain it with as little fluid as possible for the time to keep it alive while under inspection. All the other tubes are held and manipulated with the first finger in a similar manner, the rubber-tipped ones need to be pressed before placing in the water, and on release near the object suck them in, with luck.

The "live box" particularly recommended is the "Rousselet" pattern, designed by Mr. Chas. F. Rousselet, famous for his expert knowledge and work in the Rotifera (Figs. 4A and B). It has a wide circular piece fitting a lower one, sliding with an easy but exact fit within it, made of brass, both of which are screwed into a rectangular plate of brass, suitable for the stage of most microscopes. In the centre of the plate is a small glass circle about 1/25th of an inch in thickness, and in the middle over this of the sliding larger circular piece a thin cover glass. The drop of water is placed on the lower glass and the upper portion pressed gently down to it. Care is needed in this action so that the specimen is not crushed, as the two glasses come in actual contact when quite home. The advantage of the wider circle around is here apparent, allowing any superfluous water to be accommodated at this outer annular portion which is without glass. You are now ready to see the object under both transmitted light and also dark ground illumination to the best advantage. Many other troughs of varying depths to suit low powers and large objects, such as plants, larvæ, Entomostraca, etc. (Fig. 4E), are made, but for definition of smaller objects the one above described is the most generally useful. Ebony or wood slips with a central circular hole (Fig. 4C), to admit objects in water between two cover glasses upon a ledge held in place by a ring are sometimes used. This gives more freedom of movement for the condenser to focus the light, and may be used with a 1/12th″ objective up to 1·2 or 1·3 N.A. in special cases.

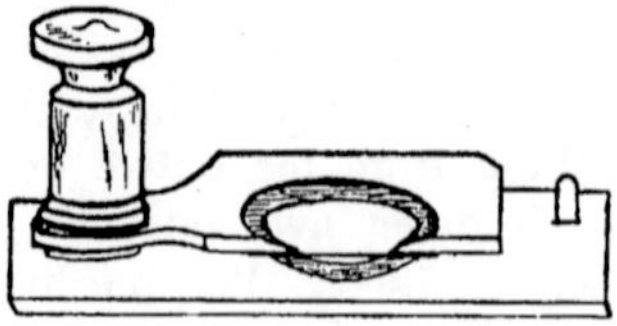

Fig. 9.

Another neat contrivance is a compressorium (Fig. 9) answering the same

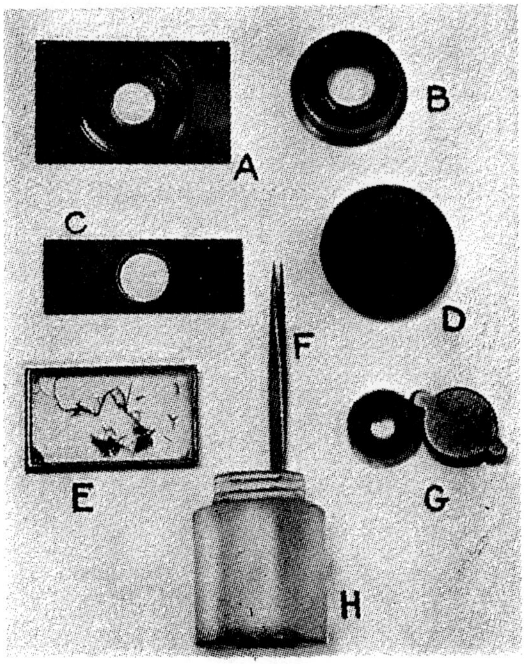

Fig. 4.

[*To face page* 16

purpose as the "live box," but having an arm carrying the cover glass, which can be turned to one side, and a portion of the surrounding brass cut away on one side to allow of chemical reagents, pigments, etc., to be run in by capillary attraction while the object is under observation. The brass plate is of a size somewhat the same as a glass slip ordinarily used, and the appliance is convenient in that the pressure is by an arm, which screws gradually upon the specimen. It is made again by the trade generally, and a catalogue inquired for of their accessories will readily bring you a number of devices as your experience grows and finds advantageous. A couple of needles (Fig. 10), secured with the eye portion in the end of wooden handles, are exceedingly useful, and a pair of fine tweezers (Figs. 4F and 3L) and a sharp, cutely shaped knife or two called "scalpels" (Fig. 11), a pair of finely pointed scissors, none of which are of great expense individually, but which will reduce the difficulties of preparing an alga or a living specimen for its easier inspection within the "live box" considerably.

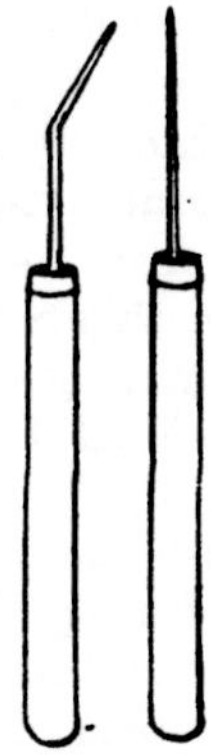

Fig. 10.

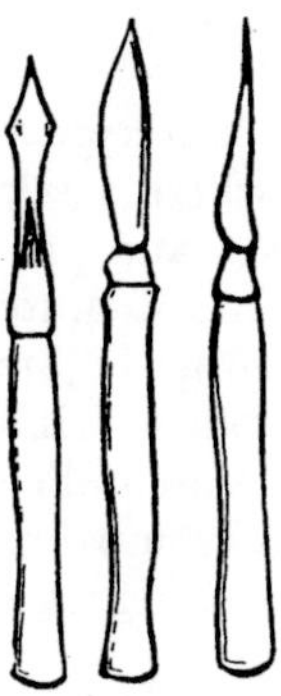

Fig. 11.

For filtering much unnecessary water from aquatic plants, algæ and the larger organisms an extra wide-mouthed glass jar (Fig. 12) is useful which will take a good-sized bung cork, in which may be adjusted two small funnels either glass or metal, preferably the latter. One is placed at one side and through the cork with its mouth outside and uppermost, and the other inside with its mouth inside; this latter has a piece of bolting cloth or cheese cloth stretched and fastened across it, and upon its outer cylindrical stem a length of rubber piping to fit is attached. This is the siphon to allow water to be drawn off, but retains the larger material within. Condensing the water and its contents in this manner gives a much more abundant number of organisms in a given dip and saves a deal of time in looking through when the "live box" is being prepared with specimens for the microscope.

Fig. 12.

As one becomes advanced a centrifuge is often used for a similar condensing purpose for exceedingly minute monads, unicellular algæ, spores, bacteria, etc., but this is, at this stage, beyond our requirements and is simply a note by the way. One of the water-tight glass troughs previously mentioned of about 3 inches by 1½ inches may be carried when visiting the water's side collecting; it comes in very useful to place a piece of branch or twig in when a likely gelatinous patch upon it is noticed, as it can then be held in water and scrutinized for Bryozoa, etc., with the pocket lens (Fig. 4G) (which may be of a power of × 10 as an average convenience), and in this manner perhaps save you the trouble and space needed of carrying carefully home only a batch of Snail eggs unconsciously.

Well then, having taken you a rather wearisome journey, through a number of appliances used for the comfort and success of "micro-fishing," we will now mention some of the likely places and methods of using them. With most descriptions throughout the book the writer has inculcated many suitable and likely places to look for the various specimens, and where in most instances he has actually obtained them himself, but it by no means follows that they will not be found in many another spot, and therefore as the tyro grows with his instructive hobby he will note many further points and situations in their collection.

In gathering Algæ, if this be loosely wrapped in waterproof paper or mackintosh cloth (or a bag of this material) where a fair quantity is taken it will save much room in the bottles, and provided it is not unduly squeezed together will convey itself quite all right until home is reached. It may then be placed either in jars or spread as suggested in white porcelain photographic dishes (Fig. 3N) and teased carefully apart with the two wooden-handled needles, and is then ready for inspection. Oftentimes a clump like this contains many varieties, some of which may be of a delicate structure, adhering to, say, a Spirogyra strand.

Vaucheria is a delicate tuft and cannot be preserved for any length of time; it loses colour and perishes very quickly generally speaking.

Plants which have grown in quickly flowing streams are apt to decay more rapidly when placed in the standing water of aquaria than those found in similar still waters. This should be remembered, and such specimens have preference of observation before their vigour is lost. With the Oscillatoria and those that adhere closely to the bottom, retain a fair amount of mud when collected; these should be dealt with by placing in a plentiful supply of water in the dish and carefully loosened so that the sand and debris may settle and leave the Alga floating freely on the top. If gathered late in the summer this will frequently be done naturally and automatically when the plants themselves, which is their habit, leave the muds and rise to the surface. They are frequently older samples then, and it is well to collect them occasionally from their primitive habitat earlier, however, and note the younger growths, their more complete life history and the differences present.

Oscillatoriæ by their nature, with time and plenty of surface and water around them, separate themselves in their slow movements, creeping apart from the earthy matters and so provide quite clean layers for the observer to select his specimens from.

Free-swimming Diatomaceæ, gathered off the surface of muds and ooze, if placed within the porcelain dish in a layer and put out in the sunshine or bright daylight, will likewise sort themselves out to the surface and then appear as a distinct browny gelatinous mass, and may be skimmed off with a spoon in almost a pure collection. These natural idiosyncrasies should be taken advantage of, as it saves a deal of time and trouble otherwise.

The stalked or stipitate diatoms may be searched for upon water plant stems, submerged tree branches, and the green algæ especially and a variety of species may frequently be found, like miniature forests standing out from the surfaces of such, the yellowy or reddish-brown tints of their tiny frustules colouring the objects and betraying their presence at once to the unaided sight.

With the Desmidaceæ the boggy heaths and swampy places, mountain tarns and moorland pools, clayey ponds with clear water, Sphagnum, mossy waters, are the ideal spots for these, and if the same procedure is carried out as for the free-swimming diatoms and are placed with the muds and ooze in a flat white dish they will sort themselves to the surface if given bright daylight, in a surprising manner and in a relatively brief period.

Washing the Sphagnum Moss in a glass vessel will also cause them to make for the light and affix themselves to the sides, and in this way greatly assist the collecting, without damage by instruments to their delicate structures.

When the moss is gathered there is a slimy gelatinous feel about it which will indicate their presence in all probability, and in taking out a tuft it should be allowed to drain a little while, but without pressing or squeezing, and when the bulk of the water has drained away, to enclose it in a waterproof cloth or bag and carefully place in the vasculum to be opened later in the porcelain dish with a quantity of the soft water brought from its habitat, if possible. On no account should tap water be used, any chalky, limy matters soon adhere to the surface coverings of these little plants and they do not survive with their colours intact for long in that unsuitable medium. To pick them up for inspection when placing in the "live box" or upon a glass slip, a wetted camel hair pencil is very convenient ; a mere touch will raise them owing to their mucous coats.

Empty frustules, that have lost their brilliant green contents, should not be discarded, the markings upon their shells is often of great value in determining the species to which they belong. Frequently these are so delicate and finely striated, punctated or knobbed that it is the only way they can possibly be seen at all.

In observing specimens upon a glass slip of the usual 3 inch by 1 inch size a ring of some waterproof cement, shellac, asphalt or other is first rung in the centre and allowed to dry hard (a few may be made beforehand in readiness), the spot of water with the objects is immersed within, and a cover glass, preferably of a slightly larger size than the ring, dropped on. With this rough-and-ready method the disadvantage is that it needs the microscope to be in an upright position, a rather tiring, straining position for long, although angle prismatic eyepieces are now made to obviate this, otherwise some arrangement must be made to keep the cover glass from slipping off. On the whole the compressorium is the most convenient for this work. It may be well to mention here that the slip and cover glass should be seen to be free from any foreign particles, grit, dust, etc., before using, also eyepieces and condenser on the microscope. Such trivial matters repay in the precision and comfort of observation and are much easier to do before than during the examination.

Desmids may be laid direct on the cover glass without immersing in water, owing to their gelatinous envelopes absorbing moisture with merely breathing upon them sufficient for a quick examination if applied to the glass slip at once. Eventually they dry up, but may again be moistened as before. Such method is merely in case a new or rare specimen is found and is desired to be carefully scrutinized before final preparation is undertaken.

With filamentous algæ like Œdogonium, Zygnema, Ulothrix, etc., these should be looked for in clear, quiet, standing waters or in trickling rivulets often attached to stones and other objects in their habitat, and allowing their wavy filaments to be gently swayed from side to side.

Enteromorpha will be seen in the shallow water as tufted clumps of green, and Spirogyra when young commences upon the mud, rising to the surface as it becomes older. Many of these may be so profuse that lifting them out by hand is quite a simple matter, and if too far out, the toothed ladle or net will come in very convenient for the purpose.

Cladophora, if recognized when gathering, should be removed

with the object it is attached to, if at all possible (often it is a small stone), otherwise it needs be gathered as close to its point of attachment as practicable. The position of its main stem and extent of the ramifications has a bearing upon the determination of the species, fragments of it being little or no use for the purpose.

Draparnaldia and the Chætophoraceæ generally need careful handling, and the use of the needles mounted in handles are advised for the placing and preparing of them for the slide prior to observation, as this entails least disturbance to their delicate filaments.

The Potamogetons, aquatic plants, sedges and all mosses should not be overlooked, and either removed completely or rummaged among with the toothed ladle or net, as many species of algæ attach themselves—Coleochæte, for instance—to such, which are far too minute to be detected by the naked eye, and so are frequently missed.

Heliozoa, Clathrulina, Actinosphærium and similar amœboid Protozoa will also be found among the mosses and plants in general, as well as Water Bears (Tardigrada), Chætonotus and the Water Mites (Hydracarina).

The "micro-fisher" will find by experience that organisms plentiful one season in a particular water may be quite absent another, and no absolute dependence therefore may be concluded one season to another on procuring them.

Here comes valuable the knowledge of other haunts wherein to pursue his "micro-fishing," some newer fields to explore, the labours of which may add considerably to his stock for observation, keep the interest alive and in the securing of some, at least, as permanent mounts through his own dexterity and skill, and in this way pleasantly augment his own collection of slides for the cabinet.

CHAPTER III

AQUATIC PLANTS

WATER MILFOIL (*Myriophyllum*) (Fig. 13A, facing p. 24)

So plentiful is this that hardly any English river is without some of it to be seen along its course. A good vantage point when in the country is looking over some picturesque bridge where one may ofttimes see this plant forming long trailing masses of light green filaments swaying this way and that, as the current decides, many feet in length. Twenty feet has been calculated from root to tip. It is not strictly correct to say root in respect of this plant; it has none, although it necessarily has a firm attachment there. It imbibes its nourishment from the whorls of finger-like threads representing its leaves. These are set at fairly regular intervals, becoming closer together the nearer the tip is approached.

They are quite smooth and flexible, a feature to be remembered when distinguishing it from Ceratophyllum, another plant somewhat similar, but whose leaves are quite stiff whorls, and which do not collapse together either when in the water or out, whilst Myriophylla will immediately they are lifted out.

It is a plant foremost among the many prolific sources of fresh water micro-organisms, and should never be forgotten when collecting. To give a list of those found upon it at one time or other would probably include all the fixed forms known.

In one species, *M. spicatum*, there is a vertical portion to the main stem which rises considerably above the water surface, and upon this a different shaped leaf will be borne, somewhat ovate, pointed and undivided, while the submerged ones always are finely separated and thread-like.

Although Myriophyllum is almost habitually a near the surface plant, yet it is found occasionally in dense patches covering the bottom where the waters have a gradual slope downwards.

WATER CROWFOOT (*Ranunculus aquatilis*)

(Figs. 13C, facing p. 24; and 14)

Where a soft, muddy surface to the bottom of the water is, there this will be found to grow profusely. It is a plant remarkable for possessing leaves of two distinct kinds, accommodating itself, according as its habit is either submerged or floating, to either of these positions. An ingenious method to ensure continuity of existence for its kind, and another instance in the struggle of the survival of the fittest. The "leaves" that are submerged are long narrow tubes, which, growing rapidly, soon form dense clumps and tangled masses of their green threads below. It is these that form the especial tit-bit for Water-fowls, and which they may be seen constantly plunging beneath to obtain. The floating leaves are three-lobed and broad, lying horizontally on the surface.

Fig. 14.—Ranunculus aquatilis.
A. Floating leaves.
B. Submerged leaves.

There are nine or ten distinct species of the Water Crowfoot recognized by botanists. *R. circinatis*, for instance, has large white flowers and no floating leaves, and the segments of its submerged ones are rigid, spreading out in one plane, while *R. intermedius* has small rather pinkish flowers with floating leaves and seldom any submerged ones. Naturally those plants with floating leaves, then, will be found principally in standing waters, whilst those in running streams have the long submerged tubular leaves and curiously the larger flowers. These latter are "the long-leaved mosses of the stream" alluded to by Tennyson in "The Lotos Eaters."

The flowers generally are a pure white with a bunch of yellow stamens in the centre, and are exceedingly pretty as they are seen floating serenely upon the swaying waters and a delight to the eye when in profusion. Very differently from those whose duty it is to look after the welfare of our streams are they viewed however, their dense bulk filling up the river beds, needing much attention to keep the water course fair against flood, and a constant cropping of their rapid growth is imperative at times.

For microscopic organisms, plants in swiftly moving waters do not offer the best advantage nor effect our purpose so well for collecting as the stiller pools and canals, etc., and so are left undisturbed to fulfil their mission alone. Most microscopic life prefers quietude for its delicate structures, and such objects are to be found in greater profusion and more vigorous life in the stillness of lakes and ponds, where too, aquatic plants are also abundant.

Upon Ranunculus may be found many of the fixed Rotifers, as Melicerta, Stentors, Vorticella, Epistylis and many others, so when we are out micro-fishing we must not forget to add a spray of it to our bottles before we leave.

BLADDERWORT (*Utricularia*) (Figs. 15, 16, 17 and 18)

About March, after lying dormant at the bottom of ditches and ponds, the tiny terminal seed-like buds of this plant begin to bestir themselves, rising to the surface to spread forth into leaves and branches. It is a very interesting plant, loving a quiet, marshy or stagnant spot for its habitat, and is of a browny-green colour (Fig. 15), floating its many finely divided thread-like segments which carry small bladders. These bladders are about the 1/5th of an inch in length when fully grown, and can be readily seen with the naked eye when the plant is lifted from the water, growing upon alternate sides of the branches. They have a rather slippery, gelatinous appearance, and were at one time thought to act simply as mechanical agents to float the plant from the bottom to the surface, to contain a gas in fact for that purpose.

Subsequent examination and investigation, however, has shown they are actually employed in obtaining and capturing food, and, strange to say, of animal food, forming at once a mouth to imbibe and a stomach to digest small larvæ, infusoria, worms, etc., which they entrap within. They are hollow and somewhat oval bodies, having at the narrow anterior end several threads pointing away from the utricle (or bladder), and any little animalcule striking against these are directed by them towards the mouth which is normally closed by a transparent cover or curtain, this it pushes aside in its movements and passes at once into the trap. This curtain is attached upon three sides of a flexible square and loose on the fourth so that it may swing in with pressure from without, but not pressed outwards from within, and it is astonishing with what certainty and power it acts. It has been observed a tail or a head may be held by it half in and half out without the little unfortunate object being able to extricate himself. Once fully

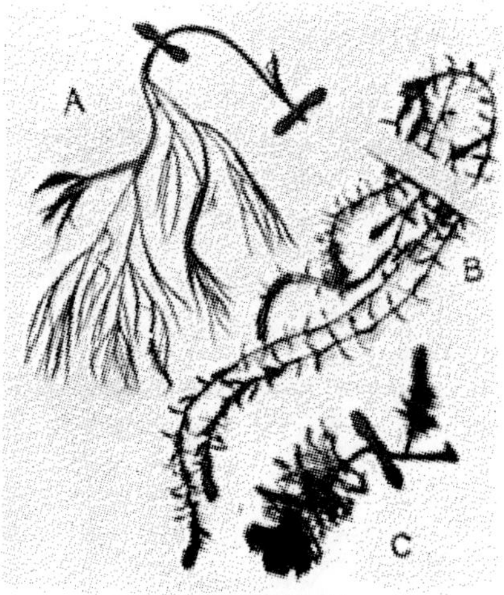

Fig. 13.

A. Myriophyllum.
B. *Elodéa canadensis.*
C. *Ranunculus aquatilis* (submerged leaves).

Fig. 15.—Utricularia.

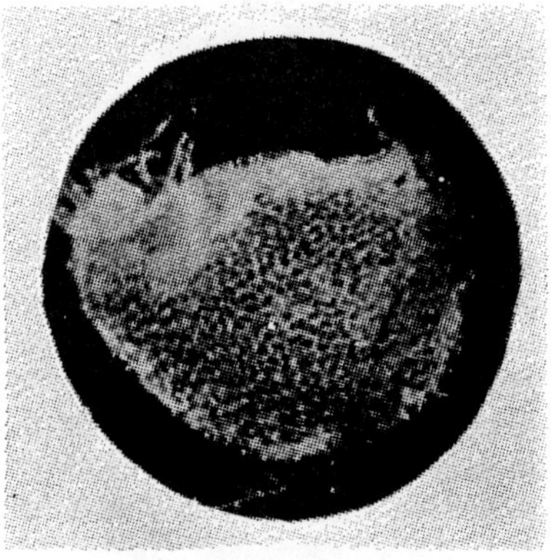

Fig. 17.—Utricularia.
Surface structure of pouches.

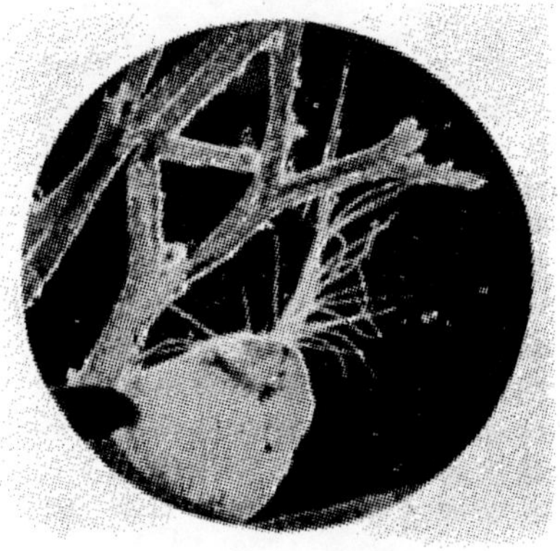

Fig. 18.—Utricularia.
Guard threads on pouch.

[*To face page* 2

inside, the captive soon becomes passive and its movements cease, and it is evident the contents of the bladder have acted in some way upon the creature and must contain a fluid with more potency than water.

Around the inside of the bladders is a lining with curious-shaped bodies (Fig. 16) like a wrist and four fingers spread out, but no thumb. These are the glands by which the absorption of the food is undertaken, and can only be seen when the utricles are teased apart and dissected, an operation few ordinary microscopists trouble to undertake.

Fig. 16.
Quadrifid processes on inner lining of pouches.

The flowers of the plant are on alternate sides of its short stem, which rises above the water with a bright yellow or orange corolla and a conical spur pointing towards it, about half as long as the lower lip. They appear usually between June and August.

The utricles are pouch-like, similar to an oval cut across the centre, and made double. When much food has been ingested they are bulged and semi-spherical; pressure is thus induced, which acts indirectly in the digestion of the contents. As the substances are absorbed the sacs become flatter and narrower, the pressure acting automatically as their normal position is assumed.

The cellular structure of their coats is rather complex and is very strongly held together by numerous stout cellulose threads joined together and thickened at their joints (Fig. 17). Upon the uppermost surface the cells are geometrically spaced, not in regular figures of hexagons, etc., but rather adapting their straight sides to meet at any angular-shaped figure. A granulated protoplasmic cover is stretched between each, and upon the inner surface of this is situated the green spore granules of varying sizes, having different duties to perform in the life and growth of the cells.

The nuclei present in all green plants can be seen and the smaller granules lying around. The utricle's coat seen with a low power shows also the further sides of the cells and gives the appearance in combination, much like a piece of Loofah, commonly used as a bath requisite, if a portion were cut out and laid flat. The strong cellulose threads join at their nodes and are often much thickened there, which still further strengthens the sac.

The spore cells as they become older are browner and wider apart, with more transparent plasm around than has the young growth. The young cells are bright green, and are packed very closely within, difficult to see through. They have also a much thicker case. With age the spores arrange along the edges of the cells in a regular file, giving a broader appearance to their outlines.

Each utricle is attached by a short, rather slender-looking, hollow stalk to the branches, having but one layer of cells to its surface, the centre being in connection with the stem and inner surface of the sac. Along the middle of the plant stems is a definite thread with stout annular rings at regular intervals supporting an inner strand or tube, and around this are bundles of longitudinal strands two or three times its width making up the whole thread. Its purpose is doubtless one for nutriment transport.

Upon the further side from that attached to the stem the utricle has many guard threads (Fig. 18) at the corner. These threads are transparent delicate branches, divided up into fairly long cells which have granulated, sticky protoplasmic pieces irregularly shaped and placed upon their outer sides. They are part of the snare which entangles small Rotifers, Water Mites, Infusorians, etc., that form the alternate food of the plant.

Utricularia essentially is a chlorophyll absorbing plant. Upon the outer surface of the utricles are many stomata, which are present in most higher plant life for the interchange and respiration of CO_2 and oxygen necessary to them. They are always situated at the nodes of the adjoining cells, and are oval or ellipsoid in shape, with a cross suture dividing them into two halves, capable of being dilated or closed as required. At the ends of the branches and along the delicate stems are very fine pointed transparent spikes which catch and hold the plant, exasperatingly sometimes, amongst Conferva and Algæ, in which it may be found. A new branch will often be seen emerging from between the forks of two branches as a half coil of young green growth before straightening itself out.

Vorticella will live in harmony near the opening of the sacs, and in a specimen before the writer there are over a dozen, busy contracting and plying their cilia with evident contentment and enjoyment. They are the long liqueur-glass shaped ones, probably Vorticella Cucullus.

Utricularia continues to grow when a branch is severed, even if this be done many times, provided the water containing it is kept fresh and similar in composition to its habitat. The writer has taken snippings from a single branch to as many as fifteen and still it flourished uninterruptedly, furnishing fresh utricles fully guarded and the stem strongly branched as at first obtained.

YELLOW WATER LILY (*Nymphæa lutea*) (Fig. 19)

This is sometimes called by its Arabic name *Nuphar lutea*. Its yellow globular flowers are well known. It is found more plentifully than the White Water Lily (*Castalia odorata*). The leaves, large, flaccid and tender, which are submerged, are among the most happy hunting grounds for finding the rare Stephanoceros and Floscules. When opportunity occurs therefore a few should be secured, kept as flat as possible in water, until they can be transferred to the porcelain dish at home. There they can be overhauled for such organisms as Plumatella,

Fig. 19.—Yellow Water Lily.

Cristatella, the Floscules and many others of great beauty, all of which find a congenial habitat upon their surfaces.

It needs a trained eye to tell a colony of Polyzoa such as Cristatella at sight, and it is often a useful plan to examine in the little glass tank carried, portions of leaf and rootlets that appear as likely, with the pocket lens on the spot, any small brown patches of plasm. This may save much disappointment later, otherwise you may find your treasured " take " on return consists chiefly of Snail eggs.

It was upon a Yellow Water Lily leaf the writer obtained his most fortunate capture of Stephanoceros. The clump of leaves was near the edge of a lake and about 2 feet down in clear water, and upon one leaf was found seven full-sized specimens, all in full plume, upon overhauling it in the tank on arrival home. The leaves were the submerged, green, tender and membranous kind, for it should be noted *N. lutea* has two quite distinct varieties belonging to each root. The floating leaf comes to the surface upon a smooth flexible and stout stem, similar to the flower stem, and is quite different, being oval and leathery, with a deep heart-shaped incision at its base.

The word Nymphæa is given to the plant from its growing in places which the " Nymphs " were supposed to haunt. In ancient mythology these were inferior divinities of nature, represented as beautiful maidens inhabiting the forests, etc., as well as the waters of the earth.

Vallisneria spiralis (Fig. 20)

This may be seen as long ribbon-like strands in slowly running streams, often of several feet in length. It is a most copious oxygen generator in sunlight, and is always useful on this account as an adjunct in the home aquarium. A perennial aquatic plant with an inconspicuous stem and grass-like leaves, it is remarkable chiefly for the manner in which its seeds are fertilized.

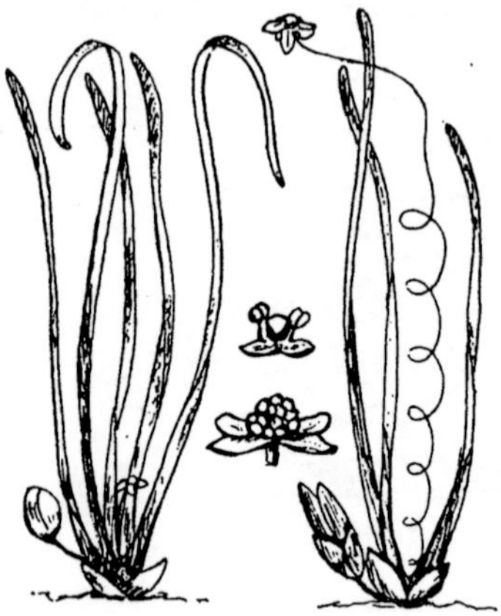

Fig. 20.—Vallisneria spiralis.

The female flower is attached to a long stem, twisted into many spiral turns, which allows the flower to rise to the surface, there keeping its level by uncoiling or coiling up its length as the height and depth varies.

The male flowers are white and globular, and as they mature, break off with but a short stalk which aids them in balancing upright and rise to the surface, floating freely among the lady blossoms hither and thither, carried by the wind and tide, there to allow their pollen grains to come in contact with the stigmas.

Fertilization thus accomplished, the long spiral stems recoil beneath the water to a short and compact cylindrical tube at the bottom, and within the cylinder so formed the seeds are ripened.

Vallisneria is especially interesting to the microscopist apart from Infusoria that attach themselves to it, in that it shows the movement of its protoplasm within the cells known as " cyclosis."

There are but half a dozen or so water plants that do this. Elodéa, Nitella and Chara are among them.

Elodéa canadensis (Figs. 13B and 21)

Formerly known as *Anacharis alsinastrum*, or American Waterweed, it was first noticed in England in 1842, and so perfectly equipped is it for extensive and rapid growth there is now hardly a stream in Britain in which it may not be found.

It is recognized by its leaves, which are set in whorls of three, each about half an inch long and of a pointed oval in shape, rising directly from the stem. The stem is readily broken, each portion being capable of growing upon its own at once. Most plants require to root in order to increase, but Elodéa is quite independent of such "trifles" and will continue growing when broken as it travels down the stream. It is a very salutary plant from a sanitary point of view, producing and freely giving out oxygen to the water.

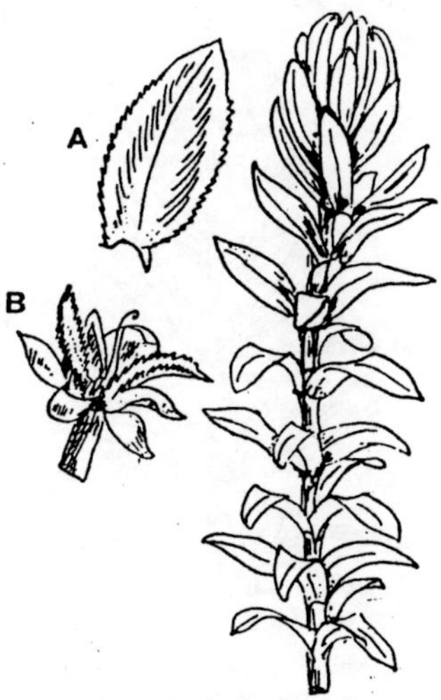

Fig. 21.—Elodéa canadensis.
A. Leaf enlarged.
B. Flower.

It is an asset to all home aquaria. Hydra have a particular partiality to it; the writer has seen as many as three upon one leaf.

The plant is also remarkable for exhibiting the rotation of the protoplasmic contents of its cells in the leaf, called cyclosis. This is most curious and interesting. It is not the sap of the plant as some have been given to think, but a flowing round and round the margins of each tiny cell of the substances within it, a streaming of the granules, the chlorophyll and protoplasm, in an apparent unceasing progression. The actual cause of this movement is not known. Cold seems to retard it and warmth to hasten the flow. No further preparation of the leaf is necessary than to cut one off near to the stem, place in warm water for preference, and transfer to the live box and a half-inch objective will reveal it "all wound up and nowhere to go."

HORNWORT (*Ceratophyllum*) (Fig. 22)

As its name implies, *Gr. keras*=a horn. Its whorl of spiky leaves are horny and stiff, and do not collapse on removal from the water. They are harsh to the touch, and their heads bushy and thickly packed together, with short leaves which expand outwards further as they become older and are left behind in the growth of the plant.

They are not smooth, moreover, each branch or filament of the leaf has its individual short and fairly sharp spines at intervals along it.

The plant is found in lakes, ponds and slowly moving waters, sometimes occurring in such abundance as to drive out all competitors, and often in company with Myriophyllum, but its tufted head, which rises perpendicular within the water, gives it a distinguishing feature from that plant apart from its roughness of touch.

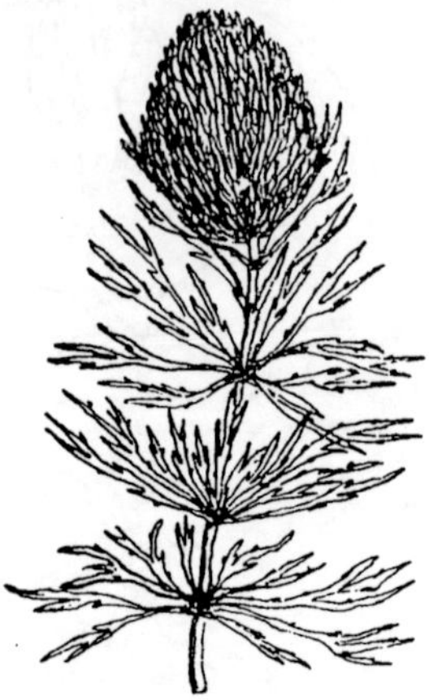

Fig. 22.—CERATOPHYLLUM DEMERSUM.

It is an entirely submerged growth in habit, the plant never coming above the surface. Even fertilization is carried out beneath, the pollen sinking on their stigmas owing to the slightly higher specific gravity of them to the water. It is entirely devoid of roots, in sharp contrast to all higher plants, and forms but a rudimentary attachment to the ooze at the bottom. It is therefore easily removed.

In the autumn the stem becomes very brittle, breaking in the mere withdrawal of it from the water. This forms its principal mode of multiplication. As the mature stage arrives it detaches short lengths of itself from the stem, and these constitute separate plants, which after resting the winter months continue growth anew. In this way it soon becomes prolific. The tufted heads are of a deeper green than the lower branches.

A French scientist, E. Rodier, in 1877, in his observations of the plant, noted a curious movement characterizing the shoots. First they would move for six hours in one direction and then return for another six, reversing for a further four hours, and returning in the opposite direction for another four. This he found took place repeatedly. The reason for this is not evident.

Lemna (Fig. 23)

This tiny plant, known to almost every country schoolboy as Duckweed, floats like a green mantle upon ponds and still waters, especially those that are replenished chiefly by the rain.

There are several species, but all are free floating, not attached in any way to the soil or to other aquatic plants. Like rafts driven hither and thither by the wind or current, they sail freely about. They have no real stems, nor real leaves, the whole plant consisting of little green oval fronds. Two or three may cohere together loosely which look like leaves.

One variety called *Spirodela polyrhiza* (*poly* and *rhiza*=many footed) has the larger fronds, and of a deeper green, that hide a

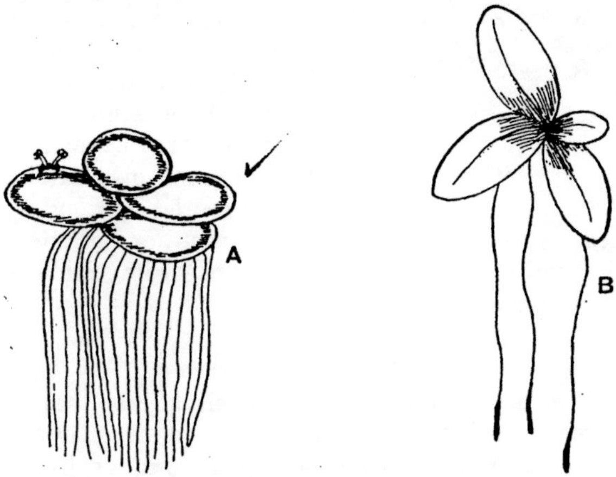

Fig. 23.—Lemna.
A. *Spirodela polyrhiza*. B. *Lemna minor*.

dozen or so tiny rootlets which hang pendant below. This is the Greater Duckweed.

Lemna grows in an ingenious manner. The wider part of a frond, on the sides where the midrib in a leaf might be, splits into two thin plates, and between these grow the secondary fronds.

Lemna minor is the most common plant, and usually has the lighter yellow-green fronds, and with but two or three little rootlets only. These rootlets have a bulbous thickening common to all the species at their extremity.

Lemna trisulca (Fig. 24)

Lemna trisulca, or Ivy-leaved Duckweed, is distinctive even to the eye on removing it from the water, appearing to grow its fronds crossway over each other, i.e. at right angles, and so spreads its rootlets wider apart. It is also more continuous and extensive in the growth of its fronds.

Tiny plants that the Lemna are, they all possess flowers, though those of the Greater Duckweed are hard to find. Two, sometimes three, tiny buds are set near the edge, on top of the frond, consisting of a little bract, enclosing two stamens and a tiny ovary. This constitutes their simple floral requirements.

Fig. 24.—Lemna trisulca.

The plant in its prolific growth, found *en masse*, becomes a shady retreat and a soft texture upon which microscopic life, such as Melicerta, Stentors, Rotifers, Vorticella, etc., will attach themselves and flourish. There is no toll or charge asked or given for this privilege, and a varied assortment may often be seen upon a single rootlet or frond. Each is thus carried by its host as wind and weather permits, and all are provided with a fresh hunting ground, from which a varied supply of food may be obtained.

MOSSES (Fig. 25)

These plants are found in so many situations that we must confine ourselves to those which are found in or near water, the edges of peaty bogs and pools, as Sphagnum, or live beneath the shady trees along the banks of rivers as Funaria, Polytricha, Marchantia, etc. Any such species will then be found to provide cover for numerous Infusoria valuable to the microscopist.

The light green Sphagnum has ever been a never-failing fount for a variety of organisms.

But the others mentioned may become covered with the rising waters at times and be found also prolific hunting grounds, and should not be overlooked. Tentacular Infusoria, Water Bears, Suctoria, Rotifers and a host of others find existence to their satisfaction among their succulent stems and leaves.

Fig. 25.
Moss. Funaria hygrometica.
A. Capsule with its hood or Calyptra.

Mosses do not possess true roots, and have a special adaptation in their metabolism for absorbing and retaining moisture during drought. This permits many tiny occupants to continue their existence when other plants are high and dry. In remembering this, one may collect from a handful by pressure over a wide-mouth bottle or phial a quantity of valuable and interesting material when unobtainable from other and more usual sources.

Chara fragilis (Fig. 26, facing p. 60)

The Characeæ generally have had a very chequered history, being bandied about from one order to another, and at present are relegated to an order by themselves called "Charales."

The plants are visible to the eye and attain several inches in length. Sometimes they are coarse and rough to the touch, the stems being constructed of several cylindrical linear bundles of cells which show nodes and internodes with whorls of cylindrical "leaves," which in turn bear "leaflets" upon them.

The whole plant from its habit of encrusting itself with carbonate of lime from the water is rendered thereby very brittle, which adds no little difficulty in obtaining perfect specimens for the collection, portions snapping off by their own weight.

The principal stem and the leaves are more or less completely covered with a layer of cells, forming the "cortex." The leaves may be six or eight or even twelve in a whorl, and each leaf with several whorls of leaflets along its length. Each leaflet generally has stipular outgrowths.

Characeæ prefer ponds and ditches with a muddy substratum where the water is clear but stationary, seldom being found in quickly flowing waters. Some of the best specimens the writer has found were obtained from the bottom of a partially sunk derelict boat that had lain in its position many years. Here the water was always shallow and warmer inside than the surroundings, and the plant apparently had found an ideal spot and the specimens were freer from encrustation.

Chara belongs to the Cryptograms, one of the two great groups of plants so named because they do not bear flowers. The other group is known as Phanerogams, which includes all plants that have flowers. It will grow from 4 to as much as 24 inches in length, and when not encrusted with lime is green and flexible. Its reproduction is by male and female organs, i.e. sexual reproduction; but the antheridia and oogonia constituting these are unlike those of any other plant.

De Bary describes the method very concisely, and says the first change taking place after fertilization is that the coarse starch granules and oil globules with which the oospores are filled begin to recede from the apical region, that part becoming filled with a light-coloured finely granulated protoplasm. A transverse septum is then formed separating the oospore into two unequal parts, the upper and smaller portion forming itself into a plano-convex lens-shaped cell called the "Nodal cell." It is from this the young plant originates.

The larger and lower portion remains as a food store of reserve nutritive material. The "Nodal cell" finally swells and bursts, the apex of the oospore splitting into five "teeth" to allow of the expansion, and the protoplasm then divides into two portions, then four, and again further, and so the new growth is produced. Each point of the crown on the oogonium is single-celled.

Chara was one of the first plants to be noticed exhibiting the peculiar protoplasmic movement called "cyclosis." The large size of the cells and its simple vegetative parts renders it easy of observation when its limy encrustation is not in great quantity.

The plant probably is an ancient type of vegetation, fossil remains of it having been found in the lower oolite beds as well as throughout the tertiary and parts of the secondary among esturial deposits and in the freestone layers.

It is an entirely submerged plant, and has a wide latitude from which its various representatives are found. As high as 69° N. in Norway to about 49° S. in the Indian Ocean, in the hot springs of the Yellowstone and in Iceland, waters that have a temperature calculated to boil an egg in from four to five minutes, it survives freely. It is therefore a most adaptable plant.

Nitella (Fig. 27)

This forms a sub-family of the order "Charales," and though in many respects similar, in its reproductive methods, it differs in its stem being simpler and composed of long single tubular cells like Vaucheria, and which do not possess a layer or "cortex" of cells surrounding. Owing to this the plant is not nearly so encrusted with carbonate of lime as in Chara, and the process of cyclosis is much easier watched and in a greater length of cell.

The stem or axis is divided into nodes or junction of cells at which normally two branches arise. These branchlets may continue simple or become divided again into fairly equal-sized lengths. Both the stem and its branchlets are of a clear translucent green when in healthy growing condition, containing abundance of chlorophyll, but as they become aged this turns to a brown colour, and the delicate tubes thin and crumple, often forming a busy hive around which Infusoria and Rotifers find many tit-bits of food to their liking.

It is noticeable with Nitella that where encrustations of lime are present upon the plant to any appreciable extent these always partake of an annular character, forming in bands around the branches of approximately the same width as the spaces between them. The explanation for this has not been forthcoming yet, but it doubtless forms a definite distinguishing peculiarity of the genus.

The propagation of the species is by antheridia, i.e. sexual reproduction. The antheridium is always terminal on the middle leaf or leaflet and situated at the junction of the branchlets, another striking characteristic of the genus. The oogonia (female cells) are also compressed, and are either single or may be several together at the points usually where the side leaflets would arise but for their presence. The leaflets project beyond the tips of the leaves which gives an appearance of being forked. The genus presents a wide range of complexity in its growing structures, varying from simple one-forked branchlets and single-celled secondary rays to numerous different forms of branchings and those in which the ultimate branchlets are many celled.

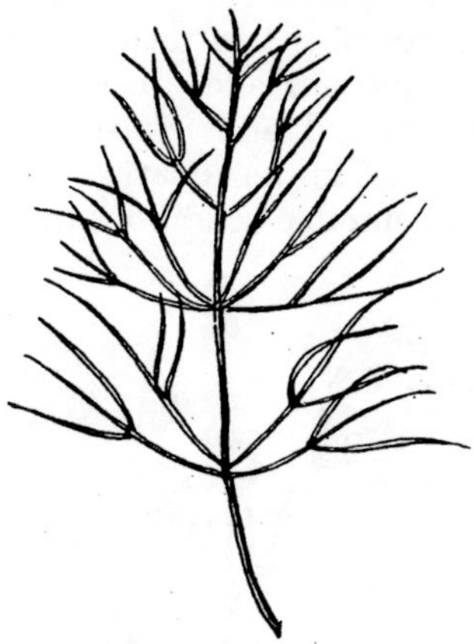

Fig. 27.—Nitella

The charm and interest chiefly to microscopists is the phenomenon within the living cells of the streaming protoplasm called " cyclosis." Each cell of whatever length (and some are quite long as seen microscopically) has many green particles of chlorophyll near its inner coat embedded in a transparent plasm. This is in constant vital action, flowing down the sides to the ends of the cell frequently in a long spiral form, carrying everything with it as it goes, apparently in an unceasing, never-ending round, a natural example of perpetual motion. Sudden changes of temperature, however, has a deterrent effect upon it, and occasionally it will slow down, but will start again if warmed. Detaching a portion of the plant also seems to paralyse the action for a time, but is only temporary. How long the procedure continues while in its natural habitat without anything untoward to disturb it is unknown, but it would seem to keep on day and night without ceasing until decay actually commences and the green chlorophyll become deteriorated. Light, too, has some slight accelerative action and quickens the movement sometimes.

There are several other plants which show the phenomenon, some of which are land plants, the hair of the Nettle, hair of Tradescantia, etc., and among the fresh water aquatics Chara, Closterium, Elodéa, and Vallisneria it is well known.

Potamogeton natans, etc. (Fig. 28)

The genus Potamogeton includes over a dozen members, all very pretty plants in their habit and all aquatic. The flowers are not conspicuous and are green, which accounts for their often being overlooked and left unconsidered, yet they form a very useful adjunct in the aquaria, giving out oxygen profusely under sunlight and suitable conditions of water supply. *P. natans* is perhaps the commonest among the species and may be readily recognized by its broad elliptical floating leaves. Like many other aquatic plants it has two kinds of leaves, the submerged ones being long, narrow and trailing. The floating ones are frequently so dense in the still backwaters of river inlets and small pools as to cover the surface during the late summer and autumn with their foliage. The colour is of an olive-green with a ruddy sheen as they grow older and decline. They are quite shiny, waterproof and sturdy-looking, and will shut out a large amount of light from any water beneath, paling the green colour of many aquatic plants. This fact will account for many organisms being absent from such dark waters where *P. natans* is profuse, but it also is a good locality often for many other species that prefer shade, Bryozoa and many of the Entomostraca, etc. The flowers have both stamens and pistil, and the parts are all in arrangement of fours: the stamens are four, perianth, four cleft, seed vessels are four.

Fig. 28.—POTAMOGETON NATANS.

Other species are *P. crispus* (Fig. 29); this has no floating leaves and is known as the Curled Pondweed (*P. perfoliatus*) (Fig. 30). The submerged leaves of this are expanded into blades, but without a leaf stalk. Sometimes it grows several feet down.

P. fluitans (Fig. 31).

P. densus (Fig 32).

Fig. 29.
POTAMOGETON CRISPUS.

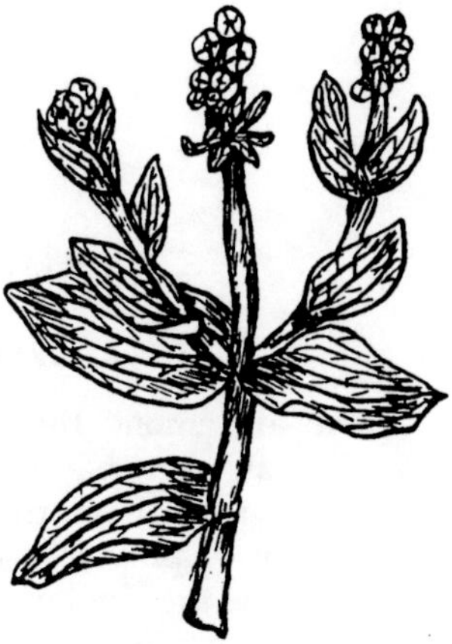

Fig. 30.
POTAMOGETON PERFOLIATUS.

Fig. 31.
POTAMOGETON FLUITANS.

Fig. 32.
POTAMOGETON DENSUS.

CHAPTER IV

LOWLY ORGANISMS:

RHIZOPODS, AMŒBA, ETC.

THESE are among the simplest and lowest forms of animal life, having an Amœba-like body, structureless in its composition, raised a stage above that organism by having acquired the power or ability to provide a covering to their naked protoplasm. The movements are similar to the Amœba, and they have neither eyes, stomach nor other organs usually accompanying animal life.

Rhizopod indicates like-a-root foot on account of their pseudopodia giving the impression of roots as the organism lies extended beneath the cover or shell. They constitute a large class, and are found in salt as well as fresh water.

Foraminifera, found in the ooze at the sea bottom, are a large constituent of the marine Rhizopods. Radiolaria are another and beautiful class of marines, the shells of which are of most exquisite shape and design in many cases. The body contents give no distinction of substance into organs and tissues and is merely a hyaline homogeneous plasm with the inherent power of contractile and extensile movements, constantly changing its form at any part, producing bud-like, stunted flowings of itself, and sometimes continuing these until they are the merest threads, projected to long distances from their source until it seems almost inconceivable how they can ever manage to withdraw them in again. This they do, however, and with it probably many granular and gelatinous objects, diatoms, etc., which form their principal food. Thus, unless they possess a shell they follow all the characteristics of an Amœba. In the animals a little further advanced shells are produced, sometimes from substances secreted by the animal itself, such as calcium carbonate, silica or chitin, or may be constituted from material particles foreign to itself found in the waters surrounding such as sand grains, diatom frustules, earth, etc. Chitinous shells are common among the fresh water forms.

Where a Rhizopod forms its shell of plates in regular forms these are often made from silica extracted from the water and deposited upon its outer envelope, taking a definite shape and fitting closely the outline of its body.

Difflugia in the formation of its shell obtains the silica in the shape of sand grains. These it passes through its body to the outer surface, and covers himself with them, aided by the albuminous nature of its substance preparing them to remain there, once they are finally on the outside. Any additional thickening is accomplished from within. They are quite formidable structures for protection, the only opening being upon the side for its body to extend itself through and gather food or to retract and use for its pseudopodia.

Simple as their whole body and edifice is, the class has a distinct power of producing their like, and a Difflugia never by any chance becomes an Arcella, for instance; there is therefore more than meets the eye connected with their simplicity.

Rhizopods to be understood require very careful observation. They should be viewed from all sides and under different modes of illumination, reflected as well as transmitted light.

The shells are of varied consistences, shapes and complexity, and these variations have been found to serve best in the classification of them into the different genera and species. The body protoplasm primarily is clear and colourless, later granules, globules and fine particles appear which diminishes the transparency and often imparts some colour from the matters taken in its food. It is a noticeable point that colouring the water while they are living will not give them colour, but let them be dead they will soon stain, like other bodies. Thus they possess ability to throw off or a power of selection not to intake coloured substances like carmine, etc., in their bodies during life when this is not desired. Iodine to some extent should give them a brown appearance, but at the expense of some of their outer protoplasm, and if too strong, to search the interior, will kill them. The axiom therefore is, if you stain the body of a Rhizopod or an Amœba he is dead, if not, or only tinted, he may be living.

Rhizopods are grouped among the Amœboid Protozoa under the general head of Sarcodina on account of their movements being temporary extensions of the protoplasm by pseudopodia.

The food substances consists usually of a vegetable nature, such as bacteria, diatoms, unicellular plants, Desmids, Algæ, etc., all of which possess plasmic gelatinous coats which the animals can readily utilize, any granular portions, shell cases or inorganic portions being ejected at any part of the sarcode, or outer cover, convenient. There are notable exceptions, however, to this, and in one species, Nebela, they are said to feed upon one another, and also to make use of closely allied species in the building up of their own shells.

Vampyrella is distinctly a parasite, boldly penetrating the soft gelatinous cells of Algæ, literally cleaning them out and absorbing the contents. This forms a striking instance of selection in so lowly an animal. There is a respiratory action performed in the body by an exchange of oxygen, received through the outer wall from the water surrounding, and of carbon dioxide, CO_2, transmitted in its place by the action of osmosis, similar in function to the human lungs in breathing.

Evidence is also adduced strongly in support of a symbiotic relationship between algæ and some sarcodina. The alga supplying oxygen and the sarcodina waste nitrogenous matters and CO_2 to it in return. There are some species which take as food only living organisms as the Reticularia, and in the Heliozoa Actinospherium captures both Infusorians and Rotifers. Plants containing chlorophyll, however, is the general staple food, and all such vegetation comes as material supply to the sarcodina.

Rhizopods, like all primitive forms, are one-celled, consisting of the cell body and a nucleus. The cell body is rich in albumen and proteid matter but poor in phosphorus, on the other hand the nucleus is rich in phosphorus bound up in a substance called " nuclein " but poor in albumen. There is also at least one vacuole which is contractile to assist digestion and expel waste matters. Sometimes there are many. In the same way there is usually only one nucleus, but again there are species with a great many, even to hundreds and thousands.

The manner of capturing food is for the false feet or pseudopodia to flow round it and link up with others coming along, and as it were creep over and cover it with their plasm and in this way to bring it into the general flow of the substance when it is absorbed and not allowed to separate until its digestible parts have been recovered and the unused and waste portions are eliminated. As the whole plasm is capable of digesting at any part of it, it matters little where the food enters, the same result awaits it.

One method of multiplying is by fission. In this the nucleus divides first and the cell body follows, each portion receiving and enclosing a part of the original nucleus. The several sections then continue separately in growth to maturity. Another manner is by conjugation, in which swarm spores are produced from either a temporary fusion or a permanent union of two individuals of the same species and an encysted period undergone before final development.

Difflugia have been observed with their mouth openings opposed, the plasma of one enveloping the oral aperture of the other, then drawn together tightly and a division into two taking place within (Fig. 42). Propagation by detaching portions of their substance as buds, similar to the Hydra, a form of gemmation, is also known. There is little difficulty, therefore, in accounting for the large numbers found in situations suited to their constitution. One of the most fruitful sources in which to find them is the Sphagnum, peaty bogs on moors, and among mossy swamps where much decayed vegetable matters abound. They have a wide altitude range occurring from the sea-level to tops of very high mountains, and are probably foremost in this respect in comparison with other microscopic organisms. The Amœba is also the oldest, in point of time, undoubtedly. The shell coverings are called "Tests," and those possessing them belong to the order "testacea." The Testacea evidence a particular ability to raise or lower themselves in the water, altering their specific gravity by inducing a vacuum within their air-tight cases or by gas vacuoles altering the amount of the heavy CO_2 contents.

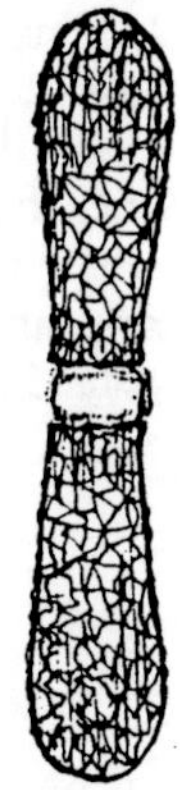

Fig. 42. DIFFLUGIA IN CONJUGATION.

In seeking Rhizopods the micro-fisher will seldom be disappointed if he lightly skims the ooze on top from out quiet, secluded pools and standing clean waters, or searches the under sides of Water Lily leaves and among Bladderwort rootlets or Myriophyllum and any kind of submerged mosses. There is no need to dip deeply into the thicker mud as this is void of any specimens. Like larger fry, they want freedom and space to live in; their weightier bodies keep them near the bottom, and substances both for their shells and food are there found in suitable quantities. This surface ooze seen under the microscope consists simply of decayed portions of vegetation, light and finely divided, intermixed with sand grains, some of which are perfectly transparent, and giving the pyramidal shape of their nature, quartz. It is an ideal land for such lowly organisms as Amœba, Rhizopods, Heliozoa, etc., and it is necessary to provide this as an environment for them, if they are wanted to be kept, in their congenial surroundings, in an aquarium at home. Mud can be obtained clean, and dirt has been defined as only matter in the wrong place. Seek out clean waters, therefore, from which to obtain your objects, they will be sturdier, healthier, and finer specimens.

AMŒBA (Fig. 33)

In gatherings from the mud in ponds and still waters, or upon the leaves of aquatic plants, these may often be found in abundance. Intrinsically they are a simple transparent substance, impervious to water, endowed with life and motion. Neither mouth, stomach, alimentary canal, eyes, nor any ordinary organs usually associated with living bodies are to be found in them. Yet they live, move, and have their being. They eat and absorb, and even select to all appearance what they will receive and what reject. They are justly considered to be the most primitive form of living matter called by

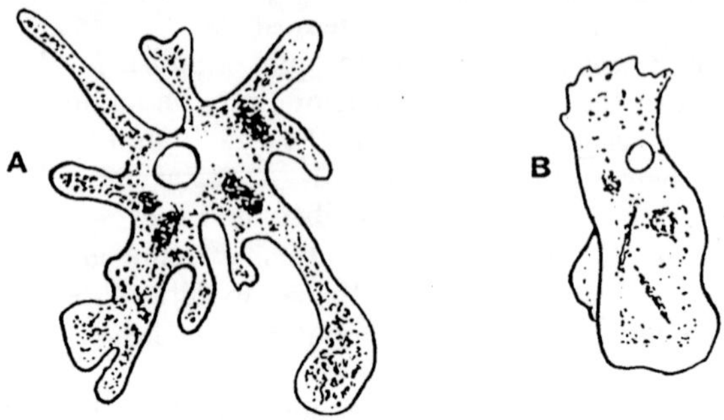

Fig. 33.
A. Amœba proteus. B. Amœba limax.

a variety of names such as The Proteus, Protoplasm, Cytoplasm, Plasm Protoplast, etc.

About 1864 protoplasm was defined as: "A diaphanous semi-liquid viscous mass, extensible but not elastic, homogeneous, that is to say, without structure, without visible organization, endowed with irritability and contractility." That seems pretty definite and leaves little to be said further.

It also has been stated that thread-like reticulations could be made out in the Amœba's hyaline substance. This, however, adds little to our interest as to their mission or what their ultimate object may be. They rid the waters of some of the mineral and vegetable constituents and more rarely animal substances. Possibly they are useful in preparing a neutral pabulum in the waters suitable for the growth and environment of all these. Gliding aimlessly upon any kind of surfaces in the water, extending first a portion of their body here and then another, they make progression whichever way random seems to take them. In eating, their method often takes the simple one of flowing quietly round the object and enclosing it within their substance. The food may be unicellular alga spores, monads, diatoms, etc., these latter are quite a usual delicacy. Quite curious at times it is to see a very small Amœba endeavouring to encompass a long diatom like a Navicula or Synedra and finally to drag it along in its quest for more.

Some Amœba have been named which possess a fairly uniform habit of outline.

Amœba limax (Fig. 33B)

This has a long Slug-like form, and though it extends its pseudopodia erratically they do not depart from an oblong shape as others may do.

Dinamœba mirabilis (Leidy) (Fig. 34)

These pseudopodia have papillæ-like little pimples projecting at their posterior extremities.

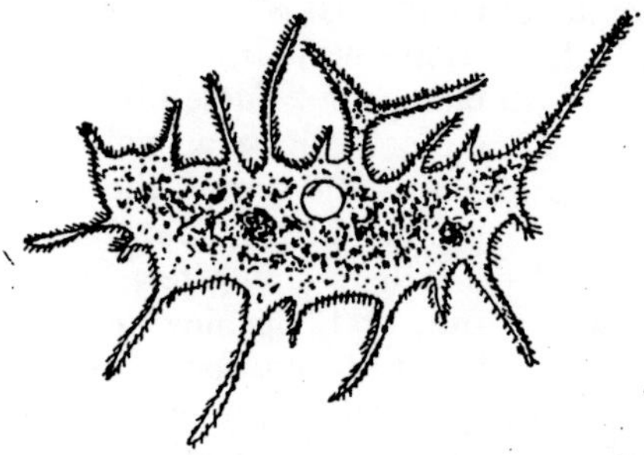

Fig. 34.—Dinamœba mirabilis.

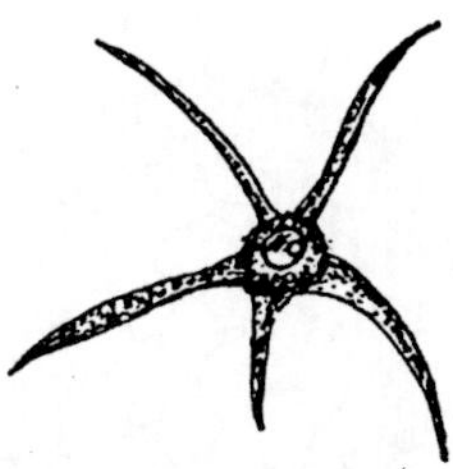

Fig. 35.—Amœba radiosa.

Amœba radiosa (Fig. 35)

Has a more or less rigid and ray like pseudopodia. Neither extended nor withdrawn save at very slowly recurring intervals. Known as radiate pseudopodia. All possess a pulsating vacuole and one or more nuclei.

Amœba may be looked upon as the base from which Rhizopods have elevated themselves sufficiently to form a covering, a house of sand grains, debris and small particles which they pass through their bodies, leaving it in the process to adhere to the outer surface and thus to protect the elementary primitive substance. They have risen in the scale of organization thus far. Amœba reproduce themselves by elongating the vital nucleus spot, dividing it in two, the portion of plasm upon each side going with each division and forming new bodies.

If a wire be passed through an Amœba it has viscosity enough to keep together and to flow round it so that no opening is presented, no nerves to harm, and so continues upon its sober way unimpeded and uninjured. Amœba have been divided into many parts without any nuclei and have still continued to " carry on."

The name Amœba is after a fabled god of that title whose power was such he could change his nature into either animal, vegetable, or mineral, a proceeding calculated to inspire awe in the early minds no doubt. In the present day he would soon be relegated by his " union " and obliged to consider more lawful habits.

Euglypha

The test of this genus is chitinous and transparent. Ovoid in shape generally, cut off sharply at the oral opening which possesses a toothed margin to it, in all of the many species spines are frequently present at the rounded bulbous end. Sometimes a few occur near to the lateral or mouth end. Its shell is composed of many interlacing plates which overlap one another like the tiles on a roof. These plates may be either oval, hexagonal, or round.

In some species the spines are produced over the entire surface. in others only a few either near the oral opening or at the widened portion. In another species they are entirely absent. The marginal teeth at the mouth distinguishes the species, often being finely dentated in themselves, giving a saw-like edge to their outline. The bulbous end is often spoken of as the " fundus." Average size is about 1/250th inch long.

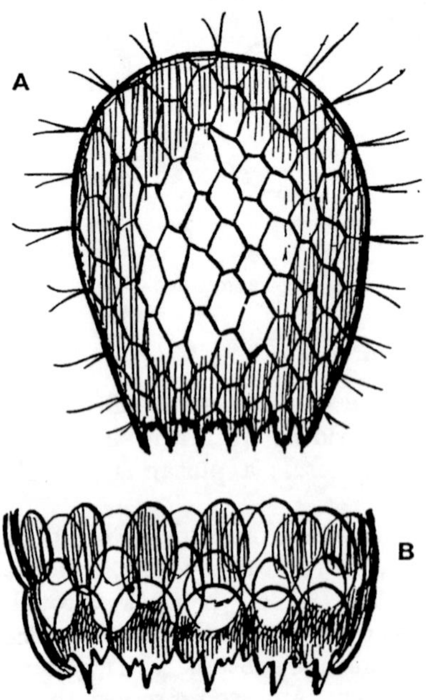

Fig. 36.—Euglypha strigosa.
A. Complete " Test."
B. Notched spines at opening (enlarged).

Euglypha strigosa (Fig. 36)

This is broadly pear-shaped, slightly compressed, being elliptical in its transverse section, although the oral opening is circular. The mouth has about twelve serrated thick scales around its margin. It has a fairly large nucleus containing small nucleoli. Around the shell many fine spines occur either singly or in pairs all over or occasionally from the "fundus" alone. This species has the notched spines around the oral opening intermediate in character between those of the shorter ones in *E. ciliata*, and the extra long ones in *E. compressa*. There are about six indentations to each spinal plate which curve inwards slightly, giving a blunter appearance to them, side view, than their reality warrants when seen end on.

The species is frequent among Sphagnum, and with the exception perhaps of *E. ciliata*, which exceeds it in numbers in some localities, is the most widely distributed. Its length is about 1/450th inch.

Euglypha ciliata (Fig. 37)

The "test" of this is more compressed than *E. strigosa* and of an elongated oval in shape. The plates forming the shell may be either oval or round and overlap each other. The spines are produced in fine straight projections, like long cilia, hence the name, over the entire surface of the shell. In some few specimens they are in line near to and around the oral opening, stouter and shorter then, and absent on the rest of it. It is a small specimen measuring but the 1/500th of an inch long or even less. It is, like its type, fond of the fronds of Sphagnum, among which it wi l sometimes be found in great numbers. It has two contractile vesicles, one each side the body near the middle outer margin. They are not easily observed on account of many food substances obscuring them, but in a fairly clear specimen can be made out with a ¼ inch objective. It rarely fills entirely the cavity of the shell, preferring the wider portion, to which it is attached at the apex.

Fig. 37. EUGLYPHA CILIATA.

The pseudopodia are fine and often spreading at their extremities into still more delicate branches. It usually presents the roof of its house to the observer, down the microscope, dragging it serenely along as it plys its protoplasmic feet, absorbing and ingesting food at the lower and open end of the shell. If the specimen is viewed between two cover glasses in an open cell so that the reverse side can be seen this forms probably the best method of getting a complete presentment of its outline. In that way the toothed denticulations are usually well seen.

Euglypha compressa (Fig. 38)

The outer cover in this " test " is composed of hexagonal plates of transparent chitin and do not overlap but border one another. The shell is much compressed, more so than *E. ciliata*. There are numerous spines which are spindle-shaped tapering towards a point at each end chiefly produced from the lateral border of the " test." The mouth aperture is lobed. The length is about 1/250th inch and is found among its genus chiefly in the Sphagnum pools and marshy waters.

Euglypha brachiata (Fig. 39)

The shell is elongated and cylindrical with but a slight convexity at the " fundus " end. The plates are oval and regularly overlapping. Its distinguishing feature beside the shape being attenuated more than in any other species of the Euglypha, it has

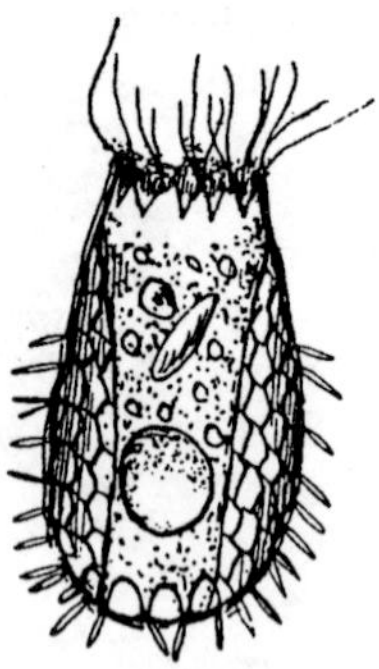

Fig. 38.
EUGLYPHA COMPRESSA.

Fig. 39.
EUGLYPHA BRACHIATA.

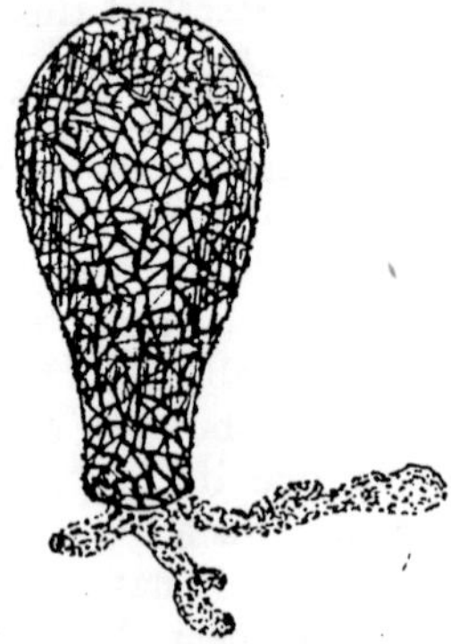

Fig. 40.
DIFFLUGIA PYRIFORMIS.

from four to six large and long curved spines which appear to be prolongations of some of the lateral plates. These spines invariably curve towards the fundus and are particularly conspicuous. Its habitat is similar to its family, among the clear upland pools in hilly places, and although found in many other localities the higher altitudes would seem to be more favourable to it, and one seldom comes away without a Euglypha from such places when hunting for Rhizopods. Its length is about 1/200th inch.

Difflugia pyriformis (Fig. 40)

This Rhizopod is also named by some authors as *D. oblonga*, considering this to be a prior name given to it by Ehrenberg. The

" test " as the shell is termed is pear-shaped with the crown arched and bulbous at that end, the sides tapering to a narrow cylindrical neck, having the opening across quite straight, forming its only entrance and exit. Through this the animal extends its pseudopodia and obtains its food.

The outer shell is covered with sand grains, diatom frustules, etc., which are taken in by the pseudopodia, passed through the body, and deposited from within upon the outer surface, not as is sometimes supposed by the Rhizopod simply rolling its body amongst such in the ooze which forms its home generally. By this means it will be seen that the grains, etc., have a mucous coat stretched around them as they are pushed to the surface, and as this is waterproof and becomes set firmly, so they become fixed and impervious, withstanding a good deal of rough usage in their existence and the adverse conditions of rapidly rising currents in times of flood transits.

The species is one of the commonest to be met with in our ponds and streams, but the forms it adopts are exceedingly variable. Some are very long, as much as the 1/50th inch, and visible to the eye unaided, others have short spines at their bulbous ends, etc.

D. pyriformis, like Vampyrella, is fond of the glutinous cells of Spirogyra when it can obtain them. The pseudopodia are rounded and thick-looking, projecting but a short distance from the tube opening. When attacking cells of Spirogyra it will perforate the sides, not breaking the joints like the Vampyrella, and boldly diffuse its pseudopodia within, draw the chlorophyll in a heap, and remove it into its body with a clean sweep *en bloc*. There is no suction action to be observed as with Vampyrella, nor finessing for any particular method in attack. Moreover, it does not appear to have that selective power its neighbour possesses. Its length of test averages the 1/125th of an inch and is often to be found among aquatic plant stems and oscillatoria filaments as well as in marshy places and quiet, shallow streams. As its food is chiefly chlorophyll its body seen through the " test " will often have a green appearance, yet its pseudopodia are invariably colourless.

Difflugia corona (Figs. 41 and 45)

The " test " is composed of large smooth sand grains which have a regular outline usually. Its aperture has several toothed processes around the circle, at least twelve, may be more occasionally, set evenly upon the margin. Its pseudopodia are fairly numerous and large with blunt, rounded ends.

To see the scallops or teeth as they are called is not particularly difficult if the cell in which the animal is viewed is tapped or tilted, giving the organism a roll over, but unless specially looked for the animal naturally has its mouth down, turned away from the observer, as the procuring of its food is its primary mission, quite unaware of a large magnified eye peering at him in his perambulations. It has a solitary nucleus, and upon the bulbous posterior end are from six to nine spines standing out covered like the rest of its body by the smooth sand grains. Found among the muds of pond water. Length about 1/125th inch.

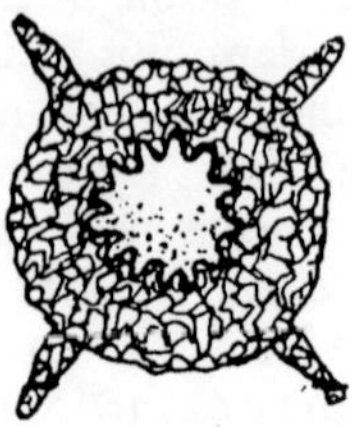

Fig. 41.
DIFFLUGIA CORONA.
(Ventral view.)

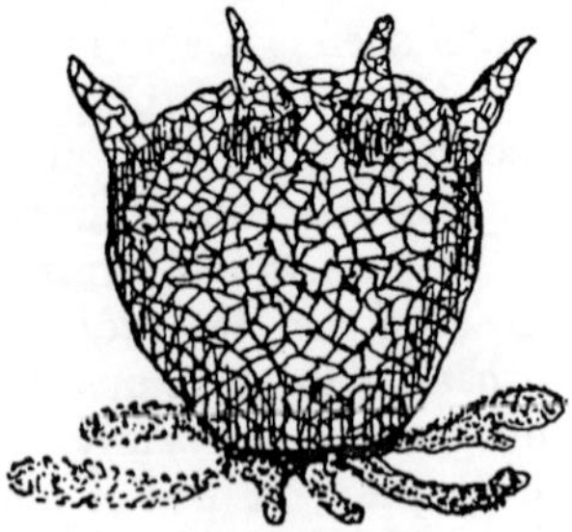

Fig. 45.
DIFFLUGIA CORONA.
(Side view.)

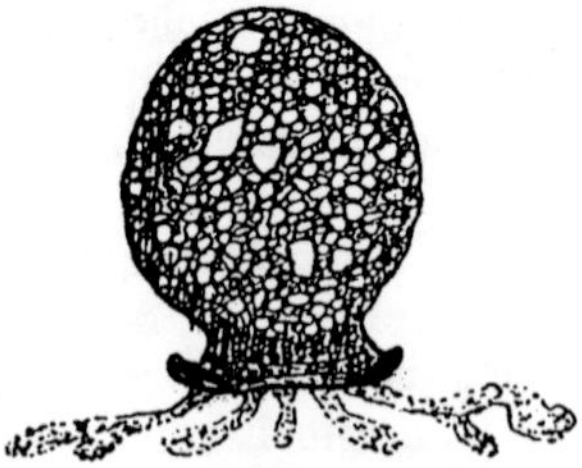

Fig. 43.
DIFFLUGIA URCEOLATA.

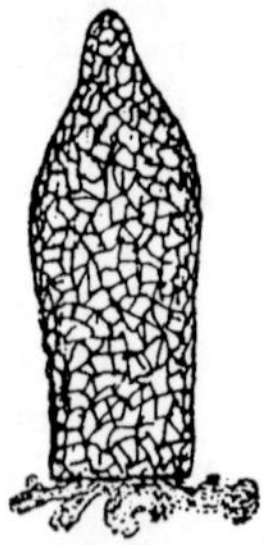

Fig. 44.
DIFFLUGIA ACUMINATA.

Difflugia urceolata (Fig. 43)

The " test " is globular with a deep attenuated constriction or neck and the mouth margin always recurved. It is without spines, but, as in other species, is covered with sand grains upon its outer coat, giving it the usual tessellated rough appearance.

In a variety of this species, *D. olla*, there are a few blunt, short spines developed at the posterior end, otherwise it is in other respects similar to *D. urceolata*. There are many pseudopodia, and also numerous small nuclei. Its outline is very distinct among the species and is found in the ooze of pond waters, etc., sometimes in large quantity.

A particular feature noticeable in observing it under the microscope is that the protoplasm always seems insufficient to fill the shell, evidently preferring a large roomy home to live in to the closely packed tenements of its neighbours. The largest forms of the species attain a length of 1/70th inch.

Difflugia acuminata (Fig. 44)

The "test" is narrowly pear-shaped, the slightly broader extremity tapering to a conical projection or even sharp point. Sometimes this ends in a knob-like process. The mouth is smooth, circular, and without teeth or lobes. It is covered with sand grains and fairly evenly distributed along with other species of the genus. Its length in full mature forms may attain the 1/100th of an inch.

Arcella vulgaris (Figs. 46 and 47)

A circular and flat Rhizopod with but a slight hump on the upper side and a small circular central opening on the under, through which it extends its pseudopodia whilst moving. Its colour is a

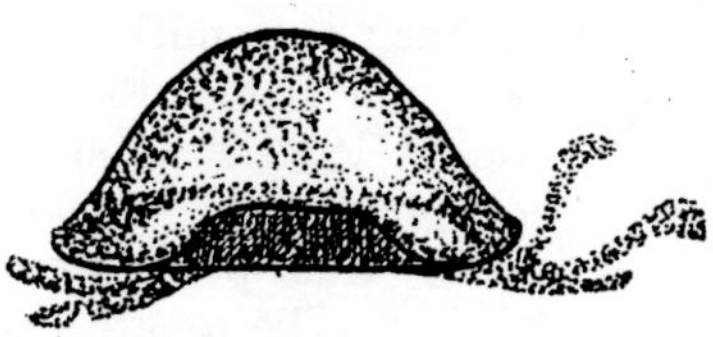

Fig. 46.
ARCELLA VULGARIS.
(Side view.)

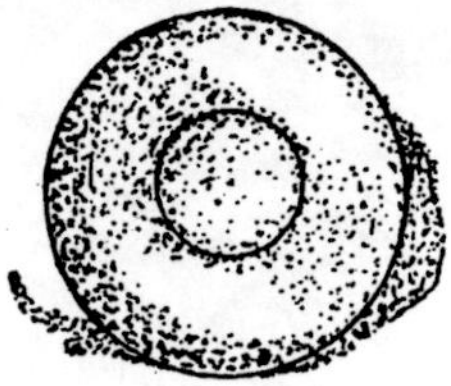

Fig. 47.
ARCELLA VULGARIS.
(Dorsal view) with central aperture showing through shell.

yellowy brown, the stronger the light impinged upon it naturally giving it a paler tint. When observing it gliding about, which it does but very slowly, one may see around its edge the wrinkled protoplasm which it protrudes and by which it carries itself, house and all along, in the usual manner of an amœboid, first extending a portion of its substance here and retracting a portion there It is the resemblance to false feet that the name pseudopodia has been given to these.

Thus it wanders aimlessly about, imbibing into its substance, through the central aperture beneath, any tasty morsel its little home happens to pass over or chance upon, never by any semblance appearing to go out of its way to obtain anything in particular. Clambering over stones, roots and leaves, or among the muddy stratum along the bottom, it evidently finds a comfortable existence, leaving us humans to wonder why he never desires to hurry through life. In the old saying, "Don't hurry and don't worry," we have here an ocular example to remind us of our own mad rush to get—nowhere. Within its structure contractile vesicles can be observed dilating and closing up, and in this way it presses and circulates, aerates and digests its food. It measures from 1/250th inch to the 1/500th inch in diameter.

Phryganella acropodia (Fig. 48)

The "test" of this genus is not unlike *Difflugia globulosa*, but is yellowish and semi-transparent and covered with irregular plates or scales having generally many foreign elements, such as large (comparatively) sand grains, diatoms, etc., giving it a rough appearance, and the pseudopodia are often ribbon-like masses which expand to finer points at their extremities. These are prolonged into variable radial lengths around the circumference. It is an active Rhizopod in its movements. The mouth or oral aperture is circular and wide, extending about two-thirds of the area across, and has a mucous pabulum usually, surrounding, of an apparently sticky nature which holds a good deal of foreign substances, large sand grains, etc., adhering to it. This may be of some protective nature to the organism. It contains a single nucleus in its transparent substance and many chlorophyll

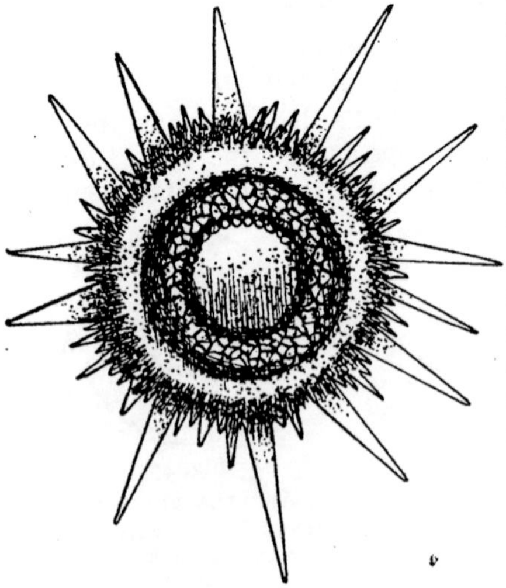

Fig. 48.—PHRYGANELLA ACROPODIA. (Ventral view.)

granules. It is about 1/600th of an inch in diameter, and is found upon the ooze, especially on the moors and bogs where Sphagnum Moss is plentiful.

The genus Phryganella was established by Penard in 1902, and all are semi-globose or ovoid with a wide symmetrical oral aperture having the characteristic deposition of large sand grains, etc., around, and with the pseudopodia issuing ultimately as finely divided points, not blunt as in *Difflugia globulosa* (Fig. 50).

Fig. 50.
DIFFLUGIA GLOBULOSA.
(Side view.)

Clathrulina elegans (Fig. 49)

This is a stalked Rhizopod and one of the most beautiful of all the class to which it belongs. Its envelope is chitinous, which is proven by its readily dissolving in sulphuric acid when heated, perforated by numerous large uneven openings, each separated by narrow bars somewhat polygonal with thickened raised borders, although the angles are always smoothly rounded. A peculiarity is the protoplasm forming the body within never entirely fills this envelope. The stalk or peduncle is hyaline and hollow with a base of root-like filaments by which it attaches itself to some support, often the fronds of Algæ or the rootlets of Lemna.

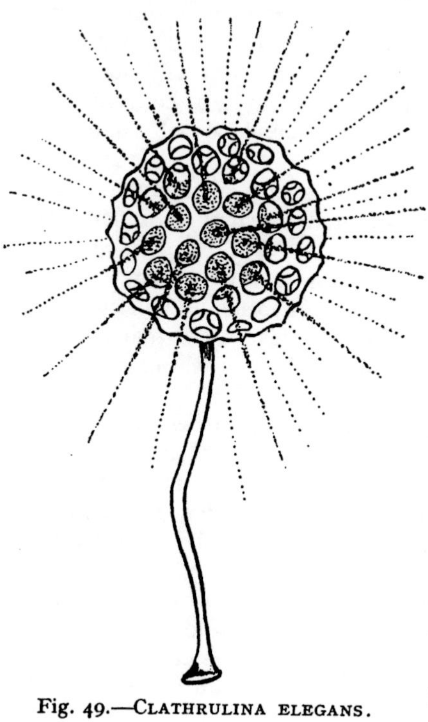

Fig. 49.—CLATHRULINA ELEGANS.

The globular head of the organism with its perforations is sometimes spoken of as the "capsule." Through the openings in the capsule it extends fine long pseudopodal rays, often two or three from a single orifice, with a common point of origin in search of its food. The pseudopodia are frequently forked and appear capable of arising from any part of the body within and devoid of any central axis.

The nucleus is single, but individuals may be found to have several if they happen to be in the process of division. There is one or more contractile vacuoles. When young the whole object is colourless, but as age advances may become yellowish or even a reddy brown.

Capsules are also found devoid of stalks which have become detached by a constriction that sometimes takes place near the junction of the stalk to the head. It then is free to move, which it does, by the aid of its pseudopodia used as feet, in a tumbling, rolling manner.

It multiplies in various ways, either by simple division of the body within into two or three, or commonly six or seven portions, the young emerging through one of the apertures in the capsule amœbæ-like. They then set about forming a stalk and a fine cover or pellicle which eventually becomes the new capsule, and they become mature with life like unto their kind. Or the embryo when out and free may encyst, in this method they simply form a spherical chitinous envelope around them, covering it with many fine projections as a protection and become dormant for a time. This has a central nucleus with the power later, at convenience, to start life again. Sometimes one such envelope will contain two cysts.

Formation by means of zoospores is also a common mode of reproduction so they are fully adapted to carry on under normal as well as adverse conditions.

Clathrulina may be found in groups of three or four together, one will attach its stem to the capsule of another and a still younger one erect itself upon the second, and in this way as many as seven or eight may be united, but independent, each enjoying their separate existence.

How the shell is formed so beautifully is rather problematical, but it has been suggested that the geometrical cells in position around the outer coat continue secreting chitin at their sutures, withdrawing the central substance to their sides, and eventually leave the opening free to the central mass of the body within to radiate pseudopodia through.

In the zoospore method of propagation a portion of the protoplasmic mass assumes a long ovoid form developing two flagella at its more pointed end and emerges from the capsule as a free-swimming zooid for a short time, similar to an infusorian. This roving existence ends in coming to rest on some object, often the capsule from whence it sprang developing a stem and finally a capsule of its own. The diameter of Clathrulina is about 1/350th inch.

Vampyrella

This is a naked plastic species characterized by a pinkish tint permeating its protoplasm, without a lorica, and capable of extending itself into extremely thin, narrow streaks of plasm with great variation. It is these mobile, very fine pseudopodia that also form an unmistakable feature in distinguishing it from Amœba when seen apart. It also has a noticeable habit of fusing separate individuals together and forming the general and larger protoplasmic mass or plasmodia. Its nucleus is unobservable, as in other genera, being diffused throughout its substance and capable of acting from any portion apparently. It has no contractile vesicle that can be recognized and the protoplasm is granular. It feeds especially upon chlorophyll extracted from alga cells and may be better outlined by describing the principal species of the genus :

Vampyrella lateritia (Fresen) (Fig. 51)

The body is normally spherical but capable of much variation in the active condition. The central mass presents the usual reddish permeating pigment. It is of an actinophrys-like shape, having many elongated filamentous pseudopodia interspersed with several thicker shorter ones, often knobbed at their extremity, contracting and expanding from its circumference. Frequently emitted from the spherical margin will also be observed lobed wave-like flowings, as shown in the illustration. Both nucleus and contractile vacuole are indistinguishable, though it is necessary both should be present in some way theoretically for its continuance.

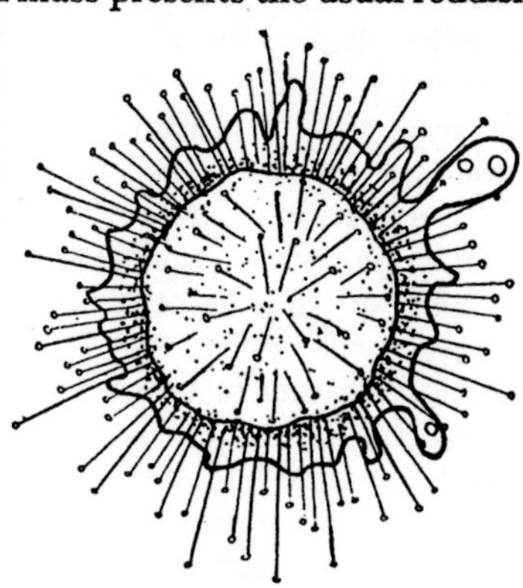

Fig. 51.
VAMPYRELLA LATERITIA.

Chlorophyll of the filamentous algæ forms its staple food, and its powers of discrimination is very pronounced, although it does not necessarily confine its attention to any particular species. It will glide through the water in the manner of an Actinophrys and even more quickly so, and upon finding an alga plant will begin investigating its outer cover, absorbing some of the superficial gelatinous substance travelling along its length, and upon arriving at a join between two cells will actually break this and pierce the cell in order to extract the contents. To watch this proceeding is most interesting and makes *Vampyrella lateritia* one of the most engaging species of its family.

Usually Amœboid animals take whatever food comes in their way, i.e. as it presents itself along their course, but this species, on the contrary, actually hunts for it and has quite a definite power of selection. Once a succulent thread of Spirogyra or similar alga is found it literally takes possession of it and seldom releases it until its full needs are satisfied. Like other plethoric animals, it then takes a nap into a resting state and will lie dormant for a considerable period.

In the method of securing the delectable chlorophyll with its encasing protoplasm, Vampyrella will make its attack in various manners according as it finds the hardness and difficulty in penetrating the cell walls or of the joints to be snapped. In the latter case it fixes itself upon a terminal cell by its longer filamentous pseudopodia, gathering them together upon one side into a heap and by a pressure or contraction difficult to observe or even understand in so apparently slender an organism the joint will give way and the plasmic fingers soon become visible within the cell, flowing forward gradually as the bands of chlorophyll are deftly, almost uncannily, absorbed and drawn into its body mass. It has been observed to sever joint after joint in this way and empty their entire contents. The broken cells were left separated and lying at right angles to one another and in varied positions after the depredations of the Vampyrella were completed. It was curiously observed that only the alternate cells were cleared of their chlorophyll in another instance, as if sufficient hold could not be obtained to exert pressure and break down the joint immediately next. It has been timed to break down the several bands of chlorophyll within a Spirogyra cell and empty it completely in five minutes, and in about thirty-five to have imbibed the contents of no less than seven cells before its appetite was sufficiently satisfied.

In a further observation it placed its pseudopodia athwart the join of two cells, pierced and assimilated the contents of both simultaneously.

Sometimes at the commencement of its resting stage, after a big meal, it will round itself in, withdrawing all pseudopodia, and encase its body within a transparent cover of its own secretion and form a cyst, attaching this to any filaments near, confervæ, lint, woollen threads, etc. Here it undergoes changes while apparently dormant, but when activity returns as many as three young Vampyrella may emerge with full vigour on a round of life akin to the species again. It is found in the ooze at the bottom of shallow pools or among confervæ and wet sphagnum and from the peaty bogs on moors. The muddy substratum at easy flowing bends in the course of hillside streams is also a favourite hunting ground.

CHAPTER V

ALGÆ

THESE are fresh-water plants which grow without roots, leaves or flowers and are found in greatest profusion where the water is still and temperate, and affording an unobstructed passage for light to penetrate. They are mostly of a delicate light green shade as seen in their habitat. Some are of a purple colour, as in Oscillatoria, and others during fruition period may assume a brownish tint. Provided there is a sufficiency of light, moisture with some species takes quite a secondary position.

Many algæ consist of only single isolated cells which, like the American steamboat, get along very well "wherever it is a little damp." The greatest number belong to the green chlorophyll secreting class and hence are styled the Chlorophyceæ. They absorb their food, which is both mineral and gaseous, by their whole surface. The oxygen necessary for their respiration; the carbon dioxide for assimilation, and the salts in solution being extracted from the surrounding water. These, aided by a plenteous supply of sunlight, supply all their simple requirements.

There are six or seven orders into which the class algæ are separated from each other based upon their differences in colour and the methods of their reproduction.

Taking first the blue-green algæ or Cyanophyceæ an example is

Nostoc commune (Fig. 52)

This is a thread-like filamentous chain of single cells, either rounded or slightly barrel-shaped, embedded within a slippery gelatinous envelope. When seen in quantity they form conspicuous clumps of a bluish green colour. They are not confined to a floating position solely, but may be found upon damp moss, stones, rocks, or wet ground.

On examining a portion under the microscope it will be found to consist of two kinds of cells, some large, some small, the larger being the cylindrical barrel-shaped spore cells and the smaller the more rounded vegetative cells. Added to these is a third cell, larger than either, apparently empty, without protoplasm.

These are the transparent "heterocysts" and characteristic of Nostoc and its near relation Anabæna. Their mission appears to be only for a breaking-off ground for the living granulated vegetative cells between to detach themselves, each piece being then capable of its own independent existence, escaping from the gelatinous mass surrounding and commencing new individual filaments and colonies of its own.

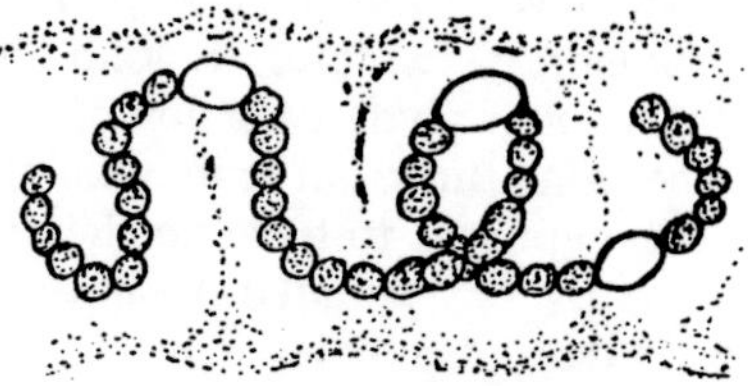

Fig. 52.—Nostoc commune.

This method of reproduction, however, is varied in more trying circumstances such as drought, when it will form what is known as resting spores direct from the vegetative cells. These swell out, accumulate more protoplasm, encase themselves with a thicker cell wall, remaining quiescent until water is obtained, when they begin to germinate and sprout seed—like a new filament from the parent spore, adding a gelatinous sheath, and continue life as before.

Nostoc, as indeed all the Cyanophyceæ, are among the lowest and most primitive members of the vegetable kingdom.

Anabæna (Fig. 53)

To briefly distinguish it from Nostoc is without the thick gelatinous envelope and generally straighter in outline, with a mere curve in place of the spiral coils of the latter. It possesses a thin outer covering and is freely floating. Its cells are rather longer than wide, with a more numerous quantity of transparent cells or heterocysts in its chain.

Fig. 53.—Anabæna.

Oscillatoria

This plant is composed of fine filaments sometimes several inches in length, and can be seen in summer and autumn rising to the surface from ponds and still waters in large clumps matted together. Temperature influences this to a great extent. As the cells become older they become emptier of granulated matters and fill with the earthy characteristic smelling gas which lessens their specific gravity and causes them to become buoyant. The threads are either green or purply in colour and divided across into narrow bands of cells. They are remarkable for their habit of swaying from side to side, hence the name Oscillatoria. They also accomplish slow creeping movements.

The newest and generally the rounded end has also another movement it executes, a kind of funnel-shaped arc or flourish, rotating on a pivot a portion of the filament, from a few cells down, at the same time that it may be gliding the rest of its length along.

Although said to have no sheath, it has a sarcode which forms a flexible overcoat similar in nature to the diatomaceæ by which it effects its movements.

Fig. 54.
OSCILLATORIA LIMAX.

Oscillatoria may be found in varied situations, an example is

Oscillatoria limax (Fig. 54)

This is often abundant in greenhouses on the wet soil, while

Oscillatoria prolifica (Fig. 55)

in some lakes and ponds occurs so profusely as to give a decided purply colour to the waters, and known as "Waterbloom."

Further, in regard to its oscillating progression, it may be noted, if a filament be obtained in which its endochrome is absent and its green colour lost it is quite usual to see this still moving about so that the endochrome has little or no influence over migration.

At the end of the threads may often be seen a short empty extension where a new cell is in course of construction. Beneath the outer coat or sarcode rows of very small spheres or spots, closely and evenly spaced, may be seen across the width and accordingly as oblique light is brought to bear upon them, lines running either longitudinally or in an X-wise manner may be made to appear. These are due to the slanting light, almost level with the spots joining them in a band and preventing the interspaces from being illuminated or seen. Within the cells themselves are small spores and granules. When a filament becomes old and worked out these little granules often appear very lively, dancing a reel of their own similar to the Brownian Movement.

Fig. 55.—OSCILLATORIA PROLIFICA.

A rather curious observation was made by the writer when

viewing a short green thread one evening. A white speck was attached on the outside and this was kept sharply and steadily in focus as it moved forward. It was noticed that when it glided from the bottom of the field to the top it always revolved spirally to the left. As it turned after a while and came down from the top to the bottom, same end first, the speck was carried to the right side. This showed the specimen in question had a uniform movement in one direction. A revolving helix to the left.

Tolypothrix (Fig. 56)

Is in general appearance similar to the Oscillatoriæ, but in method of reproduction resembles the Nostoc. It is composed of a tube of cells in single file, enclosed in a larger and transparent tube. Curiously the outer tube branches upon its own, but the inner tube of cells never join together at the branchings, although they grow

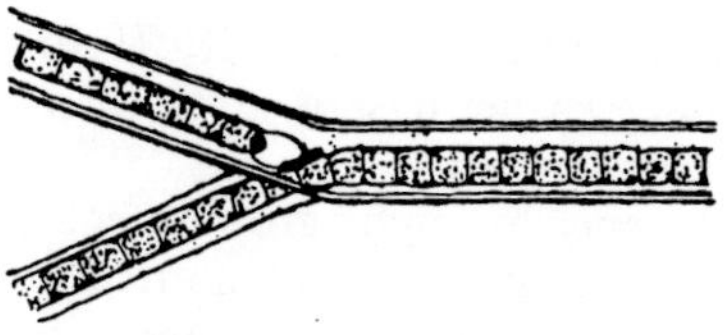

Fig. 56.—Tolypothrix.

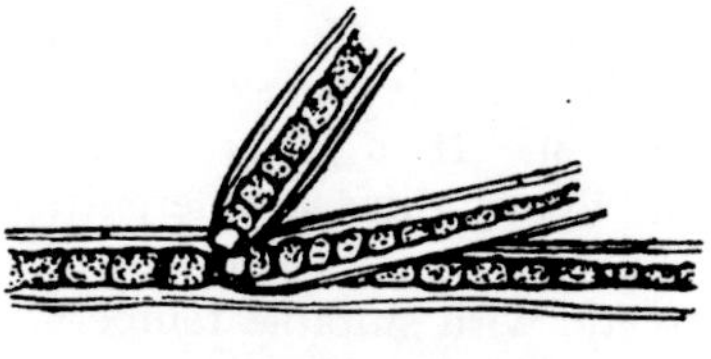

Fig. 57.—Rivularia.

within each branch. At the forks of the branches heterocysts from one to five in a row place themselves as defence, apparently a determined attempt to prevent any coalescing there.

The spore cells are slightly oval, and these in reproduction burst through the outer tube into the surrounding water, there to commence the new filaments. The plant is olive green in colour, and is found associated with Oscillatoria attached to water plants or the bottom of ponds.

Rivularia (Fig. 57)

This is more branched and tufty looking than Tolypothrix. It is found growing upon Chara, Myriophyllum, etc., and to stones in the swifter running waters of streams and waterfalls. The threads are club shaped in appearance, being thicker at their juncture or base and tapering to a thin point at the tip. As in Tolypothrix, there is a transparent cell which separates and guards any conjoining at the branch forks. There is a gelatinous substance connecting all the threads, and this forms a distinctive character when comparing other little tufts of alga. They are more definitely branched.

Pediastrum (Fig. 58)

This is a group of green chlorophyll-bearing cells arranged in a more or less circular, flattish plane and of one layer thick. The outer cells have two bluntly pointed projections from each which curve slightly towards each other and are always colourless. There is a transparent intercellular space usually seen between each of the inner cells. The inner cells here and there will be broken up and some of their contents changed in nearly all specimens. The surface of each is covered with small granules. They are angular, and the mode of reproduction is by free spores thrown off, together with the inner lining of the mother cell, in the water. These free spores, which possess two small flagella and a single nucleus, at this stage finally unite and settle down and a new colony is formed. The inner lining is used in forming the new covering or cœnobium for the whole.

The plant is free swimming and is at present classed with Hydrodictyon. It is fairly common among gatherings of algal plants, and about a dozen different species are known.

There seems no limit to which the species might grow to collectively, with suitable temperature and cultivation, but normally it is seldom very large, and on this account probably attracts but little attention or interest.

It may be noted the number of cells is usually some multiple of four when complete and which may be as many as 128 cells. This is due to repeated binary division of the endochrome of each cell.

In *P. boryanum* the diameter of these individually may be the 1/2000th inch, but generally they are less, as obtained from open waters.

Protococcus (Fig. 59)

This is very abundant at times in our gatherings. Some still pools are rendered quite thick with its green cells forming an emerald carpet upon the surface. It exists in a resting and a motile state. Its size is near 1/800th inch in diameter and even less. In the motile condition it has a transparent wall around its spherical or sometimes ovate body, and possesses two flagella. After a time it comes to rest and increases the size of its cell wall, dividing up into four complete individuals closely packed together. These continue growing, each having a nucleus and to one side a small red spot varying in size, sufficiently distinctive at times to give the aggregate cells a red appearance. Finally becoming too bulky for comfort, they burst their "happy home" and emerge fully

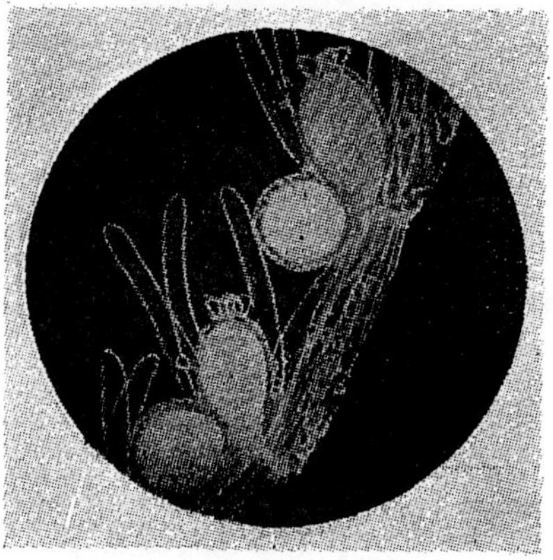

Fig. 26.—CHARA FRAGILIS.

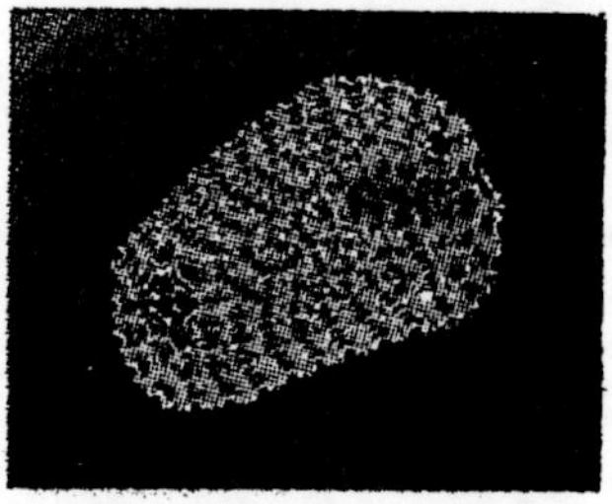

Fig. 58.—PEDIASTRUM.

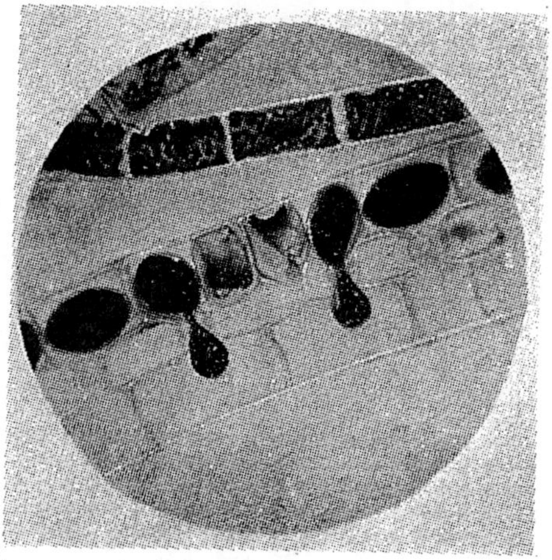

Photo by] [F.J.W.P.

Fig. 62.—SPIROGYRA IN CONJUGATION.

Fig. 64.—VAUCHERIA IN CONJUGATION.

[To face page 60

fledged with their "little increase," each furnished like their parent with two small flagella. They are then simple motile cells, free to roam and forage for themselves and to repeat with good fortune the same process over again later.

To calculate what it means in numbers, as each individual may add three more to the already countless mass, leaves small wonder at the rapidity and extent to which the numerous progeny of these minute bodies can accumulate. As Lord Dundreary says, it is prodigious!

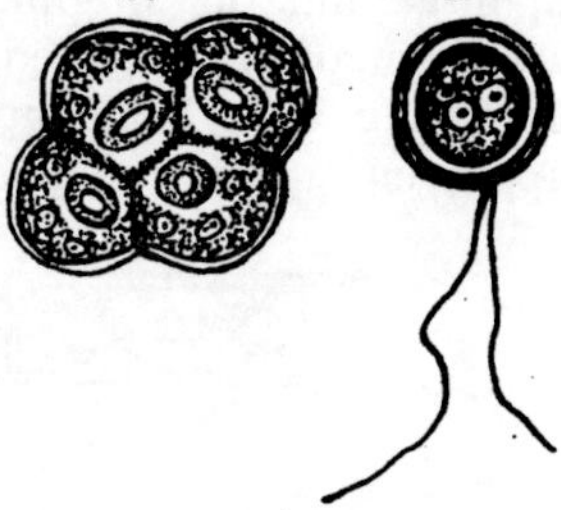

Fig. 59.—PROTOCOCCUS.
A. Resting cells.
B. Motile cell.

Protococcus has received from time to time many different names, according to its particular colour, form and the conditions under which it happens to be growing. It would need powerful microscopes to distinguish all these distinctive features.

The green covering upon the trunks of large trees is *Protococcus pluvialis*, and the soil beneath after rain is often covered with its green. Upon damp walls, fences, rockeries, etc., it can live in profusion and will withstand the drought until the rains return, then to active life, will commence again, extending by simple division and spreading in each direction, forming its green layers of minute cells until a wide area may be covered. The plant is variously called *P. viridis*, and *P. vulgaris*.

Protococcus is one of Nature's charming tints she uses to suppress those rude, colour-splashed, white-washed walls, towering gasometers, and the like, that often flaunt the peaceful countryside and so strives to tone down and harmonize with the rest of her agreeable handiwork such abrupt fabrics of man.

Zygnema insigne (Fig. 60)

This alga is found in long cylindrical filaments or threads floating near the surface of the water unattached. It is divided transversely into oblong cells rather wide and short. Within these are situated two dark green star-like chromatophores, differing slightly in shape in the various genera, enclosing one circular pyrenoid in each.

In the specimen selected the chromatophores fill nearly the whole of the space within the cell, but in some, where the length of the cell is greater, the chromatophores and their pyrenoids are surrounded by masses of transparent protoplasm.

The spores are produced about the month of April or in early summer. When a pair of cells are about to commence conjugation, which is the principal method of multiplication in Zygnema, they first become filled with green chlorophyll, embedding the pyrenoids until they are scarcely visible, finally fusing all together into one homogeneous mass provided with its own plastic cell wall, free from the inner oblong one entirely. This is the preliminary half towards the complete Zygospore from which only one individual upon germination originates.

Germination is effected in an interesting manner by a small tube beginning as a bulge upon one side of the containing cell wall appearing. This lying alongside another filament of the species

Fig. 60.—ZYGNEMA INSIGNE.

induces a similar little bulge to protrude from the side of it and eventually to meet and grow together so that a passage across is formed, producing a ladder-like resemblance termed " scalariform " with a hollow rung connecting both threads. The contents of each cell then move towards each other in the channel and effect a complete union of their mass, gradually taking on a rounded or oval shape, providing a firm outer coat to it, and as they age lose their green colouring and become usually of a dark brown. This is the finished Zygospore ready to pass through a resting stage and to burst forth on the following spring or as soon as the temperature is sufficient to call it into activity.

This in brief is the appearance conjugation has under the microscope, the why and the wherefore of affinity or the rationale for the procedure is not known.

The gelatinous nature of the outer coat keeps filaments always in touch with one another and their growth extends them alongside, but what induces the formation of the tube and its reciprocal, with another cell in readiness, or on the same mission, must form one of those points the writer has left purposely for the student of nature to fasten his enthusiasm upon and if possible elucidate. It may add zest to his hobby and labours, and if success crowns his efforts he may also have satisfaction in that it has been indicated to him as a worthy problem for solution. Various other forms of propagation are known, but conjugation is the most common.

The cell in which the zygospores are formed is known as the

female cell and the emptied one as the male cell. It is noticeable that the tube extended from the female cell is generally shorter and thicker than that from the male cell. The female cell after conjugation is frequently much smaller, indicating a fuller use of all the available plasm and nourishment within and greater energy used in contracting its bulk around the new-formed embryo spore. The cells of one filament commonly occur as all of one sex.

Hydrodictyon (Fig. 61)

The cells forming this interesting plant, commonly known as "Water Net," are arranged in a mesh or net which may become 10 or 12 inches in length individually and of great mass at times.

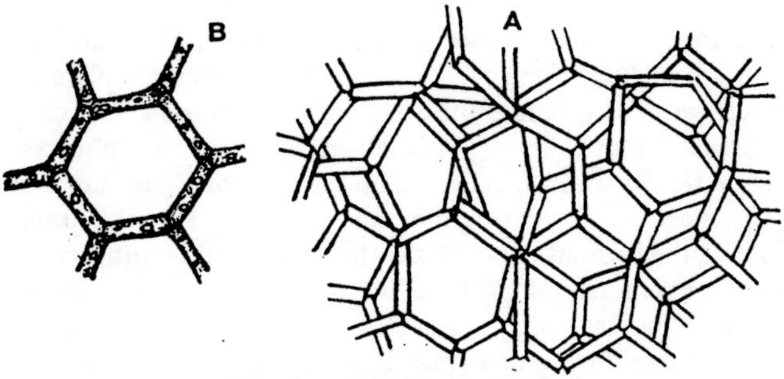

Fig. 61.—HYDRODICTYON.
A. General view. B. Single mesh (enlarged).

Each mesh is composed of five or six cylindrically shaped cells. Within each cell, attached to the inner walls, are several coloured bodies or chromatophores and numerous highly refractive granules.

There is a single nucleus in the first instance which later divides into many parts and the cell becomes multinucleate.

The new nets are all formed while within the cylindrical mother membrane and are for some time of this shape, later, however, they become broken and irregular. Environment has been shown to alter the methods of its reproduction, which generally occurs by the union of neutral four ciliated cells or gametes, which escape from the mother cells through an opening or pore at the side and finally coalesce in pairs for conjugation. After this has taken place a globose zygospore is formed which comes to rest for a time and then produces two or four fairly large double ciliated zoospores which assume various angular shapes.

The division of the contents of these polyhedral bodies and their apposition gives rise to the new net-like structure, each zoogonidia, as they are called, having one pyrenoid. The plant is found upon the surface of ponds open to sunlight, being a chlorophyll secreting organism, and can be seen as a thick light green coloured scum.

In taking specimens it is advisable to float them into the vessel while submerged and thus avoid destroying much of their delicate fabric in handling.

Swarming condition of the Zoogonidia may be accelerated or induced by the use of a reagent, consisting of a 100 cc. of 1 per cent solution of iridium chloride in 3 cc. of glacial acetic acid if necessary.

In the forming of new colonies the union of the Zoogonidia has an unfinished and rugged appearance which later becomes normal and smooth.

Spirogyra (Fig. 62, facing p. 60)

Is one of the most easily recognized and beautiful of all the algæ. Its green spiral bands are well known to all microscopists in pond life structures. The bands may consist of only one in a cell, as in *S. tenuissima*, or of three or four, as in *S. nitida*, or again as many as eight or ten in some species. The number of them has a bearing upon the species in distinguishing to which the specimen may belong. The chromatophore, as these plasmic bands with their several arrangements of pyrenoids and nuclei is called, is attached to the inner surface of each cell, i.e. parietal. The nuclei are situated amid radiating streamer-like threads of transparent protoplasm, also connected to the inner walls, and to see the plants at their best, dark ground illumination should be employed, then the brilliant green chlorophyll will stand out prominently with the many granular bodies and pyrenoids quite delightfully. Spirogyra is found in tangled masses often upon the surface of ponds, etc., and presents at such times wide patches of bright green to the eye, when it may readily be obtained and transferred direct to the wide-mouth jar for inspection. It prefers quiet waters. One of its chief attractions is its method of propagating by conjugation. This is similar to the Zygnemaceæ as described, the family to which Spirogyra also belongs, and being a rather larger species, can be better observed in its various stages if procured at the appropriate season. About June or July it is in full swing. The conjugation is ladder-like, the cell of one filament budding out upon one side and meeting or inducing another to appear upon a similar filament lying close by. On contact a short tube begins to form between the two.

It is noticeable in Spirogyra that the completed Zygospore is never found in this connecting rung or tube as is the case with

Zygnema sometimes, but always in one or other of the cells taking part. The illustration gives a good example, taken from a living specimen by the writer at the psychological moment of the intrusion of the united masses into one of the participating cells. The bands first begin to loosen themselves from the cell wall, contracting narrower and narrower until quite free. While this is going on the chlorophyll and its nuclear contents are becoming intermingled, forming an oval plasmic self-contained mass, the short bulge at the side is now lengthened, its tube widened and completed, and a passage is opened between the adjoining cells and their contents free to unite. This is the germination process. The result is a coalesced body which constricts itself and slips through to one side or the other, leaving one cell quite empty and the other containing the new body, the zygospore. This undergoes a resting period of several months until suitable temperature prevails and the season is ripe for it to burst out into new life and activity and produce a similar species to that from which it sprung.

In the process of division the nucleus is paramount and the tiny bodies embedded within the radiating globule of protoplasm are almost unseen themselves, measuring but the 1/4000th inch in diameter or even less, but though so small have a most important part to play in the life of all algæ, and are found present in every perfect cell, acting as a centre from which the vital processes always commence. In the clearer recognition of them acetic acid or other reagent is often used, changing them somewhat but making their outlines more visible. Unfortunately this kills the plant at the same time.

They are the more solid-looking portion of the transparent, radiating, protoplasmic mass. The cell ends of Spirogyra, it may be noted, are folded in, or "replicate," a point peculiar to the alga, but when the cells are separated this disappears on the release of the constriction or pressure and they are then seen as smooth, rounded, convex surfaces. This presents an elasticity in respect to their tubular composition and adds pliability. In Zygnema the divisions are straight cross sections.

The varieties of pyrenoids observed are almost as numerous as the plants themselves, some being spherical, some oval and flattened. Others appear like strings of beads, either touching close to each other in chains or separated and showing the thread-like cord intervening. Others are "staphylo," beaded and bunch-like, similar to grapes, others are seen as slender threads, or again they may be curved in worm-like convolutions, etc. In many cases they are very small and so recourse is needed in exacting diagnoses for the use of some chemical expedient to render them more easily observable.

Iodine is used to detect the starch granules around the pyrenoids and nuclei as this gives a purply colour to such substance and is of material assistance by the contrast it makes in picking them out distinctly from their surroundings.

Spirogyra often breaks off as short threads, which increase themselves by self-division. These continue their separate existence and again subdivide, and in this way, with numbers of filaments doing the same thing, a mass is soon formed. This is another method and a frequent one in propagation and without resting spores or zygospores being produced.

It should be remembered that in Spirogyra and the algæ generally although classed with plants their threads are rather an accumulation of individual cells, complete, each one in themselves, growing end to end, any one of which may function upon its own. There is no distinction of apex and base, and the cells are alike throughout the filaments. The size of the largest cells in Spirogyra attain the 1/100th of an inch in length and are easy objects for microscopic investigation. The plants give off oxygen freely in the sunlight and are kept buoyant with bubbles of this gas near to the surface, often to wide areas in extent.

The pyrenoid bodies, well seen also in Desmids generally, are the proteid bodies situated within the bands and around which starch is formed when assimilation is taking place. Each has its own coating of starch granules. In self-division the nucleus is split up by an ingrowth of the outer wall, first appearing as a faint line around the cell, gradually proceeding towards the middle until a complete disc is formed dividing the nucleus into two daughter nuclei, a portion going to either side. When the union of two cells occurs side by side ladder-like it is termed diœcious. This is the most usual method in conjugation. It may, however, take place between two cells of the same filament chain-like and is then called monœcious.

After the formation of the zygospore and during the resting stage the starch gradually disappears and it becomes a dark brown coloured body, giving place to an oily carbonaceous material which more securely withstands the rigours of winter and adverse times.

Spirogyra belongs to the order of conjugatæ placed there from its method of propagating. It was among the Spirogyra that conjugation was first observed, and it may be regarded as the primitive form of multiplication. There are about two dozen species of the genus. *S. gracilis* is the smallest, with a thickness of 10 μ or

1/2500th of an inch, while *S. crassa* is the largest, its maximum thickness being measured as 150 μ or about 1/165th inch. It has a gelatinous envelope in all the species.

Sirogonium sticticum (Fig. 63)

This is a similar, isolated, species to Spirogyra and is now called *Choaspis stictica.* It was formerly included in a genus to itself called Sirogonium, but it is doubtful whether it should be separated from its larger family. Its chief difference being in the arrangement of the chlorophyll bands which are parallel to each other instead of spiral and in the cells being of unequal sizes, and also the absence of an external mucous coat. It is the only conjugate which is not slippery to the touch. There was but the one species included in

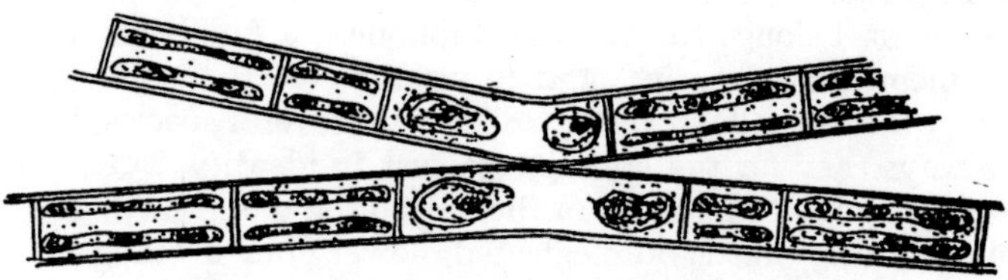

Fig. 63.—CHOASPIS STICTICA (SIROGONIUM).

the genus, the method of producing its spores differing from all others owing to the inequality of the size of the cells. The shorter cell invariably injecting its contents into the larger or female cell where the zygospore was developed. When such a union was effected there was formed a knee-like bend at the point of contact; genuflection. These touch and blend, the walls burst, and without the intervention of a tube, as in other zygnemaceæ, the contents of each cell passed through the aperture.

The zygospore is elliptical and about 1/350th inch in length. The plant is found in moor pools and ditches, often attached to stones.

Vaucheria (Figs. 64, facing p. 60; and 65)

Is readily distinguished from Spirogyra in that it has no cell walls or separate partitions dividing it. It forms one long continuous tube, and only at reproduction periods does it develop a septum. This septum specially shuts off the region in which multiplication takes place.

Vaucheria is found either on the mud of shallow streams attached by a colourless rhizoid or a number of root-like branches which serve the purpose of holding it at anchor (it would not be correct to call them roots in the accepted meaning of the term as applied to plants generally), or floating lightly above, forming thick matted masses of dark green filaments. When a filament is separated it appears as one long hollow tube growing to great lengths and branching but rarely. Its protoplasmic contents are uniform and continuous, providing a thick layer lining the external wall and containing an immense number of spherical chlorophyll granules which give it its particular deep colour. Imbedded also in the protoplasm just inside the granular layer are always numerous minute nuclei, these nuclei are a common feature and characteristic of the non-cellular algæ especially, which divide and multiply as the plant grows. At the growing apical end of the filaments the protoplasm is quite transparent, and within the rhizoid rootlets it is the same.

Vaucheria belongs to the order Siphoneæ, a family having most of its members as marine, and forms the sole British fresh water genus of this unicellular order. There are several species, however, not always easy for the lay microscopist to identify, but the genus is readily recognized. Some of them occur on damp earth and often form irritating weeds in neglected flower-pots or tangled green webs troublesome to fern cultivators. Sometimes they are found attached to stones, etc., by their slender rhizoids which give them a firm hold.

The fresh-water forms though few, differ greatly in appearance and in their methods of reproduction one from another, as in Botrydium and Phyllosiphon.

Reproduction may occur in two ways in Vaucheria, either by zoospores, i.e. animal spores (on account of their being furnished with cilia all over and swim rapidly) or by oospores, i.e. egg spores or akinetes.

Though one of the simplest of algæ in most other respects, Vaucheria ranks high as regards its reproductive aspect. The filaments are usually in the sporangial stage early in the year and are not difficult to study. In the asexual or zoospore multiplication this goes on chiefly while the plant is growing and supplied with an ndance of water.

Spores are produced in a short portion set apart at the free end of one of the threads and partitioned off for the purpose. The protoplasmic contents of the tube become dark and denser and soon assumes a firm flask-shape outline. The wider bulbous end is then separated from the rest of the thread by a transverse division or septum which now encloses the single zoospore. Here it continues to grow and clothe itself with cilia all over, bulging the cell or zoosporangium out into a wide oval. The free end finally breaks and the spore squeezes its way out, the opening always being smaller than the zoospore and changing its form somewhat in the process. It sometimes occurs that the outer portion already covered with cilia, like the rest, tries to rotate before the whole spore is exuded, twisting a portion away, and then two zoospores are formed instead of one. The cilia are in pairs around the zoospore and each pair is connected to a nucleus, it becomes in reality, therefore, a host of such, unseparated from each other.

Once out and free to swim, the spores become very active. The escape of them generally takes place in the early and cooler part of the day, evidently preferring a temperate clime during this portion of their mission. They are not long before coming to rest, however, perhaps half an hour being the normal limit, then the cilia are first to be withdrawn and the zoospore begins germinating, sending out a thread into the soil or mud, anchoring it by its tender-growing rhizoids, soon following by developing a branch, and life is in full swing once more like its parent.

This kind of reproduction is quite all right when the environments of water and weather are favourable, but with adverse times, drought and other circumstances, the alternative method of oospores is resorted to. In this two sexes, or rather sex organs, arise as offshoots upon the tubular filament. Sometimes the two organs are formed at the end of a short stalk or tubular pedicel, the oogonium being the terminal sac and the antheridium (equivalent of male) situated to one side of it immediately below. The antheridium begins as a short and narrow tubular vegetative branch containing a large amount of protoplasm, spore granules, chlorophyll and numerous small nuclei. As it grows the tube usually becomes curved, and the extremity of it is divided off by a transverse wall or septum. This forms the true antheridium. The tiny nuclei congregate towards the centre of the cell, leaving the coloured bodies and some of the protoplasm attached to the inner wall. The nuclei then break up into still more important bodies, the spermatozoids. These consist in part of nuclei and of the surrounding protoplasm and are provided with two exceedingly fine cilia, one each end.

The spermatozoids are spindle-shaped bodies, sometimes called also " antherozoids." These then are liberated from the extremity through a small opening into the water and reach openings eventually in the oogonium close by. The oogonium has assumed meanwhile a spherical form, the protoplasm developing a quantity of oil as its carbonaceous content, with many chloroplasts or pyrenoids and a number of rather large nuclei.

It is noteworthy that Vaucheria does not form starch. After a while a kind of beak is formed on the side of the oogonium and an interesting stage is now reached. From the point of this beak the protoplasm begins to withdraw, taking with it all the tiny nuclei except one, most strange to say, leaving the emptied portion as a uninucleated cell. This is cut off by a transverse wall from the remainder of the filament and becomes the true oogonium. The nuclei grows perceptibly larger and the contents adjust themselves to leave the outermost beak portion clear and free from chloroplasts. This is called then the "receptive spot." The nucleus lying near the centre of the cell is attached to this spot by a strap or band of protoplasm. The outer wall opens at the point of the beak, expelling a portion of the colourless protoplasm as an affinity or inducement to the antherozoids which swarm in at the aperture. Only one fuses, however, out of their number with the ovum, and fertilization then is accomplished. The fertilized ovum surrounds itself with a thick wall and passes into the resting stage similar to the zygospores in Spirogyra. In this condition it can endure periods of cold and drought with safety until suitable seasons permit it to germinate into a new Vaucheria plant and continue existence in a similar manner to its parent.

There is no regularity which method of multiplication make take place; this seemingly depends entirely upon the external conditions obtaining or necessitating for the continuance of the plant's survival of its place among the fittest so to do. In other abnormal situations the plant has also a third method for its safe existence should it be left stranded high and dry on the mud and the water subsides, it will divide its protoplasm into a number of portions and form individual cells surrounding these with a thick wall and so pass the time until better conditions arrive.

Before fertilization of the oogonia has taken place it is green, like other parts of the plant, but after, it changes to a brown with a red spot near the centre until germination commences. When this begins it rapidly returns to green once more.

One antheridium supplies two or sometimes more oogonia with antherozoids (Fig. 65).

Vaucheria has a common Rotifer which builds its nest or gall within the hollow stem. It is Proales Werneckii. At different places along the tube conspicuous oval sacs are produced by bulging out the sides, and in these are laid a large number of eggs, as many as one hundred being counted in one of them. The Rotifer, in fact, buries himself, until hardly visible, with the quantity surrounding him, laying almost motionless in the centre of the sac. He gives a darker brown appearance there, however, and with care can be seen the motion of this colour within his body. If it is near the hatching time of the eggs an interesting event may be witnessed. As the sac is transparent and the egg shells, or coats, are the same, the proceedings within are easily seen. The little embryo Rotifers may be observed turning and writhing about inside, their bright pink eyespot lending focus to the other portions of their structure. Presently one of them will burst its shell, then another, and in a weak and rather languid manner one will begin his meanderings round and between the other unhatched eggs still within the gall and its covering. The attachment of the gall to the main stem is narrow but hollow, and by applying his wheels of cilia and his toes, eventually caterpillar-like he hunches his way down through this outlet into the wider freedom of the tube. Here the lining is covered with delectable protoplasmic and agreeable gelatinous substances and at once knows what to do with it. Returning to the gall again another may have burst, not only his own shell but also an opening through the gall cover, and will be found freely engaged swimming near in the surrounding water.

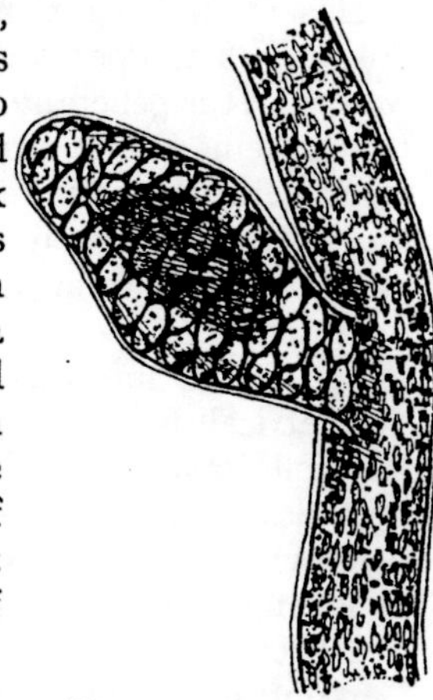

Fig. 65.—Vaucheria. With Rotifer gall.

The Rotifers that first gained access to the main stem eventually left it after many wanderings up and down, obtaining an outlet to their companions in the water by the concerted pressure of several together against an old and weak portion of the Vaucheria wall.

The use of an alga wall to form a nest for the Rotifer's offspring and thus overcoming the necessity for making a special gelatinous cover or cœnobium of its own is a step along the higher life from the usual and primitive road, and it would seem that however simple an organism may be there is a latent instinct to profit by experience, so that as time goes on heredity becomes an influence and guiding factor in the succession of the species.

These unexpected events happening in living microscopic life among the gatherings of the student lends additional interest and zest, and of all ways of seeing objects none is more generally satisfactory or more to the purpose than watching them in the medium and with the surroundings of their particular habitats as closely followed as possible. You then learn to know them and their ways more intimately.

Enteromorpha intestinalis (Figs. 66 and 67)

This at first sight, in situ, might be thought to be Vaucheria growing as it does in tufted masses upon the bottom of shallow waters. It is generally of a darker green, however, and upon lifting it out seems rather coarser and rougher to the hand. The sample from which this illustration was taken was found at the quiet bend in a small river where the water was quite clear. It looked like a rich green velvety mass beneath the water. When placed under the microscope it is seen to be a hollow tubular ulva without divisions or a septum anywhere. It is a mass of cells in many-shaped geometrical figures each having a thick cellulose wall around. The cells have three or four large spore cells within and many small granules, with one pyrenoid situated among the green chlorophyll. The pyrenoid is attached to the inner cell wall.

The method of multiplying is simple. One of the cells grows to a darker green by the extra increase of chlorophyll, and when at bursting point almost it then divides first into two and these into another two, making four, similar in method to Protococcus. A tube is now commenced and a bud is formed and continued first as a conical one-celled projection, then into a narrow tubular thread with a double row of cells. It does not increase in width yet but extends the growth lengthwise, becoming long coiled filaments which will hold to many points of an adjacent and suitable object, aquatic plant stems, algæ, or its own foster parent tubes. When so anchored against too rapid a current, etc., it continues to increase until fully mature. While this process is going on the older thread appears to possess bunches of root-like appendages to it along its length which are the younger filaments preparing for the species continuance. When the tubes attain full size they are flat and flaccid, like the intestines, hence the name, but the new growth is tubular and cylindrical. The sides of the flattened threads are not straight but irregular and limp so that upon lifting a single one upon the slide for viewing it falls flat with a tendency to adhere owing to its many new threads and its gelatinous nature.

In reproductive multiplication this is always formed in the growing vegetative cells either by sexual or asexual cells. The latter, termed zoospores, have four cilia and a coloured spot, usually red, when complete, and the sexual spores or gametes which unite to form a third body distinct from their primary

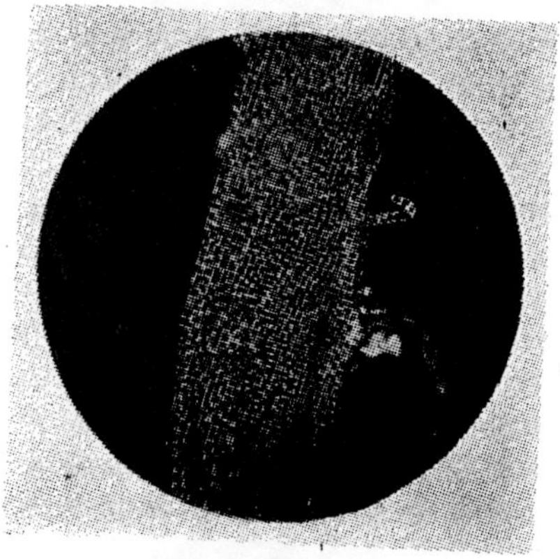

Fig. 66.—ENTEROMORPHA INTESTINALIS.

Fig. 67.—ENTEROMORPHA INTESTINALIS. With slender new growths as the older collapse and decay.

Fig. 68.—ARTHROSPIRA JENNERI.

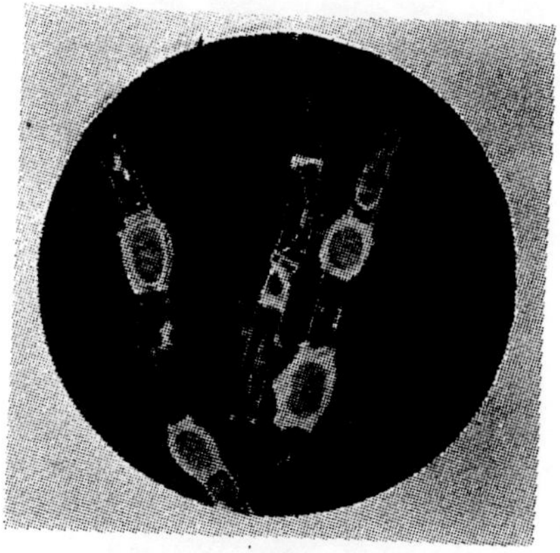

Fig. 74.—ŒDOGONIUM. Showing oogonia.

[*To face page* 72

individual selves have two cilia only. This is the winter and less frequently observed method of propagation. Fission or self-division is common in the warmer times of summer while growth of the alga is more prolific.

Arthrospira jenneri (Fig. 68)

Is a genus of blue-green algæ and may be readily distinguished by its spiral habit. The specimen from which the illustration is taken was found upon the dripping sides of an outlet or sluice from a large pond. It forms tufted slimy masses which are difficult to handle and quite as difficult to separate for mounting.

The coiled threads interlace themselves in a manner almost impossible to unravel, and occasionally two may be seen closely woven together coil for coil which gives them a rope-like appearance or a string of figure eights from top to bottom. They are not usually of great lengths, and the filaments are commonly without any sheath. Not unlike a spiral form of Oscillatoria and with which they are often associated.

The cells are numerous but would seem to be divided into thicker divisions about every two or three coils, in which smaller divisions less conspicuous are present. The cells are granulated, and one or many nuclei may be observed.

The genus differs from Spirulina, another of the blue-green algæ, by this numerous division into cells, whereas that form consists of single elongated cells of much greater length often found oscillating and rotating rapidly, which is not observed in Arthrospira. Its colour is of a dark green nature similar to many Oscillatoria, and the spirals are very regular and even in their outlines. It possesses the musty odour common to the Oscillatoria generally when taken from the water in mass, and may be looked for in any slowly running or stagnant waters where that genus is also frequent. It is capable of living singly, and an odd thread or two may sometimes be obtained in gatherings where Oscillatoria are taken in the net. Its width is about 1/1000th of an inch.

Mesocarpeæ (*Mougeotia*) (Figs. 69 and 70)

The representative of this genus, known as Mougeotia, is one of the commonest of the few species included, and except in the flattened lens-shape spore there is little difference, save size, to distinguish them one from another. The cells are much longer in proportion to their width, the single chromatophore is disposed as a flattened ribbon-like plate in which several pyrenoids are distributed along its length. The whole thread is delicate and much smaller than any of the conjugates previously described. The spore is usually formed of only a portion of the cell contents when conjugating. When the alga is in multiplication is the best time to observe the specific differences in the genera.

In Gonatonema, another species, this produces spores between two transverse membranes near the centre of the cell. The spore wall membrane being double. Its zygospore is unknown.

Gonatonema ventricosum in forming these aplanospores swell out the threads into a knee joint or bent appearance, giving the elongated cells a zigzag outline, having the spores at each of the angles. At such times there is little difficulty in distinguishing the species from Mougeotia.

Fig. 69.—MOUGEOTIA GRACILLINA.
(Chlorophyll layer seen from top.)

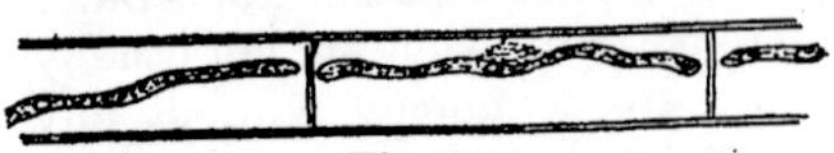

Fig. 70.
(Chlorophyll layer seen from edge.)

In generating aplanospores the entire protoplasm of the old cell ruptures its walls and develops a new and distinct cell wall around it. This process occurs frequently in the algæ when zoospores are formed. In other species the spore is produced by conjugating cells united by a tube and the zygospore developed within this channel. The spore is almost circular, of a greenish brown colour in the species, and of a bright green and oval in another.

Spore formation occurs about September as the autumn is approaching. The plant frequents ditches and slowly running streams.

Sphæroplea annulina (Figs. 71 and 72)

This is one of the Confervoid family and the only species of its genus consisting of thread-like cylindrical cells, unbranched, free-swimming, and more or less gelatinous. The cells are very long and tapering, averaging fifteen times their width. The chlorophyll is attached to the cell walls in streamer-like bands, leaving large and conspicuous vacuoles surrounding. There are numerous pyrenoids embedded in the chlorophyll.

When oospores are about to be formed some of the vegetative cells group themselves into rounded masses while others separate their contents and at the same time turn a reddish yellowy colour, finally breaking into elongated pear-shaped bodies furnished with two little lashes or cilia at the pointed end of each. During this division process small holes appear at the sides of the cell walls both in these producing the oogonia or female cells and those producing the antherozoids or male cells. Through these openings in some unaccountable way the antherozoids escape into the water and then find their path to the holes in which the oogonia are forming, with which they unite, using their cilia as the means of locomotion and as antennæ. This accomplished, the green colour of the oospores is soon changed to a bright brownish red and a gelatinous hyaline cover produced, having a folded, furrowed appearance. The cell wall breaks up and the oospores are free and take on a resting period. At the end of this they split up into a number of small oval cells each possessing two tiny cilia; these are the final zoospores, and it is from them that the new plants propagate.

Fig. 71.—SPHÆROPLEA ANNULINA. (Young growth.)

In the preliminary stage of this growth the zoospores change to a spindle form, then the cilia disappear and the red pigment turns to green chlorophyll arranged into the streamer-like bands as in their parent cells. As the tapering cells grow the points swell out to the same thickness as the main stem, and the length of the filaments is increased by the repeated division of the cells.

Fig. 72.—SPHÆROPLEA ANNULINA. In full growth (points widened out).

Sphæroplea annulina is rather scarce in Britain but is often found in the pools of quarries and pits in other parts of Europe. On that account it should be treasured as a prize worthy of study in its very interesting life history whenever it comes within the " net " of the micro-fisher.

Œdogonium (Figs. 73 and 74, facing p. 72)

This genus is composed of a great many species, and, if the water is shallow and exposed to the rays of the sun, may usually be found in ponds and streams or where the current is slow. In cisterns and drinking tanks used by cattle in the open it is also frequently obtained. It is thread-like, and the filaments are barely distinguished by the unaided sight. Its habitat is attached to stones, water-plants and submerged objects by a flat disc-shaped foot in its early stages. When adult, most of the species float freely on the surface.

The main outline consists of a single row of cylindrical cells, some of which exhibit transverse striation at their junction. These striations are produced by the insertion of a new cell wall piece, as the mother cell divides in propagating and the daughter cells continue to grow. The cells near the attached end are generally clearer than the others and more free of chlorophyll.

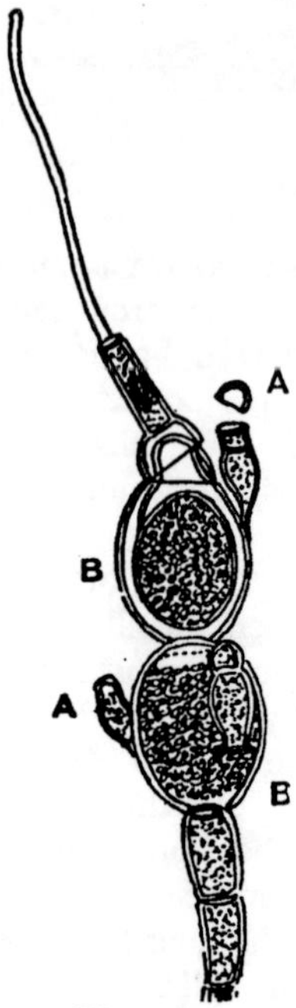

Fig. 73.
ŒDOGONIUM (after Pringsheim).
A. Male cell.
B. Female cell.

The tip of each filament is not continued as a growing point, but has a long transparent hair-like cell which does not divide, and appears to be but an ornamental finish to the growth.

The plant may have its reproductive cells on the one stem, or they may be female on the one and male on another. When the latter case obtains the filaments bearing antheridia (male cells) may be normal filaments, or tiny growths of separate cells called "Dwarf males" which are attached near the female cells (or oogonia). The dwarf cells originate from special development of the spores called androspores. Thus the life history is made complicate to follow in detail, but there are two distinct methods of multiplication, one of which is asexual, that is propagation without a fertilized cell, and the other sexual, or with egg cells and spermatozoids being formed. The asexual serves the purpose for rapid growth during the summer or when conditions are especially favourable, and the sexual method in producing resting spores to weather the cold during winter, to become rejuvenated into active life as the following spring approaches.

The chromatophore within the normal vegetative cells is sometimes linear or with irregular, net-like perforations, and contains several pyrenoids.

Along the filament's length various shaped cells will be observed, some large, bulbous and oval, others long and striated, some clearer and finely granular, others at times having the little "dwarf males" bent alongside the much-expanded oogonia or female cells.

Numerous zoospores with their circlet of cilia at the smaller and clearer apex may be seen swarming around the filaments, little caps or operculum-like lids floating near by, which have been thrown off as ripening and maturity is advanced, with many oogonia cells, coloured orange-red or brown in contrast to the green of the chlorophyll ones. The whole plant at such periods during multiplication being a busy hive of silent industry in the continuation of its species.

Bulbochæte intermedia (Fig. 75, facing p. 78)

forms the only other allied genus of the Œdogoniaceæ represented in the British Isles. It differs but little from Œdogonium except in having branched instead of single filaments.

Bulbochæte intermedia is a further example of the Confervoid family, but almost any simple filamentous algæ is spoken of as Conferoid impartially owing to them having been included by Linnæus in the old genus of Conferva. This has been so divided and subdivided, however, since his day as now to be without any distinctive meaning, and as species have become better understood they have been allocated into more definite classes.

Bulbochæte is not so common as Œdogonium but can occasionally be found attached to Myriophyllum growing profusely. The cells bear long transparent hairs or bristly hollow spines growing from their apex and always so from the growing cell at the tip of the branch. Sometimes two will be seen there. Not every cell in the filament and branches have them, but so frequently do they occur that the specimen will have them interlaced and extended about it so profusely, they form almost the first view of the plant, and in preparing for the microscope life-slide; need attention the soonest to straighten their tangled meshes out.

Their length is exceptional, comparatively to the average size of the vegetative cells, and their use is problematical, taking no part either in reproduction or division. Sometimes they will exceed an inch or more in extent. The appearance on an aquatic plant is one of a downy and feathery semblance, and if this is observed a portion should be secured and not improbably will turn out to be Bulbochæte.

Reproduction is similar to Œdogonium either by asexual or sexual propagation, and in the case of the latter the plant provides dwarf male attached cells very frequently so that it forms a good example for studying this interesting method of multiplication.

Broken branches may occasionally be found among other alga in a gathering, as it grows quite as well free as in its normal way attached.

The cell cases are very stout and are of a chitinous nature. There is a very fine structure to be observed upon the outer side of them for those who like "tests" for their objectives and apparatus and to try their skill in manipulation upon. It consists of very finely pointed short hairs studded evenly over the whole surface. They are conical shaped, i.e. wide based and very transparent, and being situated on an equally transparent coat are not readily observed. They may even be made to appear as holes, but this is an incorrect view.

These short hairs or spines are not an uncommon accompaniment on many algæ and may be found on Volvox between the little cells and their cilia and also upon many species of Desmids. In Bulbochæte an empty cell or the parts usually transparent is the best place to look for them successfully.

Ulothrix zonata (Fig. 76)

Another genus of Conferva, extremely common, growing preferably in running brooks and streams, and on this account probably does not appear in the microscopist's gatherings so often as its abundance would suggest. It is a long filamentous alga, having its cells in a single row closely set together, being little longer than wide and very slender. The width is rarely greater than 1/1000th inch, so that it does not catch the eye save by its tangled masses.

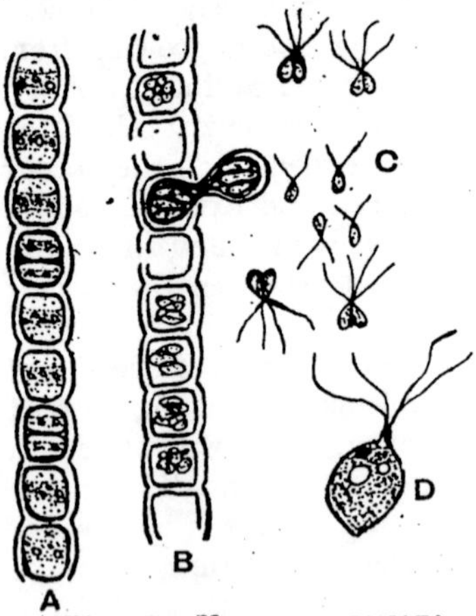

Fig. 76.—Ulothrix zonata.
A. Vegetative filament.
B. Cells in various stages of reproduction.
C. Isogametes.
D. Macrozoospore with four cilia.

The cells in a thread will count to many hundreds, each one with its chromatophore lining the inner wall, either entirely so or only in part and with one pyrenoid embedded within and a single nucleus. The cells are often seen of two separate sizes, the larger ones, those rather longer than broad, being the spore-producing cells and the smaller ones, the vegetative, often little more than half as long as broad.

The chlorophyll gives the filaments *en masse* a bright green colour to the eye and affords that delightful shade, when viewed in brilliant light under the microscope, common to all the growing algæ.

Its method of propagation is either by the formation of zoospores (and for distinction known as macrozoospores owing to their larger

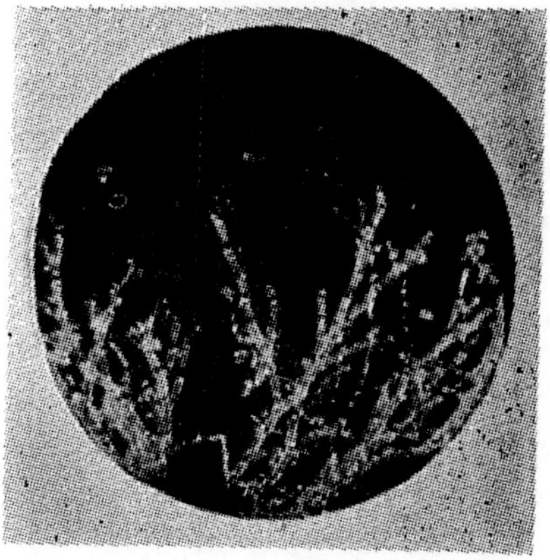

Fig. 75.—BULBOCHÆTE INTERMEDIA.

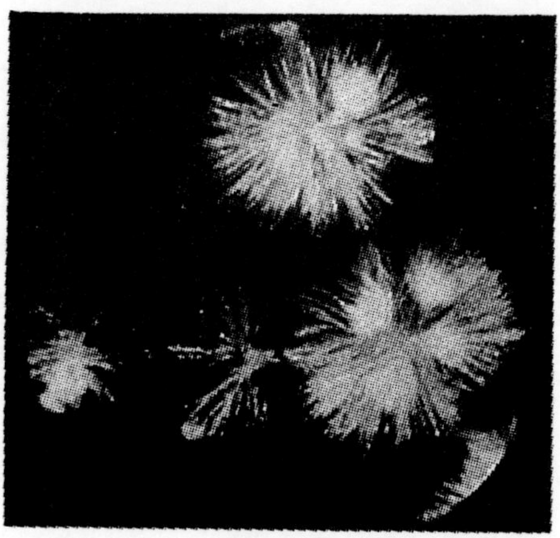

Fig. 78.—BATRACHOSPERMUM MONILIFORME.

[*To face page* 78

size and having four cilia for their locomotion) or by isogametes, i.e. neutral cells that do not exhibit sexual or other differentiation, the gametes of which are smaller and with only two cilia, these are capable of germinating without uniting and are termed the micro-zoospores. These latter, after swimming about freely, unite in pairs and become zygospores, taking on a spherical shape, with a stout transparent coat covering their green chlorophyll interiors. It is the green portion which finally buds out after the resting period into new filaments.

In the macrozoospores the mother cell may produce from one to as many as eight of these from its contents. In the commencement of this process the whole of the protoplasm of the cell is first formed into one mass withdrawing from the inner walls towards the centre, and if only two macrozoospores are to be produced will divide in half, each portion becoming a separate zoospore. If it be four then a longitudinal as well as a transverse one will succeed this, if eight a further division at right angles in the twos will partition the contents. The separated portions become pear-shaped containing one nucleus, the chromatophore occupying the bulbous part of the cell and the pointed end being left with a clear protoplasm. In this clear portion the four cilia arise, near to which is a red pigment spot or " eyespot."

Each zoospore possesses a pulsating vacuole alternately contracting and expanding once in every twelve or fourteen seconds, the presence of this vacuole being the difficult problem opened as to the classifying of them with animal or vegetable organisms. Their movement precisely resembles some unicellular animals and may be considered relatively rapid. A zoospore can travel two or three times its own length in a second, and comparing the fastest ship, this takes ten to fifteen seconds to traverse the same. Of course the microscope magnifies the speed as well as the size, but the relationship between their dimensions and movements are proportionate. The actual speed is about a yard an hour because their size is so small.

Light has a great influence upon the reproduction of zoospores. It seems that in darkness they cannot reproduce, wandering aimlessly about in any direction, but as soon as daylight begins or a bright illuminant is brought near, this at once exercises a directive effect upon the zoospore's movements. If the light be too bright they will recede from it, however, so that natural conditions of dawn is evidenced in their sensitiveness.

Ulothrix yields readily to cultivation and has had much experimental work carried out upon it.

Draparnaldia plumosa (Fig. 77)

The main stem or axis of this plant is always the largest, and the side branches are formed in tufts and whorls of half its size or less, each terminal cell ending in a long colourless hair. These form its distinguishing features.

The cells for the most part are clear throughout the plant, and only a light green chlorophyll band is present occupying the centre portion of each cell.

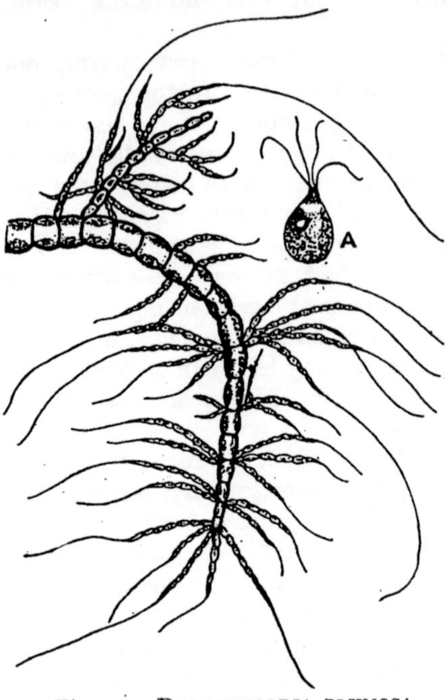

Fig. 77.—DRAPARNALDIA PLUMOSA.
A. Zoospore with four cilia and pigment spot.

Its method of multiplication is by the production of zoospores, as in other confervoideæ, each having a pear-shaped outline with a transparent protoplasmic portion at their apex and furnished with four cilia for their locomotion.

There is a red pigment spot near the base of these in each zoospore, and, like Ulothrix, show at this stage decided unicellular animalcule kind of motion in their existence.

It is in the tufted branches that multiplication proceeds from, principally owing no doubt to the main axis being so deficient in its chlorophyll, and from which the sustenance and interchange of its food supply is also established.

The plant attaches itself by a cellular disc to aquatic plants generally, Mosses and Liverworts being a favourite habitat, while the writer has frequently obtained it upon Myriophyllum in clear streams where the current was not very rapid.

The main axial cells are somewhat rounded and slightly incurved towards each division, which gives a beaded appearance, the tufted ones being even more so, inclining to an oval shape.

It is an alga distinctly pleasing under dark ground illumination; the amount of cell structure free from chlorophyll and other obstructing matters allows the bright illuminant to reflect itself in silver-like beams from the clear outlines of their chitinous transparent nature, and habitually attracting admiration from interested onlookers when a clean specimen is carefully set up for inspection.

Batrachospermum moniliforme (Figs. 78, facing p. 78; 79 and 80)

This is a rather large plant to be included, but its details are microscopic and coming within the catch occasionally, warrants a brief description. It belongs to the order Florideæ whose forms mostly inhabit sea waters, a few of which, however, are represented in fresh waters. It is commonly found attached to stones in clear, shallow streams and to the eye looks like a purplish green or browny thread of gelatinous spheres swaying about with the current. It is extremely slippery and difficult to take out of the water intact owing to its mucous covering, and in its resemblance to Frog spawn was allotted its name, derived from two Latin words meaning *Batrachos* = Frog, *spermum* = spawn. It is sometimes called Bead Moss. At one time it was thought to have some affinity to the Bryophyta or mosses. But in Florideæ the fruit is always in organic connection with the sexual plant and scarcely therefore to be considered as a new generation, while in the mosses the sporogonia or male cells are always distinct from the oophytes or female cells although dependent upon one another.

It is a much-branched plant and often grows abundantly, the branches being formed of numerous short cells end to end and set closely together in whorls about the main stem, and it is these clumps that give it the semblance of a string of beads or balls. It is very tender, and pieces are almost sure to break off in the manipulation of handling, often by their own weight. Sometimes the current of the stream will do this if rapidly increased, which forms no doubt one mode of distributing the species at different points along its course, there to affix anew and establish itself afresh or scatter its spores as it is carried along.

The cells are somewhat oval or cylindrical and grow smaller as the ends of the threads are approached where they are exceedingly small comparatively. Some are contained in radiating needle-like threads. The main axis or stem is composed of a different kind of cell set end to end in single file and in separate rows of various-sized cells, each row being quite independent of the other. At the centre is a row of large cells, and around these several rows of smaller ones. These latter it will be observed do not bring their ends quite in contact and form merely a kind of skin or cover to the others, and it is this skin which marks Batrachospermum as a higher plant from all other algæ and forms a distinguishing feature of it, approaching a kind of protective bark absent in alga generally.

In most other parts of the plant the cells are independent and convertible to any form required, while the axial and larger cells keep distinct to their purpose and are never interchanged. The side branches develop from a smaller stem in a similar way to the main one, but on a smaller scale, becoming crowded together and less distinct to the unaided eye.

In the life history of the plant the beads or clusters around the stem enclose within them, almost hidden, smaller and denser balls of threads on still shorter stalks. These are the sporangia which form the fruiting spores. In their development one of the cells at the end of a normal thread commences to swell out, not evenly, but at its lower end more rapidly than the other, producing shortly a flask-shaped cell, the neck portion being at the end of the row. From this is grown, with the aid of other special cells producing antheridia, a separate body which bursts forth as a roundish lump of protoplasm destitute of cell wall and without cilia. This is the antherozoid free to be carried with the current or, as it is observed, to attach themselves to the flask-shaped necks and finally to invade and disappear within them. So soon as this is completed the neck becomes separated from the bulb by a stout partition and withers away. The bulbous portion being now fertilized remains to emit a number of the shorter filaments from its surface and complete a protective covering for the now denser cluster within.

These on becoming ripe burst their envelope and are free either to sink to the bottom or to be carried to rest by the stream as small naked spores of protoplasm, having neither cilia nor cell walls, where they remain until ready to germinate. These delicate morsels of protoplasm form, so singularly provided, tasty tit-bits for the little fishes and others to delight their palates upon, still in Nature's wonderful profundity of seeds and germs freed, to obtain but one specimen to grow and thrive, there is a sufficiency for all contingencies allowed and so the balance is kept.

Before actual germination of the protoplasmic spores takes place, however, a cell wall is grown which upon adhering to stones and other surfaces commences to form a long thread of cells end to end lying close against its support, and from these a number of delicate branches are thrown upwards and so continues into adult life like to its species.

Fig. 79.—Main Stem.
B. moniliforme.

Fig. 80.—B. moniliforme.
With sporangia.

Fig. 81.—Amphitetras ornata.
A salt-water specimen showing the transparent covers, or "balloon cells," as collars around base of apices.

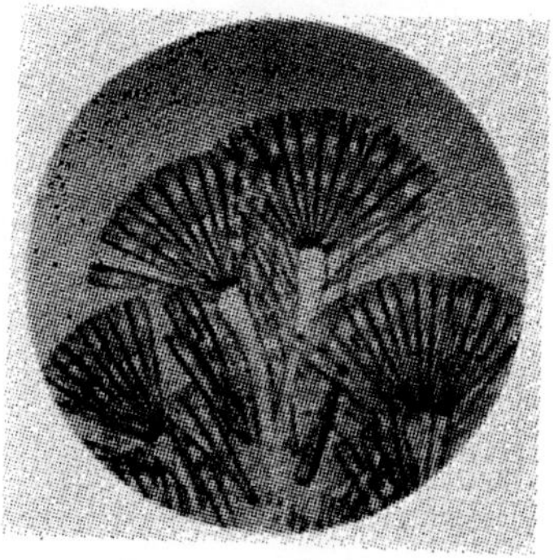

Fig. 89.—Licmophora flabellata.

[*To face page* 82

CHAPTER VI

DIATOMS

ARE single-celled organisms of great variety and shape found in all kinds of open-air waters, salt, fresh, or brackish. Hardly a dip of the collector comes up without one or more being among the gathering. First classed by Ehrenberg among the Bacillaria, they are there now, having been at one time considered animals, then plants, backward and forward many times. They consist of a frustule having two parts called valves and a third piece, the girdle. Complete, a circular or discoid Diatom is very like a pill-box in shape. The shapes, however, may be like a boat, naviculate, winged or alate, keeled or carinate, tubular, ramose or branched, sigmoid or S-shape, twisted or saddle-shaped, angled with horny processes called "feet," undulate, etc., etc. They may occur isolated and free-swimming, or attached to plants, etc., by stalks or stipes or adherent in chains curiously cemented by their extreme corners, or again fan-like at the extremity of a pedicel and radiate; or frondose, living within a gelatinous, irregular-shaped tube lying in single file or side by side in pairs, or again, branched. Some specimens are spinous.

Not all of these may be found in fresh water, but the reader will obtain a large variety there and most that are upon a stalk. The remainder are salt water and fossil forms, and their number is legion on account of the universal extent of their production. Mountains in different parts of the world are composed in the main of their frustules, and in volcanic eruptions they have been showered from the air in the dust discharged and found at great distances away (Fig. 81).

The interior of the cell contains during active life a soft mucilaginous substance having numerous granules within it, and is usually of a yellowy brown colour called the endochrome. This is attached upon the valves inner surfaces either side, leaving the central portion more or less clear and free. There are also several larger globules.

The valves and girdle differ in chemical composition from other plants, being of a flinty or siliceous nature, which withstands being boiled in strong acids and alkalies without being reduced.

The Diatoms one buys in the opticians' shops have been thus cleaned and mounted in a suitable refractive medium the better to observe their structures. Some may be mounted dry and sealed, while yet others will have them prepared to crack away some of the various laminæ in which their valves are encased and then mounted. All these methods are merely done for the purpose of observing the beautiful sculpturings with which they are embossed to the best advantage. The sculpturings take the form of lines, pits, bosses, spines, hexagons, etc., and it is these that have endeared the Diatom to lovers of the beautiful when thus presented to the microscopist.

As observed living and in the water of their habitat, the markings are not seen at their best and must not be looked for under those conditions. Several reasons may be adduced to account for this, chief among them being the fact that they are encased in a clear mucous membrane above their siliceous coat which hides it, and which is, in the case of the free-swimming, constantly on the move and blurring the under outlines.

Again, the refractive index of water is insufficient to see the very fine lines, being only 1·3 compared to air which is 1·0 and silex 1·5. It is the siliceous coat that presents them, and this needs proper preparation. Finally, the difference between the refractive index of the Diatom and the numerical aperture of the objective used is all important.

Diatoms present one of the greatest perplexities of all organisms to the microscopist, but incidentally in the probing and studying of them we owe largely the excellence of our objectives of to-day. They move, but how? They have no apparent organs of locomotion of any description. True, all kinds of suggestions have been made to provide them such. Some have said cilia, others protoplasmic extrusions equivalent to feet, some again to the extraneous hyaline threads that grow parasitic about their outer coats. These have been averred the motive organs, others of a fluid stream pulsating and emitted from the ends, a kind of breathing or dilation and contraction of the frustule as the cause. All of these have been surveyed and found wanting.

Finally, it is considered the outer protoplasmic coat possesses an inherent spasmodic twitching movement that is continually going on during active life, and this upon touching any surface gives sufficient impulse to move them and to provide ample power to execute the forward, jerky movements of their very light mass.

As the outer coat, caterpillar-like, makes a flow in either direction, so the Diatom may follow, more or less erratically according as it completely or otherwise touches the surface, either forward or backward. This is consistent with what is observed. The outer coat is of an elastic nature and also an adhesive, sticky one. Owing to this the Diatom will sometimes erect itself on end while touching a solid surface in the water.

Again, it may often be seen emerging from a mound of debris with one extremity stuck to a portion of the mass, stretching the outer coat as it proceeds, and this, after an appreciable and easily recognizable distance, will be snapped in, drawing its obstruction or anchorage with it at the same time. This shows definitely the nature of the outer membrane as well as its existence.

Diatoms are not perfectly symmetrical one side with the other, the striæ and markings are affected by circumstances of growth, age, locality, etc., and may count unevenly. Some are concave one side and convex the other, especially those fitting top to bottom, the connecting membrane not always being an essential segment of a frustule but an after development connected with the process of self-division.

Reproduction is by subdivision lengthwise, but upon reaching a certain size this ceases and they group themselves in pairs, becoming surrounded by a gelatinous substance, and the contents of the two cells merge together to form a third and larger body called an auxospore. Often two of these are produced at a time, which eventually become the parents and continue the process similarly.

The girdle around the side of the Diatom in mounted specimens may be found detached and is often striated and sculptured and will always bear inspection, and should not be mistaken for extraneous bodies owing to its curvature, even when broken. Such multitudes of the frustules with their brown endochromes are produced, that streams, lakes, and all bodies of fresh water are more or less coloured with their presence. Rapidly running waters are no exception in this case, and the waving plants lining the sides, the lock walls and gates, the boat sides will be coated with their brown gelatinous fronds.

They are of immense hygienic value in keeping the sanitary condition of our waters good, giving off as they do large amounts of oxygen in sunlight, bubbles of which may be seen arising under its influence.

Drinking waters contain many broken or even entire frustules of their harmless selves occasionally, and cattle using open-air sources must imbibe large quantities of their tiny skeletons without being any the worse.

Many forms are but the 1/2000th inch across and even less, which may account for their prolific numbers in small compass and also for the appeal so interestingly made to enthusiastic microscopists in the investigation of Nature's minute life. They are so plentiful and ready to hand, their forms so varied, there is little wonder they have become endeared and installed as favourites.

In describing some of the genera it may be as well to group them into tube-dwelling forms, stalked forms, chain forms and free-swimming forms, all of which are to be found in fresh waters. Mounted specimens, fossil and ocean-going forms, which make up the larger bulk on the globe, have to be left out for the purposes of our present work, and indeed the classification and nomenclature of such is of so stupendous a task that a lifetime of study would not exhaust the subject, and there are many at work endeavouring to unravel their structures and to classify them in some order at the present time, which it is hoped will bear fruit in some standard volumes in the future.

It will be necessary to say that for the purposes of this work the following definitions are adhered to. By the front view is meant the side usually presented to the observer having the striæ and principal details upon it. The side view may also appropriately be called the girdle view in those thick pill-box Diatoms like the Disci and others where the two valves are sutured and a broad quadrilateral or other surface is presented.

The Raphe is the median line upon the front view; it may be either straight, plainly curved, sigmoid or waved. The nodules are the terminals of the median line in general and also the larger boss at the centre, usually being smooth and clear as in the naviculæ, where the raphe end within it. The edges of the front view and the edges of the side view then explain themselves. The frustule is the complete diatom, the valve the half diatom, and usually the front view. The girdle, when separated from the two side valves of the frustule and complete is either oval or circular with few or no markings upon it and in descriptions is generally ignored. It probably forms part of a tubular case which splits away as the diatom is completed and so is entirely absent in some specimens the microscopist will encounter. In such thin and flat species as the Navicula, Cocconeis, Amphipleura, etc., it is obscure and unseen.

Encyonema (Fig. 82)

This is one of the tube-dwelling frondose genera of the Diatomaceæ. It lies generally aslant in single file, and has a somewhat similar outline to Cymbella.

The raphe may be straight or slightly curved. The back is arched, and the ends are more or less pinched in or irregularly contracted, so that the outline is not curved uniformly. There is a central nodule fairly large and oval leaving a clear space there and one smaller one at either end through which the extremity of the raphe passes and continues as a curved line a little farther before coming to the end. The tubes are seldom branched, and are of a gelatinous, semi-membranous character. They adhere to aquatic plants, and one tube may contain as many as a hundred or even more frustules.

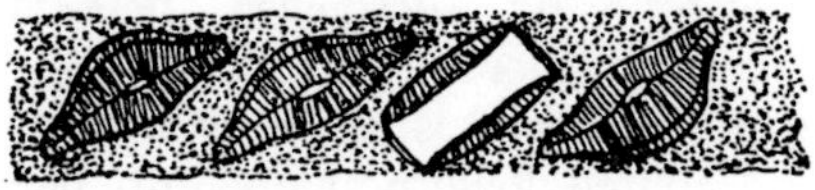

Fig. 82.—ENCYONEMA.

Seen as a side view, the surfaces are quadrangular, the ends then appearing as straight and the whole pill-box figure. From the raphe or medium line to the edges the valves are striated about fifteen of these in 1/1000th inch. The length is about 1/500th inch and often less.

Encyonema at times may be found in great profusion, and at the present moment in a fair-sized lake near the writer, upon which boats ply for hire, there is the larger portion of an Osier or Willow tree broken down by the wind during the winter, lying in the water, the slender branches of which are literally coated with them, the whole presenting a brown, gelatinous mass around every twig to be seen near the surface.

Such a quantity must number individually many millions of the tiny frustules; what their countless striæ and still more minute details upon them amount to is beyond calculation, and each is so perfectly fashioned when seen with the highest powers of the microscope. Truly nature is lavish and prodigious under favourable conditions of sunlight with her energy and substance.

Colletonema, Schizonema and Micromega (Figs. 83, 84 and 85)

These are forms of Navicula which have assumed the frondose form and live within gelatinous tubes somewhat similar to each other.

Colletonema is usually found arranged in single file or again in crowded clusters in a tube thread-like in form and considerably branched, almost tuft-like at times.

In Schizonema the frustules have a more flaccid and delicate structure and are encased within a single tube, having even, parallel margins which also is very slender.

Fig. 83.—Colletonema.

In Micromega there is generally a series of files of the Navicula within the tube, and the frond is generally stouter, also several tubes may be found united longitudinally into one compound frond which gives it a slightly distinguishing feature from Schizonema. All the forms, however, are capable of assuming these various habits, and it is difficult to divide them by the shape of their tubes or envelope alone. If specimens are separated out they are true Navicula. On the other hand, as they have taken upon themselves to live and grow within distinct gelatinous tubes they must be counted as distinguished from the free Navicula form while so encased. As we know more of the life history of our simple, primitive organisms no doubt better nomenclature and classification will arise.

Fig. 84.—Schizonema.

Fig. 85.—Micromega.

The Navicula in all the frondose tubes are very small and will average little more than 1/1000th inch in length and the striæ count fifty to the 1/1000th inch.

There is the little nodule in the centre and at the ends of each raphe, while some of the shapes will be sigmoid as well as boat-shaped, and are quite as perfectly fashioned, seen individually, as their larger brethren. Nature has moulded them and that suffices.

Gomphonema (Figs. 86 and 87)

This is one of the stalked diatoms. The frustules in this genus are attached by fairly long, slender stalks or stipes of a transparent, silicious nature. They are flexible to a slight degree. The outline of the diatom from the girdle or side view is wedge-shaped, the edges being straight. On a front view the appearance is quite different, and the two sides often lead to quite erroneous notions as to the diatom they belong to when seen apart.

The front side is gracefully curved on either edge longitudinally, first widening, then an incurve, then a larger curve, again incurving more and again bulging out at its middle to gradually taper down to a rounded end of the same width as at the other extremity. The median line is straight with a central nodule only, and from which to the edges, radiate the lines of striæ containing small bosses, so small as to be deceptive whether to be looked upon as punctures or pimples. The latter is the correct view.

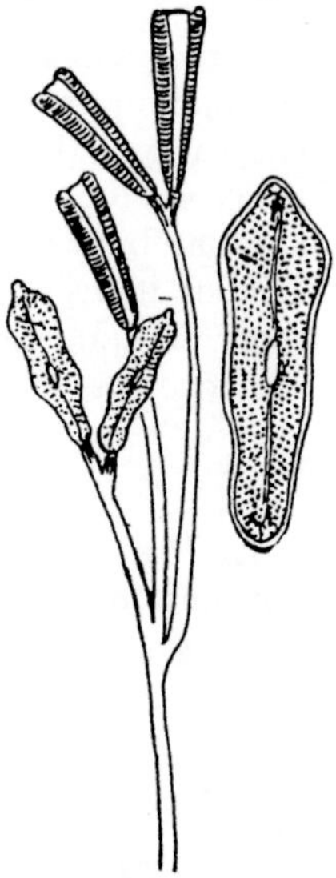

Fig. 86.
GOMPHONEMA ACUMINATUM.

Gomphonema acuminatum illustrated is a slender specimen tapering below into its stalk-like base and to which its stipe is affixed and continued down to its own support either upon some aquatic stem, a stone, a tree branch submerged, or other object. Its length is about 1/600th of an inch.

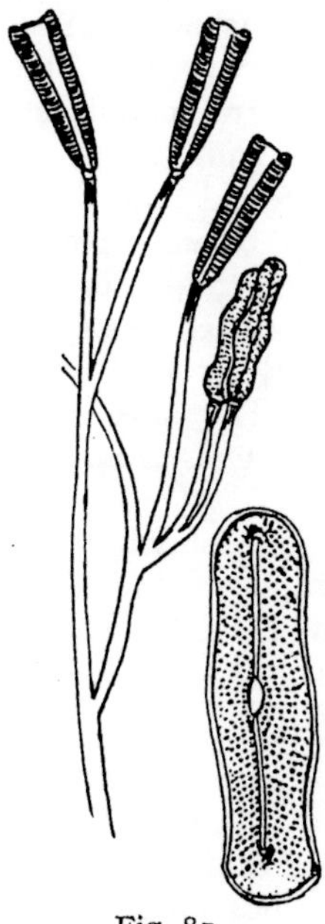

Fig. 87.
GOMPHONEMA GEMINATUM.

In *G. geminatum* the stipes are frequently branched and bear a single frustule on the summit of each. It is a larger specimen than the foregoing. The outer end of the frustule is round on front view, but has the square end wedge-shape upon side view. Often the narrow end is prolonged by a short stipe which meets another frustule upon a similar stipe and the two unite at the attachment to the main stalk.

When a young cell derived from an auxospore finds its resting place the thinner end secretes a stream of gelatinous pabulum which soon sets solid and becomes the stalk for the frustule. By and by the frustule commences to divide longitudinally and the stipe substance follows suit, so now we get two streams, while yet the diatom may not be completely severed. This they do very shortly, and so the branching goes on, eventually producing quite a tree-like appearance of pellucid branches with the diatom at each extremity. Occasionally two stipes will be seen running close beside one another and the frustules unattached.

The striæ in this species are wide and more conspicuous than in *G. acuminatum.* It is found plentifully in waters where rocks and solid stone banks form the margin, attaching itself in large tufted, gelatinous masses to these, but is also common in our ponds and streams, clothing the stouter plants with its brown, velvety covering. Its length is about 1/480th inch and sometimes more.

Cocconema lanceolatum (Fig. 88)

Frustule arched or cymbiform, i.e. in a boat-shaped manner, and the outline follows very much the Cymbella genus. It is attached to some firm, submerged support at one end by a stalk which is often much branched and wavy.

Fig. 88. COCCONEMA LANCEOLATUM.

On the front view the valve is slightly crescent, and the median line is not central but slightly towards the concave side. The raphe has a rather small central nodule as well as a smaller one at either end.

The frustules are often found detached, and the genus to which they belong then is difficult to decipher so much do they resemble the free Cymbella. The arched margin, however, is not so frequently or decisively convex as with Cymbella. The side view shows a tapering towards the ends from the centre and is lanceolate.

The striæ are transverse, simple and verrucose on the front view, running from the raphe to the edges the whole length. The length of frustule is about 1/150th inch.

Licmophora flabellata (Fig. 89, facing p. 82)

This genus is essentially marine, but as it is occasionally found in our brackish waters and the saline marshes of the East Coast, and is so strikingly beautiful a specimen of stalked diatoms, I offer no further apology for its inclusion. The writer has found it attached to seaweeds along the coast and upon the rocky beaches of the south-west, among the teeming wealth of vegetation and miniature life to be found in the rock pools and slopes of weather-beaten, wave-washed boulders in that locality.

The frustules are wedge-shaped on a front view, closely united in a spreading fan-like manner upon the end of a fairly stout stipe or pedicel. The pedicel is irregularly branched, and becomes thicker as it nears the attachment to the frustules. The genus has much similarity to Synedra save that the latter has its frustules more nearly straight. Both, however, are attached to stalks.

Licmophora frustules are fairly long and narrow, tapering to almost a point at their base and attachment, where many specimens are seen so regularly formed, there presenting a perfect section of a circle as they are splayed around. The frustules do not seem so firmly silicious as Synedra, which accounts for their specific name, flabellata. They may often be found detached from their stalk and become difficult to decipher at such times to which species they belong.

The front view shows a striated though smooth surface longitudinally, not ribbed in any way. At times the fan is in two or more portions, and as many as a score frustules may unite in its construction. Under dark ground illumination it is a picturesque sight and specimens may be found where the simile to a fan is followed completely, handle as well.

Within the frustules each has two rounded masses of brown protoplasm, the endochrome, one near the wider end, the other near the centre. The outer circumference is frequently split at the sutures, giving a dentate margin. This occurs with age when growth becomes more or less longitudinal before finally ceasing.

The stipes as well as the endochrome will admit of staining owing to their sub-membranous nature. Methylene blue among other stains being readily absorbed, and when desired in mounting specimens for preservation this can be done before placing in position with satisfactory effect.

The length of frustules are about 1/250th inch, the stipes being of varied lengths up to an inch. Each stipe produces one or more fans, and at intervals along the stalk others may bud out from the side. The ones at the tip are usually the finest.

Synedra splendens (Fig. 90)

This forms a rather frequent species in fresh water, often covering the fronds of Lemna with its brown coating. It is also obtained from other aquatic plants, twigs and rootlets near the surface in quiet sunlit situations. Only during its earlier stages, however, can it be considered stalked, then the frustules, somewhat cylindrical and elongated, are aggregated in clusters upon a cushion of gelatinous matter which may be floating thus freely or attached to their stipes fastened to the leaf or stem of some water plant.

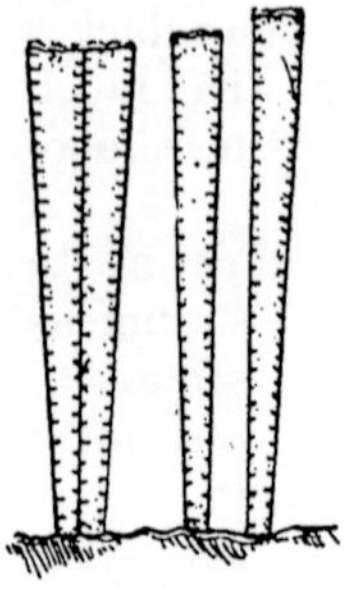

Fig. 90.
SYNEDRA SPLENDENS.

On a side view they are narrow, more so than upon a front view, and are slightly broader at both ends as well as at the centre. The front view shows a narrow line down the centre with fine transverse striæ crossing. The outer margin is slightly wider than the inner and attached end.

Several small masses of the endochrome are seen distributed along its length. The length of the frustule is about 1/75th of an inch.

As the diatom ages it loses its hold upon the point of attachment and the frustules break away in pairs and singles and eventually become free to move in the same way as those of the species which are never so attached. Their movement, however, does not seem to be so constant or regular as the true free-swimming species.

It is extremely difficult to distinguish the various species at such times, and many have probably been thought distinct and a new name given to them that in reality are but another phase or condition of ones already catalogued.

Fragilaria capucina (Fig. 91)

This diatom forms a long band of frustules lying closely side by side. You do not therefore look at the valve or front view, but upon the side of it, much as a pile of books grown top-heavy and have fallen over for more stable equilibrium might be situated. It abounds in reservoirs, ponds, lakes, and in almost all fresh waters open to the sunlight with no swift currents coursing along.

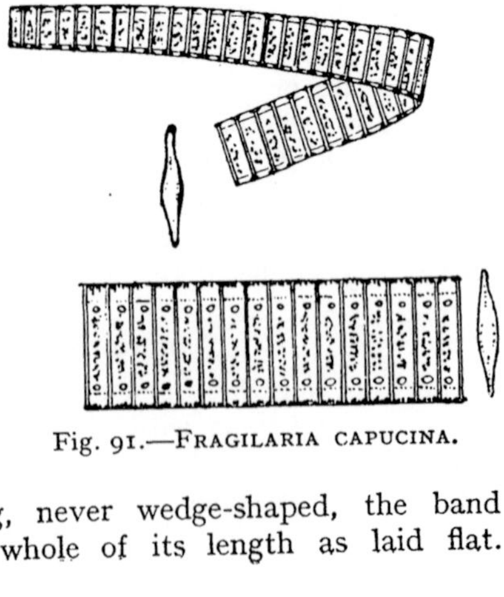

Fig. 91.—FRAGILARIA CAPUCINA.

The bands may be an inch long or more, but the breadth is little greater than 1/700th of an inch wide. The frustules, seen from their side, ribbon-wise, are narrow and oblong, never wedge-shaped, the band presenting parallel sides the whole of its length as laid flat.

Sometimes the bands group themselves together and grow twined around leaves and stems of aquatic plants, or spread as a brown velvety covering upon stones and other objects in the water.

The diatom is so compressed and thin as regards depth it is not possible to obtain a front view of it except from a detached specimen. When this occurs the valve or front view shows the frustule to be rather canoe-shaped and not unlike a narrow species of Navicula save that there is no nodule at the centre, nor any raphe down the middle. There are extremely fine striæ crossing the width transversely when seen under high powers.

The endochrome forms into yellowy brown masses along the sides of the frustule, but this varies considerably. Sometimes it occurs as a more or less rounded patch in the centre, and when this is so and the ribbon is flat, lengthways towards you, it gives the appearance of a single yellow line running along the middle. Sometimes there will be two patches, one near each end of the valve. This will give a double line, altering the appearance in a variety of ways according to the positions it occurs in.

Out of water in a more or less dry condition the diatom presents quite a silvery object, due to its transparent silicious nature reflecting the light, glass-like, which falls obliquely upon its surface.

Himantidium pectinale (Fig. 92)

Frustules of this genus are united into either ribbons or tables. When in ribbon form they are readily recognized, but should they be separated, then occurs a difficulty in distinguishing them from the free species of Eunotia.

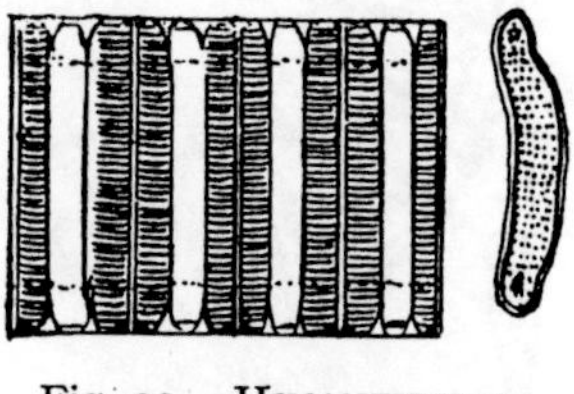

Fig. 92.—HIMANTIDIUM PECTINALE.

The striæ in Himantidium on the front view are parallel, while in Eunotia they are radiate, but this is partly optical on account of the upper margin in the latter genus being strongly convex.

On viewing diatoms arranged in bands it is necessary to remember that the individuals comprising them are growing from the ends, after the manner of a pile of equal-sized books that have grown so top-heavy in height they have fallen over and assumed a more stable position. Thus you look at the side of the separate diatoms as they lay so joined together. The front view is only obtained when a single specimen is detached. There should be no difficulty with the sparse number that take this form, therefore, when dealing with the host of diatoms that are free, in recognizing the valve front. Upon this the minute striæ and markings are always presented, while with the band forms only the clearer suture and often a small portion of the front markings continued round that edge are visible. The appearance, too, is usually oblong, whatever the front view may be like.

In *Himantidium pectinale* the frustules are much wider than in Fragilaria, and the front view shows one side to be straight while the other is arcuate or waved with a concave depression central, then a hump either side, tapering to narrow, rounded and obtuse ends.

The striæ are transverse and count about twenty-five to thirty in the 1/1000th inch. It is found in marshy ground, and though not frequent is often profuse when its habitat is discovered. It occurs in all the four continents, Europe, Asia, Africa and America, and so cannot be said to be rare.

Meridion constrictum (Fig. 93)

The frustules of this genus are all wedge-shaped, both upon the valve or front side and also on the girdle or side view. The girdle view is the one normally seen, as with the ribbon variety of diatoms generally.

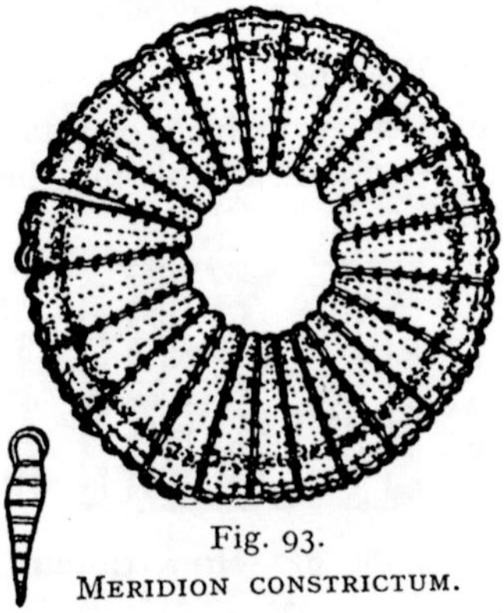

Fig. 93.
MERIDION CONSTRICTUM.

Owing to the cuneate, wedge-shaped growth of the frustules they naturally curl around an imaginary axis, forming very beautiful spirals of themselves, and so exact are they in their growth that a central circular space is apparent by the time one circuit is made. They rarely make very lengthy spirals, however, possibly on account of their fragile nature, being unable to adhere for any great distance. They may be met with in isolated frustules or among algæ in quite large masses at times, and are fairly common.

Save for their coiled shape the individual frustule differs little from Diatoma, another genus. On a front view when a separated frustule can be found there are transverse partitions or septa to be seen which are fairly broad and conspicuous. The wider end of the valve is rounded, then a constriction occurs, then a bulging out, tapering to a narrow, rounded point at the opposite end.

Between the partitions are fine striæ crossing the valve, these pass round the edges of the front view and are seen endwise as the diatom lays normally on its girdle for a short distance to the suture of the frustule. They give a crenated or wrinkled outline to the sutures. Their habitat is generally in shallow streams, forming a gelatinous brown stratum on stones, leaves of aquatic plants, algæ, etc. The length of the frustule is about 1/350th of an inch.

When the diatom is young a gelatinous cushion may be observed which suffices to adhere it to the stone or other object, forming a starting-point for it, but otherwise there is no stalk or attachment at any period of its growth. The endochrome runs along the length of the frustule either in bands or in separate, rounded masses and of a yellowish brown.

Diatoma elongatum (Fig. 94)

The frustules of the Diatoma generally are found in zigzag chains adhering to one another by a slender portion at one of their corners. The valve or front side shows transverse costæ or striæ, rather thick and consisting of a beaded line without an interruption, crossing in the centre longitudinally. There is no raphe in fact nor any nodules at either centre or ends as in the Naviculæ. The striæ show around the valve edge a little way, and on the girdle side present a readily defined margin of beads along its outline, so that the counting of these beads determines the number of striæ on the valve front.

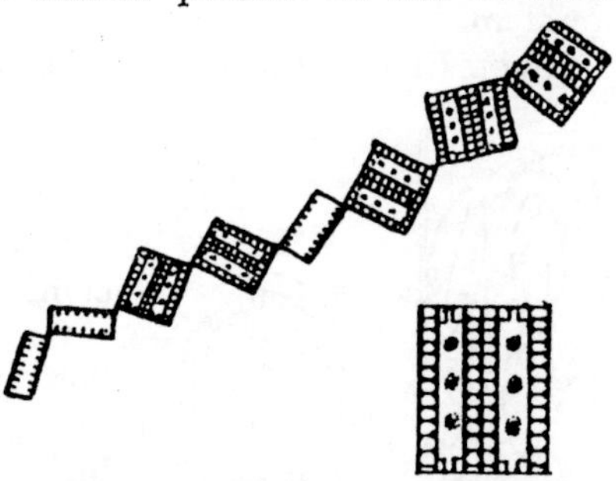

Fig. 94.—Diatoma elongatum.

When the diatom is young and the frustules unseparated it will grow some length as a ribbon-like band, later, however, they open out and remain as shown, attached by their corners only, and not always by the same side corner particularly. The attachment is not a large piece, barely is the portion seen, yet it is very tough, and will bear a good deal of rough treatment at times before parting. It is, moreover, somewhat flexible in a length of chain. The diatom is found appended to aquatic plants, and the specimen used in characterizing and illustrating this account is from a tangled mass of the frustules attached to the alga Cladophora.

The ends of the valve fronts are slightly inflated, which gives the side view a double-concave attenuated appearance in the middle.

The common species *D. vulgare* on side view is oblong, the four corner angles being right angles, and the costæ are three times finer than those of *D. elongatum* or as six is to eighteen per thousand.

Frequently a mass of Diatoma contains numbers of frustules differing slightly from one another, and in these massed growths it is with difficulty one can get conclusive evidence of their specific names in accordance with the schedules.

Some costæ are distinctly beaded, some but slightly if at all. Some girdle views show the frustules concave, others straight, or again waved at their sutural joints. The microscopist must therefore be prepared to find several variations.

Cocconeis pediculus (Fig. 95)

This is another very plentiful species growing upon aquatic plants, sometimes so closely together as to literally cover the leaf or stem.

Upon Elodéa, Cladophora, Lemna and Myriophyllum the writer has frequently found it in such masses. It is an oval diatom of the pill-box form, which can be readily seen as it is situated around the circumference of a stem.

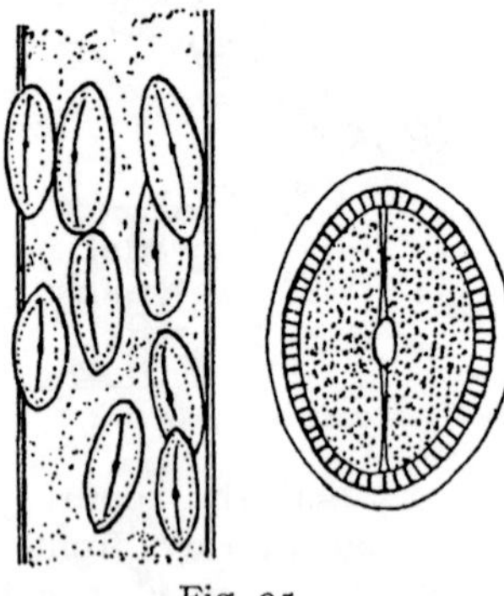

Fig. 95.
COCCONEIS PEDICULUS.

On the front view it shows a median line and a central nodule, and if a favourable one is before you, exceedingly fine and delicate striæ can be observed running transversely. These consist of fine granules, and by oblique light can be made to show themselves running longitudinally as well, so closely set are they together and evenly separated.

Near the outer margin a second band following it is situated, having evenly spaced costæ across it which gives a very pretty appearance to the diatom as a whole. There are no end nodules as in the Naviculæ, and the frustule in the side view is curved; this can be best seen as a specimen is rounding a plant stem to which it is closely attached. It is very small, the diameter being about 1/850th of an inch, the longer axis about 1/700th inch, but many still smaller are to be found. Its habit appears to be to attach one of its broader surfaces to the plant upon which it is growing completely, so that if it is a slender thread like the rootlets of Lemna the curve is greater than if found on an Elodéa stem, and in all such cases an end-on view shows the curve as well as the girdle or side view. Upon a flat surface it is not seen at all and is insufficient in itself therefore to form any circumstantial evidence of its species.

Asterionella formosa (Fig. 96)

This is a distinct method of diatom grouping from either stalks or chains or ribbons. It is in the form of a star, *L. asteron* = a star, and with its spreading rays of flashing light, as it is brightly illuminated, reflecting from its long and narrow radiating frustules it is aptly named.

The frustules are generally broader at the base, tapering slightly to a rounded knob at their apex. There is a central mucous cushion to which they are affixed, and when an entire specimen is present quite a circular hole will be seen there, not unlike the hub of a wheel where the axle has been removed. As the wheel is presented, flat, it is the girdle or side view that is seen, and upon this the costæ or cell divisions may be observed along the frustule.

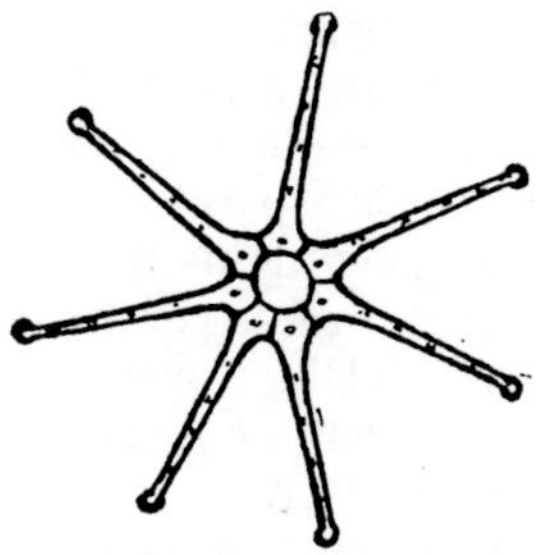

Fig. 96.
ASTERIONELLA FORMOSA.

On the valve front where a specimen is detached delicate striations may be noted under slightly higher powers. Unfortunately it is often difficult to find a specimen with all the rays complete, usually one or more will be found missing. When circumstances are fortunate the diatom gives a very beautiful appearance and adds yet another instance of the manifold ways Nature has in ordering her materials and of displaying their simple artistic charms.

It is frequent in lakes and fairly large and clear ponds. The writer had an abundance from the reservoirs at Staines, near London, in which were many complete specimens. It seems to prefer waters of evident purity, and there it will flourish luxuriously.

The number of frustules forming the star is usually four or eight, but there seems nothing to prevent them forming in any other figure except that the method of growth is by division, and one would therefore expect the number to be an even number. The length of frustule is about 1/450th of an inch, sometimes more.

Asterionella has been considered to be a form of Diatoma nearly allied to *D. tenue*, and, excepting for its method of arrangement, there is a similarity in its outline and general markings. It is frequently found near or on the surface of the water, probably on account of its spreading frustules giving it a buoyant nature. For this reason it is common in the plankton or gatherings from the surface nets in suitable broad stretches of water.

Naviculaceæ

This genus of diatoms constitute by far the most numerous family of the Diatomaceæ, abounding in all latitudes and at all altitudes from alpine springs and mountain lakes down to the bed of the ocean, where water is perfectly fresh or of every degree of saline impregnation. The marine species, however, keep to their habitat and are not found with the fresh water ones and vice versa, this is so constant as to form the rule. There are exceptions where it has been disregarded, but principally in waters that have been found to be brackish and intermediate between the two in its constituents.

The frustules are free, solitary, either straight or curved, elongated and boat-shaped, with the extremities similar, either end. The valve or separated half is more or less convex and has a median line longitudinally with distinct central and terminal nodules. The surface is covered partially or completely with smaller nodules arranged in transverse, oblique or longitudinal lines called striæ, which may be situated closely together and so present the appearance collectively of ridges or "costæ," or separated sparsely as to be visibly segregated, this under powers of from five hundred to a thousand diameters.

With extremely high magnifications beyond this, various views of the frustules' structure present themselves, according to their mounting medium, the arrangement of the light impinging, the quality of the objectives used, and last but not least to the training received from an acute and practised eye that is *au fait* with the many optical illusions light can trick the unwary with when dealing with the substance of pure transparent silica and approaching a half wave length of usable illumination.

Small wonder some of our greatest microscopists in the past have fallen astray under such difficult conditions in elucidating the very minutest details which have had to be encountered in their structures. The march, however, has been a forward one, and the benefits derived optically and mechanically from their labours incalculable.

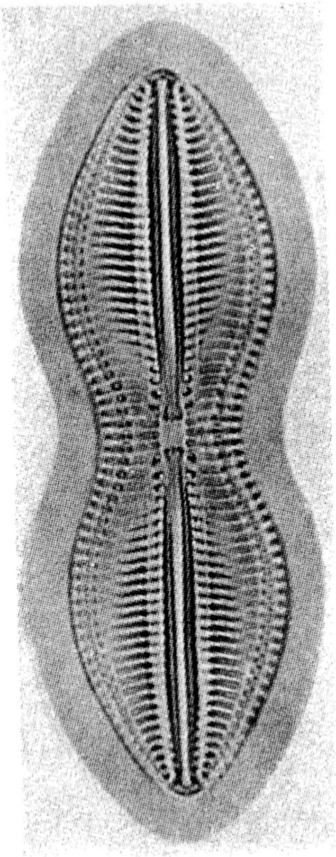

Fig. 97.—NAVICULA CRABRO.

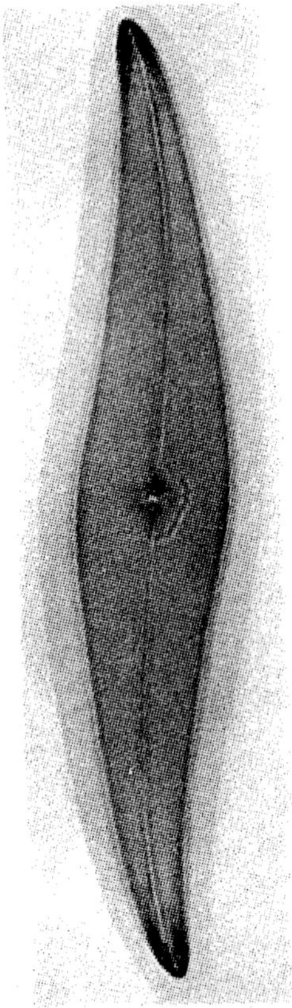

Fig. 100.—PLEUROSIGMA ATTENUATUM.

[To face page 99

Navicula crabro (Fig. 97)

The frustule of this is large and usually described as panduriform, i.e. fiddle-shaped, or having an inward curve about the middle either side approaching a division of the diatom into two ovals longitudinally. The apices come to rounded ovate extremities. The striæ are transverse and costate, or finely and obscurely granular, and interrupted longitudinally by a broad straight groove, along each side of which is a single row of conspicuous spherical nodules.

This diatom has had various names at different times, which points to the great difficulty in attaching definite specific titles. Nature loves diversion; no two blades of grass grow exactly alike, so one can only satisfy himself with resemblances. It has had, for instance, the appellations *N. pandura*, *N. nitida*, *N. didyma*, *Pinnularia crabro*, *P. pandura* and even *Diploneis crabro*. It is one of the many diatoms that show considerable variations both in size and outline. In some the striæ appear finally costate, in others there is an appearance of the granule structure, whilst in the more robust, and under the finest instruments and manipulation, they are definite and distinct tiny nodules. Microscopists now call them rosettes because they are typical of the larger nodules which can be further detailed into smaller markings of a rose-like pattern.

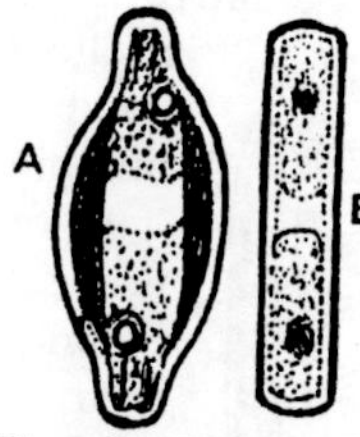

Fig. 98.—Navicula amphisbæna.
A. Front view.
B. Side view.

The costæ in themselves are fairly coarse and average about 11 to the 1/1000th of an inch. Specimens have been found in the North of England and in Scotland and are prevalent in places across Europe and Asia.

The photographic illustration was taken with a very special lens by the late Mr. Max Poser, and in the original shows faintly the transverse lines of fine granules within the costæ near the centre of the frustule. It was presented to the writer many years ago and is exceedingly fine. The diatom is convex in each oval portion, which requires depth as well as aperture to delineate it satisfactorily as a whole. Reproduction cannot do it full justice.

Navicula amphisbæna (Fig. 98)

Is a very common species in the ooze from shallow streams and similar situations, where it can jerk along its erratic course in touch with something firm enough for its purpose. It is a small specimen with narrow capitate ends bulging out widely about the middle when seen from the valve front view. Its measurement is anywhere from 1/1500th of an inch to as much as 1/300th, but the specimens usually found by the writer have been about the 1/750th inch, and a very active little object it is getting along at a good speed across the field of view. To watch its energetic movement as it points in its course first to the right and then to the left one might fancy it could see something interesting in those directions. When it comes against a mound of debris or other obstruction it rarely waits long, but soon alters its course or backs away in the reverse direction without turning round. It has its light brown endochrome along either side in a band breaking short of the narrowed apices. There is a fairly large central nodule and a smaller one each end of the raphe or median line, and generally one or more good-sized transparent globules at the commencement of the bulging portion from the narrowing ends.

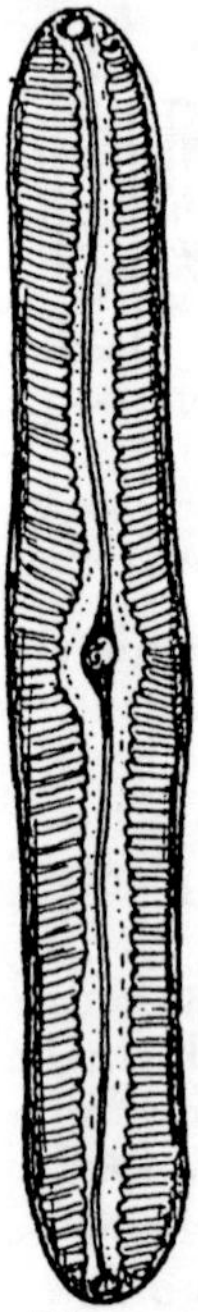

Fig. 99. NAVICULA MAJOR.

Upon a side or girdle view the diatom is typically navicula-shaped, box-like, with squarish ends, giving an oblong outline. The clear space in the centre is well seen in this view also, when the endochrome is present. With a clear frustule, delicate close striæ may be observed on the valve, running from the raphe each side to the edge transversely. A similar diatom, but with more robust, conspicuous striæ to this, is *Pinnularia amphisbæna*, which in all probability is but a variation or state of the same species.

Navicula major (Fig. 99)

Known also as *Pinnularia major*, *Frustulia major*, *Navicula viridis*, *Bacillaria fulva*, *Pinnularia viridis*, etc. There is nothing fundamentally different in its structure, moreover, to separate it from *Navicula nobilis*, *Pinnularia nobilis*, *Navicula gentilis*, *Navicula gigas*, *Pinnularia mesogongyla*, *Pinnularia gigas*, and quite a number of others. The writer mentions these names primarily to show in what a confused state diatom nomenclature has finally arrived at, owing no doubt to the varied quality of apparatus used at different times, and to the age, locality and perfection of growth to which specimens observed have attained and the particular skill or otherwise in the manipulation of the microscope and its various accessories, screens, illuminants, etc., that go to make up the outfit of the microscopist at work.

Diatoms multiply by subdivision. A new valve is developed to fit in each of the old ones, the inside valve of the old cell becoming the outside valve of one of the two new ones, and in this way must gradually grow smaller and vary the size and probably the shape as well slightly. The most notable and substantial difference in the appearance of this diatom from other of its species is the broad costæ upon either side the raphe and the large size to which it will attain, which makes it also a very suitable and easy specimen to observe for its more minute details. Its girdle view is box-like and oblong, typical of Navicula generally.

Specimens named "*Viridis*" are usually illustrated smaller, and so called, as someone aptly remarked, because its endochrome is brown, not green, and the median line may take a slightly varied curve or wave as it recedes from the central nodule. There may be eighteen instead of thirteen striæ to the thousand, or the ends of the frustule appear a little squarer than its neighbour. Such like natural idiosyncrasies have been enough to allot new names to the struggling growths, perhaps under very adverse surroundings. The diatom has been misjudged whilst doing its best.

These observational remarks apply with equal force to the Desmid families. It is equally certain also that under cultivation many variations can be obtained. Specific names therefore must not be implicitly relied upon with the Diatomaceæ or the Desmidiaceæ until a more scientific and accurate method can be inaugurated.

The silicious covering of the frustules of diatoms are altered by accretions from the surrounding water after their embryo formation with age, partially obliterating, in some cases, their primary details, and in the preparation and mounting of them for preservation as slides this also affects their natural appearances.

Navicula major is bulged around its central nodule, and the costæ there have a convergence, and often they are set slightly oblique on the valve front and not perfectly transverse. The length is about 1/80th inch in some of its largest forms and of all sizes downwards. Its habitat is frequently among the ooze at the bottom of well-lighted, shallow streams, where a good supply of decayed vegetable matter is found alongside. It moves very stately along, and is very prone to roll on account of its rounded valves and bulging centre.

Pleurosigma attenuatum (Fig. 100, facing p. 99)

The frustule of this diatom is of the usual Navicula, box-shape type, with both raphe and valve outline sigmoid or S-shaped. It is very finely marked on the front and is a free swimmer. There is a conspicuous nodule in the centre of the valve, but the end ones are obscure, the median line running through to the apices.

Upwards of two hundred variations of this genus have been tabulated owing to some slight deviation from the normal, but all may be readily recognized as belonging to it by the particular curved sides and raphe upon the valve front. On the girdle or side view the frustule is seen as an elongated oval showing none of the sigmoid curve, and in this position may be difficult to decipher the proper class to which it belongs. The brown endochrome in the living frustule extends along the sides, seen in front view, to within a short distance of the ends, leaving a clear space there and also down the centre longitudinally. On the girdle side, however, it appears to fill the central as well as the sides, leaving clear spaces either side the central nodule, where the thickened nature of this intervenes, indicating its outline and substantial character, and also at either apex of the diatom as in the front view.

Like most of the commoner diatoms, it has had numerous names appended to it, both of species and genus. It was separated from Navicula first under the genus of Gyrosigma and again styled Sigmatella and so on. It is at present under the Naviculaceæ, which is defined as constituting " valves not arched or keeled and usually symmetrical with reference to a straight or a sigmoid raphe."

The striæ transversely count about forty in 1/1000th inch, and longitudinally slightly less, thirty to the 1/1000th inch. This is on account of the minute rosettes or nodules being set slightly wider apart on the valve front from side to side than they are from end to end.

The frustule consists of more than one layer of silica, the upper one may be likened to glass marbles, round and transparent, set in a more impure heterogeneous mixture, resembling alabaster and of a softer texture, in the living state. This looks as if it could be cut with a knife, cheese-like, while the nodules are crystalline and brittle. The under layer, if a dry, mounted specimen is viewed, may often be observed where the upper has peeled apart, and is of a smooth surface, upon which an imprint or embossed surface of the upper is left upon it.

Each tiny nodule is made up of smaller details of a cellular nature and peep out through the layer surrounding, and can be made to show either black or white according to focus and the arrangement of the light. The writer suggests the microscopist endeavours, in viewing them, to stop out central axial cones with a small stop in his condenser and to view them as white rosettes with the intermediary substance as black lines to obtain the most correct view when using achromatic objectives and upon a specimen that is mounted dry. A mere touch then of the fine adjustment will reverse the view and give black little columns with clear spaces between. Sometimes along the curve of the sides of the valve these lines of black dots will show branching points to a line beside it, like the two prongs of a fork, and this a little farther may again fork out, this method filling in the greater surface upon the curve so that the whole front shall have its regular quota for the extra space.

A Zeiss E objective on a 160 mm. (about 6½ inch) tube and an eyepiece of ×10 shows these points nicely upon a dry specimen, and there is no necessity for using an oil immersion for either condenser or objective. This arrangement gives a magnification of about six hundred diameters. In mentioning this the writer does not imply that a lower power will not succeed, but that as an achromatic combination it has been used by him successfully and comfortably, and that an air angle of illumination is amply sufficient for the purpose.

Surirella splendida (Fig. 101, facing p. 114)

The frustules of this species are large and widely distributed. It is frequently found in the surface sweepings of reservoirs, lakes and broad stretches of open water. The empty frustules are light and buoyant and its habitat therefore favours positions where there is little or no current, yet a free, open stretch of water prevails which receives plenty of sunlight and replenishment.

The front and valve view is egg-shaped with a broad, rounded base tapering to a narrow curved apex. The costæ are stout and distinct, somewhat irregular, reaching the median line, also distinct. The spaces between them appearing lower than the edges of the valve which gives a winged algæ-like appearance at the margins. In the girdle view these wave-like edges are absent, the frustule is broadly wedge-shaped, which readily distinguishes it from Cymatopleura, a somewhat similar diatom in which such outline is conspicuous. The light brown endochrome in the living frustule follows the outline of the edges on a front view. There is a clear, rounded space at either end, but no central nodule nor any raphe as in the Naviculæ. The whole diatom forms a box-shape structure, however, and between the transverse costæ fine radiating lines of small rosettes may be seen. These will require careful manipulation in some cases to show themselves here and there upon the wave-like surface of the valve. They are very regularly placed side by side, but a much wider space intervenes between the rows from top to bottom.

The length of the frustules attain as much as the 1/100th to the 1/200th of an inch. They form a very beautiful species, and in mounted specimens their pearly structural details and waved contours glisten under bright illumination in a very charming way.

The greater number of species are marine, and *S. gemma*, which is often used as a test object, is found commonly in marine marshes and tidal basins, harbours, etc.

Stauroneis acuta (Fig. 102)

An elongated, slender, diamond-shaped, or lanceolate diatom, having a distinct stauros or cross formed by its longitudinal raphe and a conspicuously plain, central, clear band, fairly broad and dilating outwards as it reaches the margin of the valve, front view. It will be readily recognized by this distinguishing feature. The apices are also plain, as though thickened at these points. The striæ are oblique and the valve highest along the median line.

Fig. 102. Stauroneis acuta.

There are many variations of the stauros; some are quite straight across, some show but little more than the median line longitudinally, others have a marginal line within the edges around the valve and the cross falling short of the edge transversely. All will, however, leave no room for doubt about the genus to which they belong when the cross is present. In some the raphe and median line is sigmoid, as in Pleurosigma, and with a broad plain band across, while again both raphe and transverse band may be narrow but unmistakable. The apices too will sometimes be knobbed and slightly capitate in their outline with a narrowing upon the outer edges towards the centre.

In *S. acuta*, here illustrated, the diatom is widest at the centre and tapers gradually towards the ends. The striæ count about thirty in the 1/1000th of an inch and the length of the diatom longitudinally about 1/500th inch. It is found principally among the ooze at the bottom of lakes and the larger-sized ponds in many parts of Britain, and in such places is fairly common. Its girdle side is box-like and oblong. It is not dredged for so frequently by microscopists perhaps owing to its habitat, and only stray specimens are observed as they float upon some object to higher levels and present themselves, but even in this accidental manner one will often come across a specimen of Stauroneis of one species or another. It is a very pretty diatom, with its brown endochrome situated along each side of the valve.

CHAPTER VII

DESMIDS

THESE are exclusively fresh-water plants, not a single species being found in the sea. They are a very large family, all single-celled. True, in a few species they exist end to end ribbon-like within a gelatinous sheath, but each is competent upon its own to produce its kind and may therefore be looked upon as a colony of individual single cells. The lovely emerald tints and sculptured contours, the diversity of embellishments and design, have justly endeared them to all microscopists for their meritorious and intrinsic beauty.

They are to be found in quiet streams upon the ooze in slowly running waters in positions similarly favoured by Euglena. Upon the light green Sphagnum Moss around the peaty bogs and pools on hills and moors they are often in great abundance. Sunlight is attractive to them, and when gathering, the time of day employed has much influence upon the success of the undertaking. If seeking them in the stream upon the soft mud let the sun be high and shining if possible, then a gentle disturbance of this will bring them floating to the surface and in that way a cleaner collection may be obtained and in a simple manner by skimming the surface and placing this in the bottle and allowing to settle. Some of the water may then be poured off and a fresh gathering added. Should the stream be thought too swift for this method, take a portion of ooze out gently and pass through a linen handkerchief or cloth and when most of the water and mud has drained off dip this into the dish or wide bottle and so wash out, carefully, a cleaner and more fruitful catch at each operation. Bolting cloth of varied meshes may be obtained for such purposes.

When searching Sphagnum their likely presence is indicated by the gelatinous, slippery feel it gives to the touch, portions of which may be gently detached and floated into the vessel while this is in the water. Clayey subsoil in clear, still pools that are open to sunlight is another congenial habitat. Stems of sedges, rushes, and many fixed aquatic plants are also favoured spots. These may be gently scraped of the mucus upon them and will often be prolific of Micrasterias, Cosmarium and the flatter species especially.

Unlike the diatoms, they have no silica in their composition and are therefore more delicate. This must be remembered in the handling of them. Neither heat nor acids should come in contact with them. To remove single specimens from one place to another a fine hair brush or pencil may be employed; they readily adhere to these and can be thus easily manipulated. They may be allowed to dry upon glass, and with the application of moisture they once more swell out and assume their former proportions. This applies more especially to the flatter specimens.

Their mode of reproduction is twofold. It can be accomplished either asexually by the cell dividing into two halves budding out upon the broken sides a new half-cell portion, fellow to the old, and thus becoming complete once more, or by conjugation, that is a more complex procedure, two cells coming into contact side by side and from the middle portion of each by a union of the two a third distinct and separate cell body called a zygospore is produced. It is evident, however, that this latter method is not to the advantage of the family as the destruction of two plants for the formation of one must eventually arrive at the elimination of the species. Its purpose appears to be chiefly for their preservation during the cold and rigours of winter.

If desiring to preserve specimens, one very old and simple method may be used, known as the " 1, 2, 3," i.e. 1 part glycerine, 2 parts distilled water, and 3 parts of pure alcohol. The specimens, if to be mounted, are taken up on the hair pencil and laid on a clean slip, a drop of distilled water touched upon each, and then the above solution laid over. This has been well tried and allows of the form of the plants and the colouring of the endochrome to be kept very faithfully for a number of years. The sealing down of glycerine mounts being the chief bugbear, but as this is here in very thin solution, with care has been generally overcome.

Another method devised by M. Petit, a French microscopist, which is also considered an excellent medium is: Camphorated water and distilled water, of each take 50 grammes, glacial acetic acid 0·5 gramme, crystallized chloride of copper 2 grammes, and crystallized nitrate of copper 2 grammes, dissolve and filter. Immersed in this they keep their colour well and the cell contents are not prone to shrink perceptibly for a considerable period. Seal, if mounting, with shellac.

Scenedesmus quadricauda (Fig. 103)

From the pools on the Welsh and Yorkshire moors, where peaty bogs abound, this curious little plant is very plentiful. It is a union of cells more or less oval in shape placed side by side, having a curved hook or horn projecting from each corner. Curious to note the cells are generally found in some even number together, very rarely an odd, usually four, but may be either two, four, eight or sixteen. A central chromatophore or colour cell is plainly seen amid the otherwise granular material. It is a hardy plant, and once established will survive where most other alga fail in unfavourable conditions. It increases in size by the cells evenly dividing transversely into two halves, forming two groups, attaching themselves while still within the parent cell side by side, and finally bursting the outer case where there is now two quartettes to roam freely.

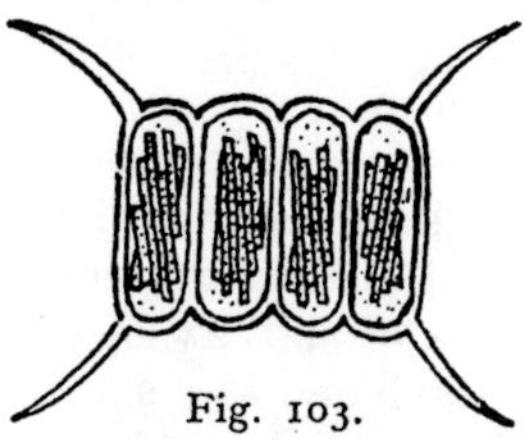

Fig. 103.
SCENEDESMUS QUADRICAUDA.
Showing segmentation of cell contents.

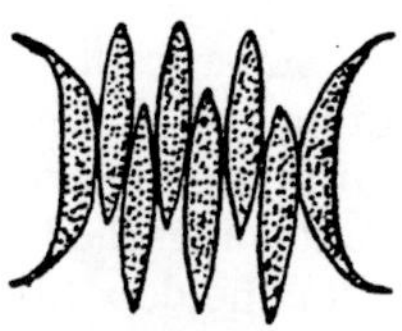

Fig. 104.
SCENEDESMUS OBLIQUUS.

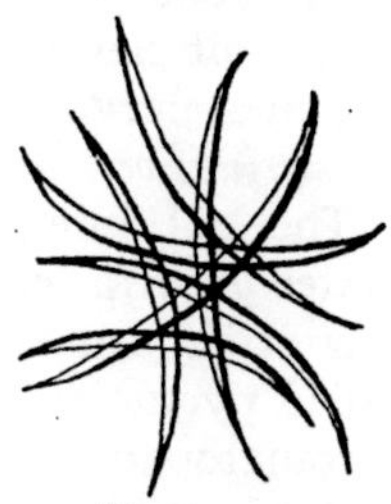

Fig. 105.
ANKISTRODESMUS FALCATUS.

It is an alga which readily adapts itself to artificial cultivation, and many curious forms of it have been produced upon a medium of agar and glucose. Agar is a product of a Japanese seaweed and is procurable at most large optician houses.

Scenedesmus obliquus (Fig. 104) has its eight cells placed alternately above and below the median transverse axis and has been observed to possess the property of liquefying gelatine. It will also accommodate itself to various media and temperatures, and remarkable deformities can be obtained by such experiments with it.

Polymorphism (i.e. many forms) occurs in the families of most alga, however, and some possess different vegetative forms, as in Pleurococcus and Botrydium.

Ankistrodesmus falcatus (Fig. 105)

Found either in single cells or in fascicula, i.e. faggot-like bundles. The separate cell is needle-like, attenuated or often acutely curved, the length always exceeding many times the diameter. Their reproduction is by oblique division. It is common in all fresh waters.

A bundle may have as many as sixteen cells, crossing like boomerangs over one another in their centres, the ends being arranged apart as much as possible. Their length may be the 1/500th inch and the width as little as 1/7000th inch.

Closterium

This is an elongated and usually more or less crescent-shaped cell, having a narrow constriction transversely across its centre and enclosed in a transparent cellulose case. Owing to its long body proportionately to its width and the influence of light principally, it never manages to grow uniformly all round and thus becomes curved. In *Closterium lineatum* this is slight, but in *Closterium venus* it attains almost to a complete ring.

The cell interior has superficially five or six radially placed plates or strips constituting the chromatophore in which is a single row of circular bodies of a denser structure and proteid character called pyrenoids. These are regarded as centres of activity of the chromatophore, serving for the deposition of starch and as a source of food supply. In short, a synthetic food factory. At each end is a clear space in which is situated a single vacuole containing many highly refractive granules in constant motion, dancing away their merry lives with a Brownian, jigging movement whilst many others are passing by in the protoplasmic stream, encircling their home, giving a pause, a look on, but never mingling within. Each is composed of two or more still smaller ovoid or circular particles and all enclosed in their hyaline coats. Their purpose is not definitely known.

Between the cell or chromatophore and its covering is the stratum of streaming protoplasm constantly gliding along here and there, round and over, the whole cell. This contains small rounded bodies amongst it, which may be seen to hurry along so far, then to stop and return a little way, then to find a hollow in which another little friend may be, when both career around together. Separating, one may meet an avenue amongst numbers of others where it will, with a few ups and downs, continue the whole length of the cell. At some points, as if sliding downhill, they will make quite a rapid dash, and so the busy round goes on, each one continuing its switchback course incessantly. This is called "cyclosis," and Closterium is the only Desmid in which it can be seen to such perfection, if at all. Their method of increase has already been explained, and is either by dividing in halves and budding new parts to the broken edges or by forming a sporangium or zygospore and developing anew from this specially formed body.

Upon the outer cellulose coating with good lenses a very fine series of lines may be seen; these striations are part of the actual structure of the membrane, and run lengthwise along the whole cover of the cell. When the plant is living and the green chlorophyll is within, they are seen with some difficulty, but upon empty cell cases they are easier. They are probably thickenings of its transparent medium.

Closterium has such a numerous progeny that only a few examples can be given. Most if not all of these are figured in W. and G. S. West's "*Monograph of British Desmidaceæ,*" which contains very full and lucid details.

Unfortunately, line drawings give no adequate idea of the intrinsic beauty of these objects; the brilliant greens and flashing, transparent membranes must be seen to be realized, whilst in their living, natural state beneath the microscope.

Closterium aciculare (Fig. 106)

This is a very long, thin specimen almost straight, and while its width is only 1/500th inch, its length is 1/5th inch, so that it is merely a needle in appearance. It is very slightly wider in the

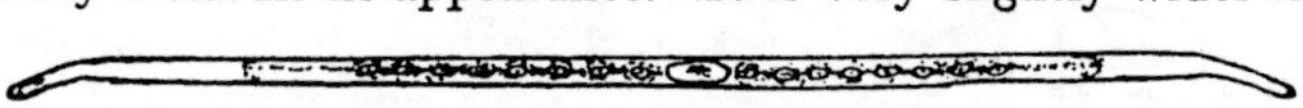

Fig. 106.—CLOSTERIUM ACICULARE.

centre, with a smooth cell wall tapering to very narrow and slightly incurved ends. Has twelve or fourteen rounded pyrenoids amidst its green chromatophore, and in rather long, transparent vacuoles at each end, just one or at most two moving, dancing granules.

Closterium lineatum (Fig. 107)

More crescent-shaped than the last, the centre cylindrical and more full, tapering to broader and rounded ends. The cell wall

Fig. 107.—CLOSTERIUM LINEATUM.

is striped with about fifteen longitudinal striæ, the number of striæ always being representative of those upon the actually visible part and not all round the specimen. Its length is about 1/50th inch and breadth 1/1000th inch.

The chromatophore has five or six ridges of darker green bands in which, along the length, are about twenty pyrenoids. In the vacuoles at the ends are clusters of small granules. The centre is oval and transparent and the fine division line clearly seen.

Closterium strigosum (Fig. 108)

This is about 1/80th inch long and 1/1600th inch wide and the extremities of the cell incurved and acutely pointed. It has three

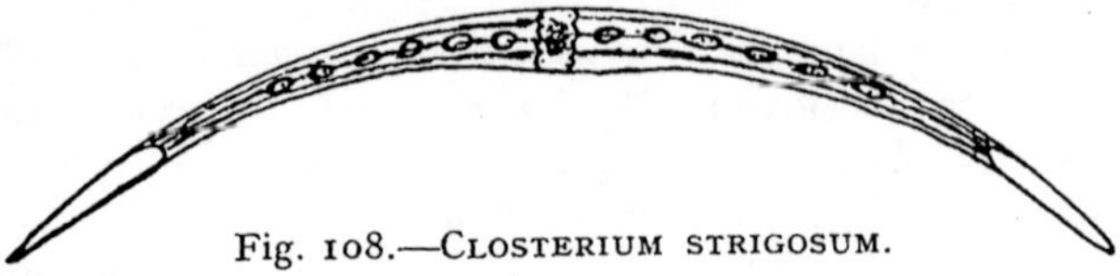

Fig. 108.—CLOSTERIUM STRIGOSUM.

bands of green along the length and fourteen or fifteen pyrenoids, its vacuoles at either ends being tapering and with several moving granules.

Closterium cynthia (Fig. 109)

This is exceptionally curved, almost to a circle. It is larger than *C. venus*, although still very small, being but 1/200th inch in widest part across and 1/1600th inch in central diameter. It is not swelled in the centre, and narrows very gradually to an obtusely apice each end.

Fig. 109.
CLOSTERIUM CYNTHIA.

There are six pyrenoids in the light green chromatophore, and the terminal vacuoles have but one granule in each. This differentiates it from *C. venus*, which has several. The cell wall is also finely striated, having fourteen striæ visible, while *C. venus* is smooth. The empty cell membrane is of a yellowy brown in colour.

Closterium dianæ (Fig. 110)

Strongly curved; centre slightly bulged; chromatophore obscurely ridged and containing twelve or fourteen

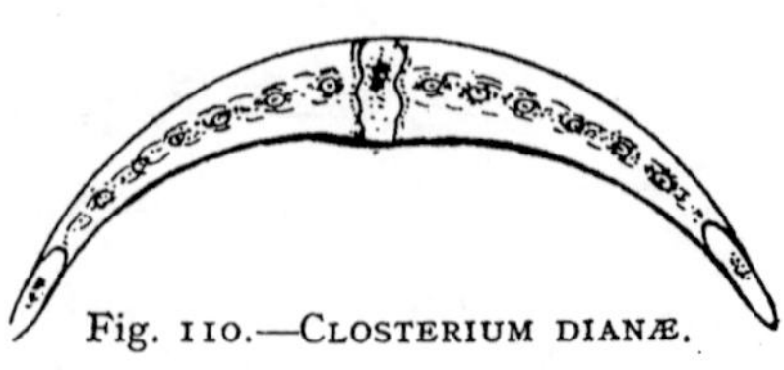

Fig. 110.—CLOSTERIUM DIANÆ.

pyrenoids in the longitudinal median line; terminal vacuoles with many granules. Length 1/80th inch, breadth 1/800th inch.

Closterium lunula (Fig. 111)

Broad and almost straight, slightly thicker in the middle. The ends are recurved somewhat and obtusely rounded. There are five broad thick bands lengthways in the chromatophore and many

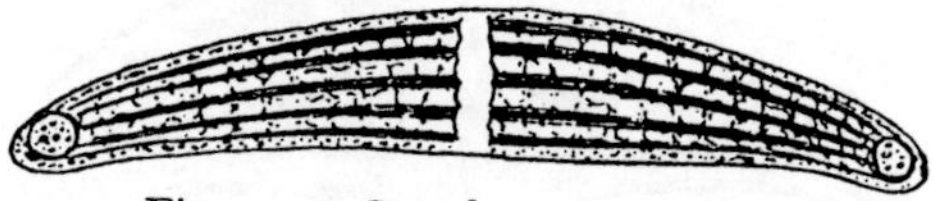

Fig. 111.—CLOSTERIUM LUNULA.

small scattered pyrenoids. Large clusters of granules in the terminal vacuoles. The length 1/50th inch and breadth 1/250th inch. The cell wall is devoid of colour and smooth.

Closterium didymotocum (Fig. 112)

Cells large, outer margin but slightly curved, inner almost straight. Middle portion of cell has sub-parallel sides. Chromatophore with twelve to fourteen pyrenoids. Many moving granules

Fig. 112.—CLOSTERIUM DIDYMOTOCUM.

in vacuoles, and usually an annular thickening of a dark brown colour at each apex. Cell wall is reddish brown, rarely striated, and the ends bluntly cut away. The zygospore is not known.

Closterium leibleinii (Fig. 113)

Very crescent in outline, the outer margin sometimes as much as 190° of arc and the inner strongly concave. At the middle it is bulged. There are six large pyrenoids along the length of the chromatophore. In the end vacuoles, numerous moving granules. Its length is 1/165th inch and breadth 1/1000th inch. The cell wall is smooth and transparent when seen apart from its green chlorophyll. Rarely it is a pale straw colour.

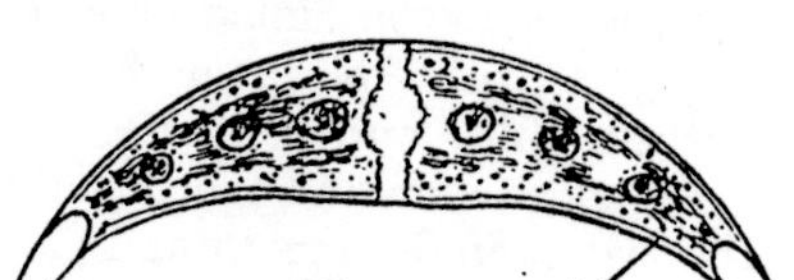

Fig. 113.
CLOSTERIUM LEIBLEINII.

Closterium ehrenbergii (Fig. 114)

Large and stout, not quite so crescent as *C. Leibleinii*. The inner concave, and is inflated in the central part. The chromatophore having eight or ten bands of darker green, and the pyrenoids

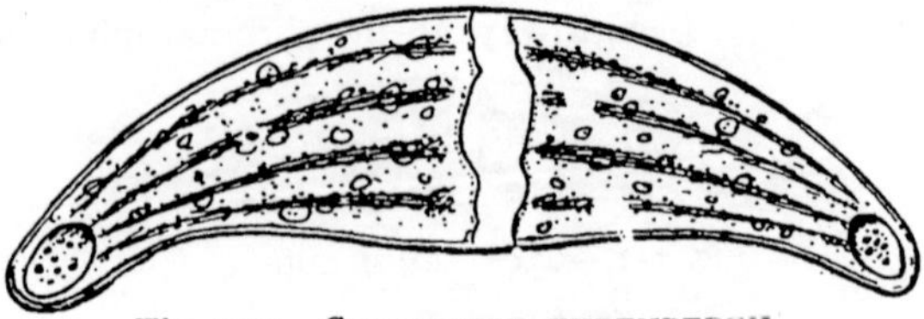

Fig. 114.—CLOSTERIUM EHRENBERGII.

smaller, more numerous, and scattered over the whole cell. Vacuoles with many small granules. The zygospore of this is smooth and globose with a mucous lamellose coat.

Penium

This genus is a somewhat artificial one, differing little from Closterium, save that its species are of a straighter and more cylindrical outline, with blunt apices, and includes species as at present arranged of widely differing affinities.

In *P. margaritacium* (Fig. 115) the cells are cylindrical, tapering slightly to rounded obtuse ends, and furnished with about ten rows of granules in longitudinal plates or ridges.

The cell walls are of a reddish brown when seen apart from the

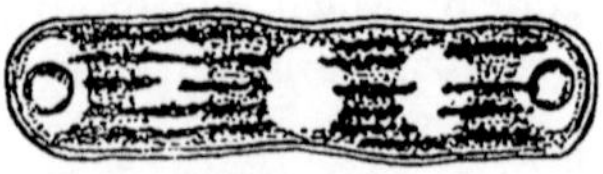

Fig. 115.

PENIUM MARGARITACIUM.

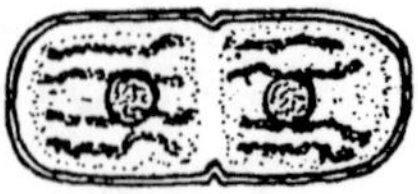

Fig. 116.

PENIUM CURCUBITINUM.

chlorophyll. There is a slight but distinct constriction in the middle of the frond with frequently a terminal vacuole at the extremities each containing moving granules. The bands are of a darker green than the main chlorophyll, which is attached to the inner cell walls, and contain several pyrenoids. The length is about 1/250th inch and the breadth 1/1000th inch.

The zygospore is smooth and globose and about 1/550th inch in diameter. Moss and peaty bogs is a frequent habitat. It is fairly general, but never in any large quantity together.

Penium curcubitinum (Fig. 116)

Cells slightly constricted in the middle and broadly rounded at either end; about six times longer than broad, with a smooth outer wall which may occasionally be punctated or striped longitudinally.

The chromatophores are placed radially in each cell half, and enclose one fairly large pyrenoid in each. There are no vacuoles nor moving granules in this species at the apices.

It is a small, chubby specimen.

Spirotænia

This is similar in outline to Penium, rather more spindle-shaped, however, and has a spiral band of endochrome running from end to end. It is cylindrical, with rounded ends and without any constriction in the middle. The band contains several small pyrenoids. There is no clear space at the extremity and no moving granules. It multiplies by division, severing the cells obliquely, not transversely, across the centre. This is *S. condensata* (Fig. 117).

Fig. 117. Spirotænia condensata.

The gelatinous envelope in all the species is very thick and readily apparent. The length of the desmid is about 1/200th inch and the width 1/1000th inch; it varies with age and environment slightly.

In *S. obscura* (Ralfs) (Fig. 118) the frond has several spiral bands slenderly lining the inner wall with two or three helices to a cell, and will sometimes leave a clear space at the extremities in which may be a single free granule in motion.

Fig. 118. Spirotænia obscura.

Micrasterias

Is broadly oval, almost circular, divided nearly into two halves at the centre by deep fissures. It is flat and in sharp distinction from Closterium. The edge is indented in most species, with many longer and shorter clefts around its margin. Outwards from this bright green margin is a transparent band, sometimes having a few moving, dancing granules within it.

This desmid forms an excellent example to show the method of self-division and its mode of growth in forming a new half upon the old one.

Some of the species are widely open at the margins of the central fissure and might be thought to constitute two cells, so slender is their attachment at the centre.

In the process of conjugation the cell as usual divides at the narrow central bridge, and the contents of both then run out through the apertures, blending together to form a new body, the zygospore. This becomes circular and covered with a thick envelope protected

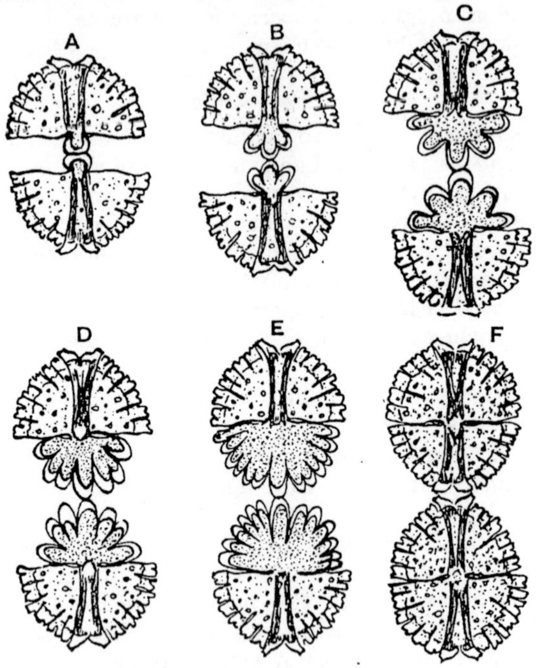

Fig. 119.—MICRASTERIAS ROTATA.
In self-division.

by several branching spines, cleft at the extremities and projecting on all sides. In that condition it rests until a suitable season arrives to break open and commence active life, finally assuming the form and shape of its particular species.

In the sketches (Fig. 119) self-division is shown in various stages. First the bridge is broken along the line of central fissure and upon this part on both sides a small disc begins to swell out, plain-edged, and having a short hyaline envelope around. These remain touching each other, but making a wider space between the older cell halves. Presently indentations appear around the disc as it grows larger.

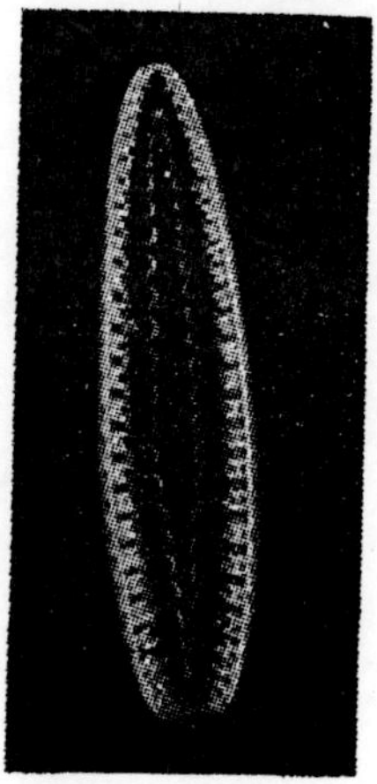

Fig. 101.
SURIRELLA SPLENDIDA.[1]

Fig. 120.—MICRASTERIAS ROTATA.

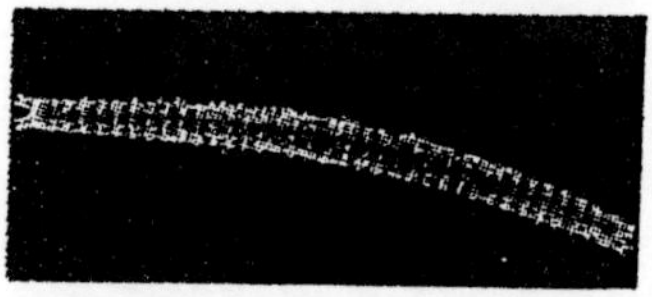

Fig. 130.—TRIPLOCERAS VERTICILLIATUM.

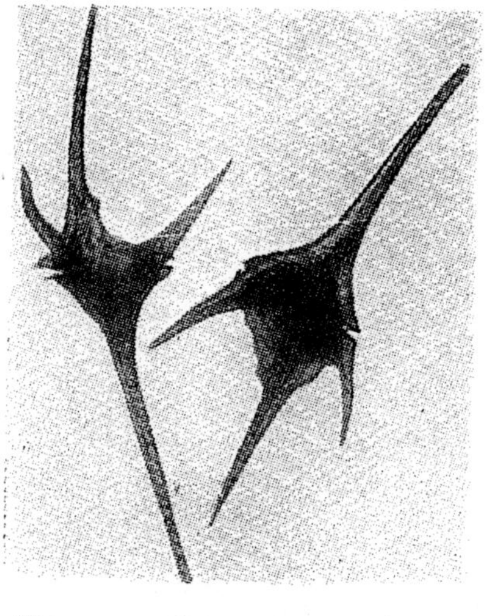

Fig. 179 A.—CERATIUM LONGICORNE.

[*To face page* 114

It continues in contact, however, until the new halves are almost completed. The deeper clefts appear first and the shallower next, the very small ones being last of all.

If the season is a warm and suitable one the new halves will attain almost full size before the clefts are fully formed, while if their growth is retarded or becomes sluggish the indentations take place much earlier.

Like many of the desmids, Micrasterias favours peaty pools that are little disturbed from one year to another.

Micrasterias rotata (Fig. 120)

This has its larger polar lobes, triangular, with slightly concave sides projecting farther outwards than the others, the base corners being slightly protruded into two lateral lobes. The side view of it is ovate, with an inflated base. The apex somewhat broadly conical. Its cell wall is finely spotted. It is said to rotate, hence its name. The dancing granules within its hyaline margin, sometimes seen when old, may give it gentle movement, but the writer in witnessing many hundreds has never yet seen a specimen make even one complete rotation.

Its zygospore is circular and beset with spiky protuberances. Its length is about 1/85th inch and breadth 1/120th inch. The zygospore, without its spines, 1/225th inch.

Micrasterias denticulata (Fig. 121)

Is broadly oval, with deep indentures. The smaller, saw-tooth edge is not so pronounced as in *M. rotata*, and its polar end lobes more dilated at the apex. The cell wall is sometimes coarsely spotted or punctate.

It is a very common specimen among desmid gatherings. Its length is 1/90th inch and breadth 1/110th inch.

Fig. 121.
MICRASTERIAS DENTICULATA.

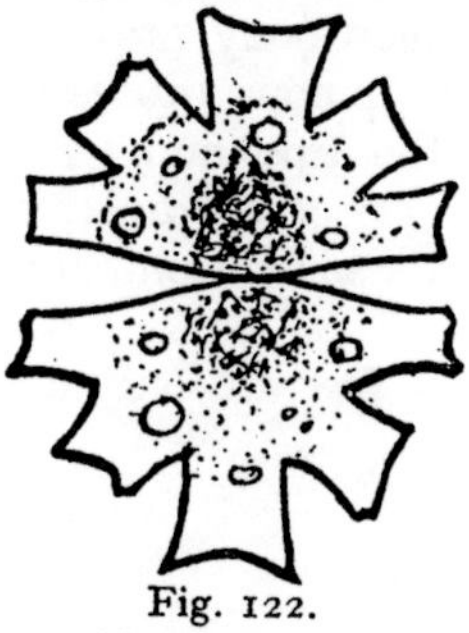

Fig. 122.
MICRASTERIAS APICULATA.

Micrasterias apiculata (Fig. 122)

Has very deep constriction across its centre with wide outer openings. Its semi-cell is five-lobed.

Length 1/100th inch, breadth 1/110th inch.

Its polar lobe projects slightly, and its denticulation less evidenced.

Micrasterias truncata (Fig. 123)

The indentures at either pole of this are absent. It is barrel-shaped, with a deep fissure either side of its centre. Seen in profile, the side view is broadly ovate.

Its length is 1/250th inch and breadth 1/300th inch.

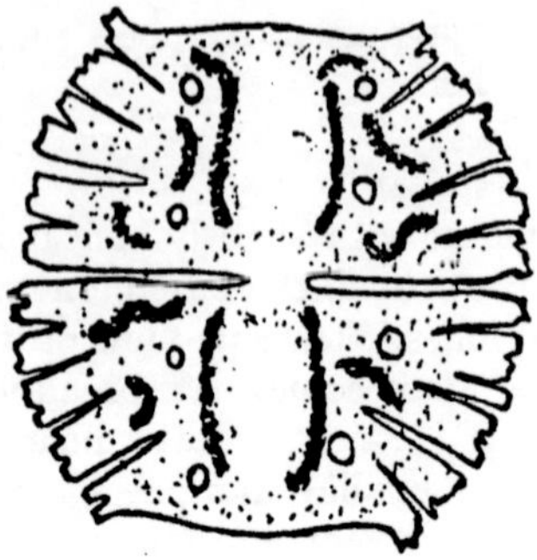

Fig. 123.
MICRASTERIAS TRUNCATA.

Cosmarium

This desmid seen at first glance might be confused with Micrasterias, its outline being somewhat similar in shape but devoid of any indentations save its central sinus. It may, however, be promptly distinguished from it if an end-on view is obtained. From such an angle it is oval in shape and not a mere line more or less as with Micrasterias.

If an empty cell can be found this will also differentiate it, rendering its characteristic knobs upon the surface distinct and the thicker nature of its body becomes then more clearly recognized.

These warty prominences in Cosmarium are arranged in decreasing semicircles from the edge to the bridge of each half cell. This bridge is narrow, amounting to little more than a short tube connecting either half.

In the multiplication of its species it breaks in two, allowing the contents of each part to exude, unite, and form a new body, growing a wall of its own, circular in shape, and with spinous processes around it. These processes are indented at the tips, and have three sharp points. This altogether forms a perfect sporangium or zygospore. It eventually turns brown, and its size is comparatively large to the size of the plant.

There are many scores of species and only a few are outlined here as representative of the variety.

Cosmarium orthostichum (Fig. 124)

This is rather small and deeply constricted at its centre, the sinus having a slight dilated extremity. The half cells are evenly semicircular, and in side view become almost circular in outline.

The upper surface is knobbed with relatively large protuberances in three convex rows toward the margins of each semi-cell. They are fairly distant and distinct, varying sometimes in their relative sizes.

It is not a common species, but plentiful in old and deep Sphagnum bogs. Its length is 1/700th inch and breadth 1/800th inch, about.

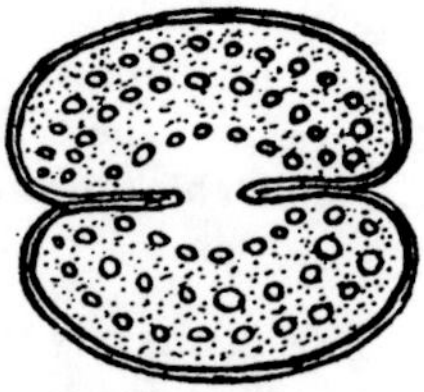

Fig. 124.
COSMARIUM ORTHOSTICHUM.

Cosmarium taxichondriforme (Fig. 125)

Another small specimen, somewhat hexagonal, with a deep fissure at its centre, slightly undulating in outline and more open than is usual with most other species. Within its chromatophore are two circular pyrenoids in each semi-cell.

Its length is 1/700th inch and breadth 1/750th inch.

Cosmarium botrytis (Fig. 126)

This is circular or elliptical in its contour and with an undulating margin or sheath. The central constriction is deep and straight.

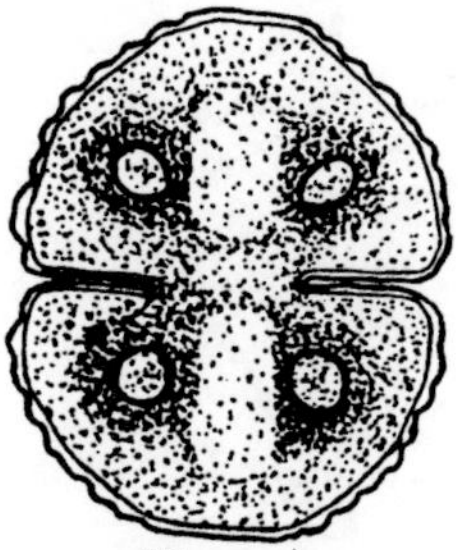

Fig. 125.
COSMARIUM TAXICHONDRIFORME.

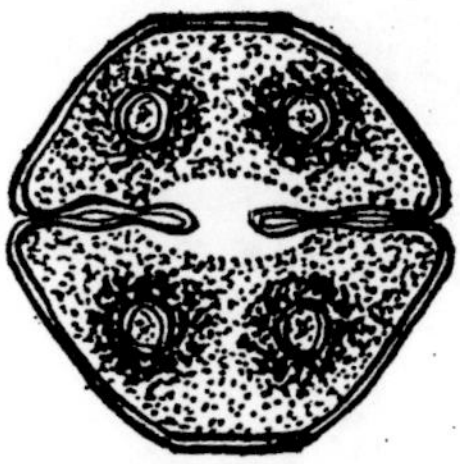

Fig. 126.
COSMARIUM BOTRYTIS.

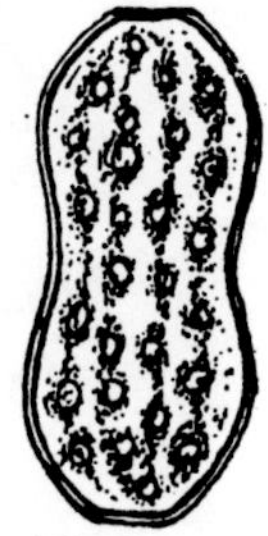

Fig. 127.
COSMARIUM TURGIDUM.

It is a thick specimen in relation to its upper view, the end-on aspect being almost circular.

The chromatophore in each semi-cell consists of thin radiating layers having two pyrenoids situated within. It is usually punctated with many minute knobs.

Cosmarium turgidum (Fig. 127)

The cells are very large and but slightly constricted in the centre. Each semi-cell is ovate, the sinus consisting of a mere rounded notch. The vertical view is circular and the cell wall minutely pitted or depressed. Chromatophores are in bands attached to the walls of the cell and contain about twenty-four pyrenoids along the four chloroplast strips.

Euastrum

This genus has an elongated, oval outline with the ends truncated generally. Its form is plainer than Micrasterias. It has a central fissure, and is waved along the margin, never sharply pointed. It is usually filled with the endochrome, and the surface does not show its markings well, but if an empty cell be found it will be seen to have numerous minute knobs standing out upon it.

Euastrum verrucosum (Fig. 128)

Has three lobes in each half cell, the central one somewhat square and each with a concave outline. The points are rounded and without indentations.

The side view of each semi-cell is widely inflated on the lower part

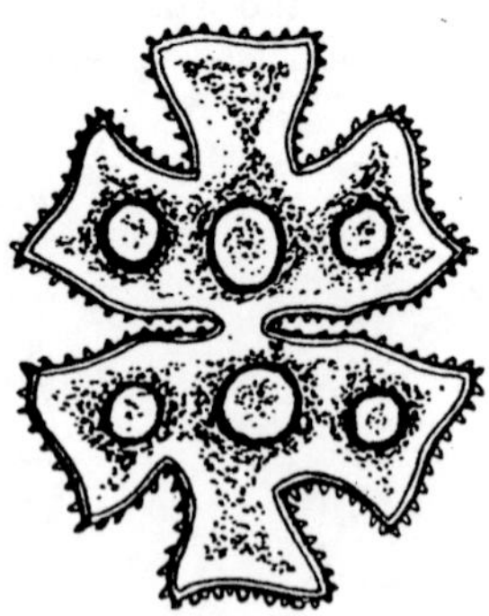

Fig. 128.
EUASTRUM VERRUCOSUM.

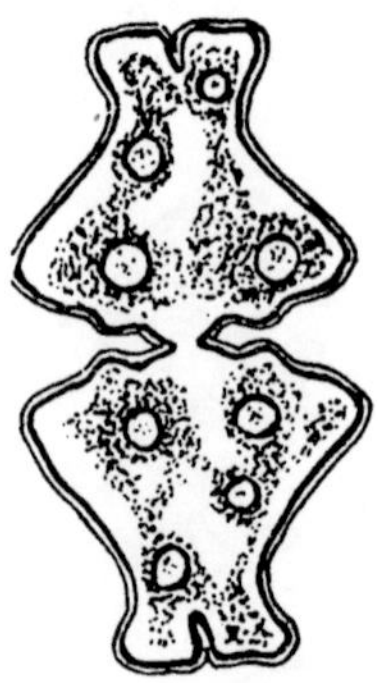

Fig. 129.
EUASTRUM INSIGNE.

owing to the central protuberances showing at each side. Hence the name verrucose or warty. It then narrows, forming a neck, the apical portion dilating.

It reproduces in the same way as Micrasterias. Length 1/250th inch, breadth 1/320th inch.

Euastrum insigne (Fig. 129)

The cell is deeply constricted in the middle, outer edges widely dilated. The semi-cells are broadest near the centre sinus, tapering with concave sides to a square-cut extremity, having one central short indent.

The outline of *E. insigne* varies considerably, some are very narrow and some much wider in their stems. The chromatophore contains four or five pyrenoids in each semi-cell.

Length 1/210th inch, breadth 1/400th inch.

Triploceras verticilliatum (Fig. 130, facing p. 114)

This is an elongated species not unlike a Docidium and was formerly classed with it, erroneously. It is about fifteen or sixteen times as long as broad, having two bidentate projections at each end. Along its length are numerous rings of pyrenoids, the outline being circular, and above these are fairly sharp-pointed spines standing off from its cylindrical outline. There are as many as ten or twelve pyrenoids in each row around and about thirty-six rows.

If the specimen is seen mounted the outer coat sinks in between the projections and gives longitudinal rows of spines in appearance, but seen as a fresh object in its natural habitat, growing, it is cylindrical with a slight tapering towards each end from the middle and a fine sinuous line dividing the two halves at that part. Sometimes the points are finely notched.

The endochrome terminates at each end in a rounded, clear space near to and within the projections in which may usually be seen active granules dancing their time away merrily. There is no clear space at the central suture as in many of the Closterium species, nor rim or inflation there as in Docidium.

Being long and somewhat soft in texture it occasionally is observed with a slight curve to one side, giving an arched appearance, this is a distortion mechanically produced in removing or mounting it, and does not constitute its natural outline.

Staurastrum

It is particularly necessary in deciphering the outline of this desmid that it should be seen from both the superficial view and also an end-on or side view, otherwise it might in some species be taken for a Cosmarium. Should there be difficulty in effecting this while in water and under the microscope, gentle pressure upon one side of the cover glass will often cause it to roll over, if the thickness of water be not too shallow, and afford the necessary angle of observation.

Unlike Cosmarium which upon a side view is circular or broadly oval, Staurastrum is either triangular or star shaped in the greater number of species and so can be readily recognized. Some present a dumb-bell or hour-glass shape in this position.

Specimens may be smooth on their outer surface or with knobs or spines. The endochrome rarely fills the whole cell but congregates about its central part, leaving the triangular points clear and transparent.

It is common in peaty pools and in the shallow waters upon the moors. It favours the surface at the bottom of such places, and a skimming from there will usually provide some representative of the genera.

Staurastrum grande (Figs. 131 and 132)

In this the cells are deeply fissured in the centre and acutely angled, giving the semi-cells an elliptical outline. From an end-on view they are triangular with blunt or slightly rounded apices.

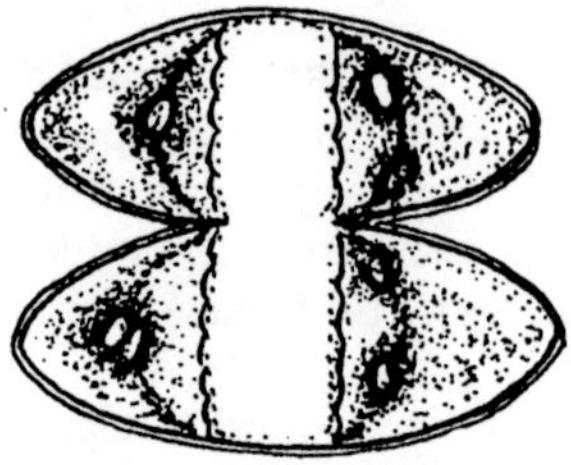

Fig. 131.
STAURASTRUM GRANDE.

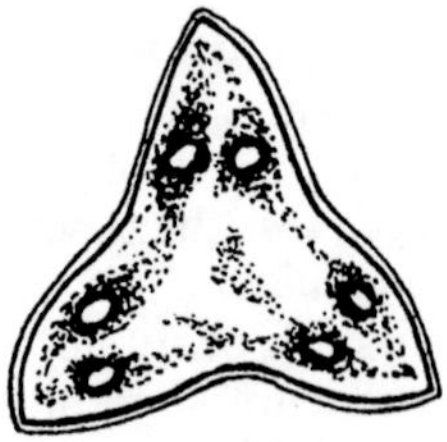

Fig. 132.
STAURASTRUM GRANDE.
(Vertical view.)

The cell wall is covered with small depressions. The chromatophore is attached to the cell walls and contains three pyrenoids usually in each half cell.

Its length is 1/250th inch and breadth 1/300th inch. It is devoid of spines.

Staurastrum longispinum
(Fig. 133)

These are rather large cells with an acute sinus in the centre, opening out widely. End view shows the margin convex and the apices extended into two stout spines

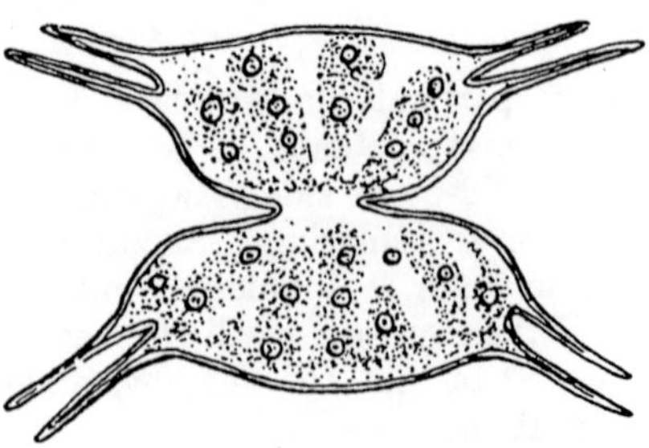

Fig. 133.—STAURASTRUM LONGISPINUM.

which vary slightly in their lengths. These project obliquely outwards.

The chromatophore is disposed into several bands attached to the cell walls longitudinally and contains numerous scattered pyrenoids.

The length is 1/250th inch, and breadth, without spines, 1/300th inch.

Xanthidium

This is not unlike a Cosmarium with spines at each angle. The cells are circular or broadly oval with a deep central fissure, the end-on view being somewhat elliptical, the outer membrane having two rows of stout horn-like spines.

Xanthidium Smithii (Fig. 134)

Cells are small with a widely open sinus at the middle and acutely angled. The sides are slightly concave, and at each corner are a

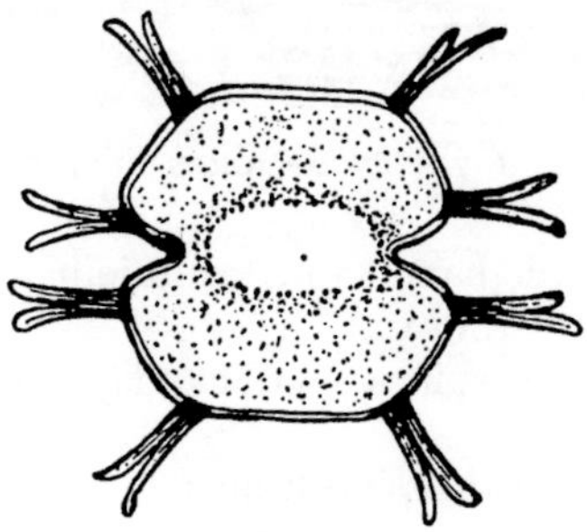

Fig. 134.
XANTHIDIUM SMITHII.

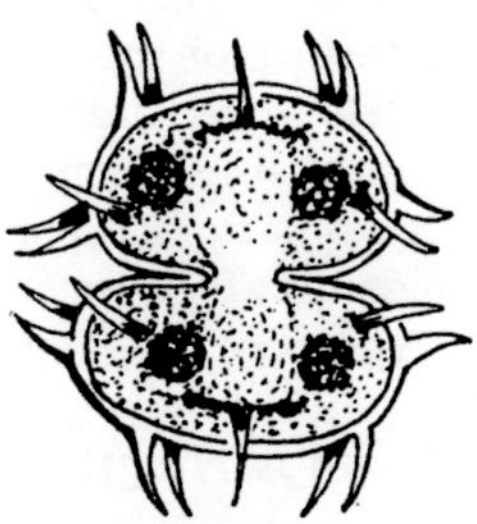

Fig. 135.
XANTHIDIUM FASCICULATUM.

pair of straight and fairly long spines. It has a small thickened area in each semi-cell. The side view of each half cell is circular.

Its zygospore is globular with sharp acute spines arising from their comparatively wide bases; about seven are visible at one time.

Its length is 1/1000th inch and breadth about the same.

Xanthidium fasciculatum (Fig. 135)

Cells are almost circular with a deep narrow sinus, the end view elliptical and often with convex sides.

Endochrome is attached to the cell wall and is somewhat divided, containing several pyrenoids.

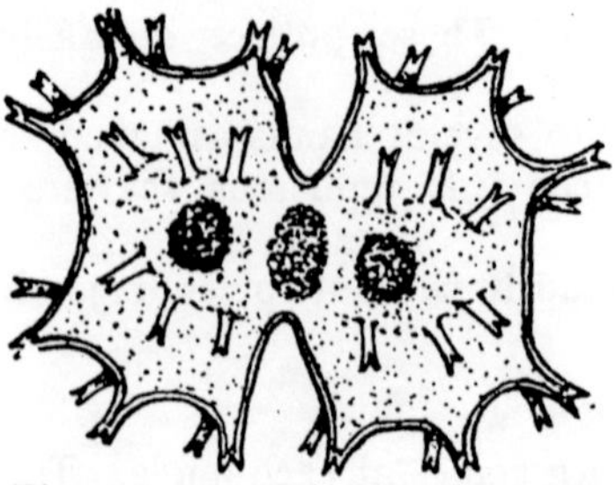

Fig. 136.—Xanthidium armatum.

Xanthidium armatum
(Fig. 136)

This has only short spines, their ends being notched, constituting one of the very few having this peculiarity.

The whole cell is about twice as long as broad, and has several rings of small raised knobs.

Tetmemorus brebissonii (Fig. 137)

The cell is sub-cylindrical, having a noticeable constriction at the centre. The semi-cells are slightly tapering towards the apices, which are broadly rounded and with a deep incision or notch in its middle. This is a characteristic feature of the genus.

Fig. 137.—Tetmemorus brebissonii.

The cell walls are minutely pitted in distinct longitudinal lines, chromatophore in a central band and containing five pyrenoids usually in each semi-cell within it.

Its zygospore is globular, without spines, and having a thick cell wall.

Its length is 1/125th inch and breadth 1/600th inch.

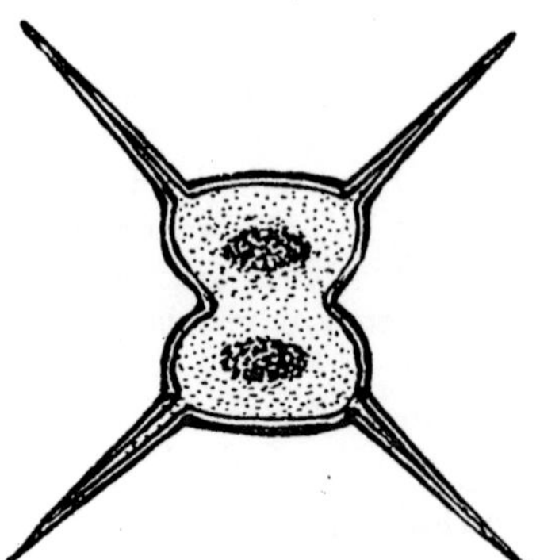

Fig. 138.—Arthrodesmus incus.
(Var. inflatus.)

Arthrodesmus incus (Fig. 138)

Rather a thick-celled species, the semi-cells being much expanded with the apices elliptical and angular, appearing as if cut off at the corners.

Spines are set at each angle slightly divergent and very stiff. Sinus in the centre short and somewhat rounded.

Length 1/1000th inch and breadth about the same.

Desmidium cylindricum (Fig. 139)

This takes the form of a long thread or cylinder with each cell placed side by side within. It is usually seen enclosed in a gelatinous sheath as in other species, Hyalotheca, etc.

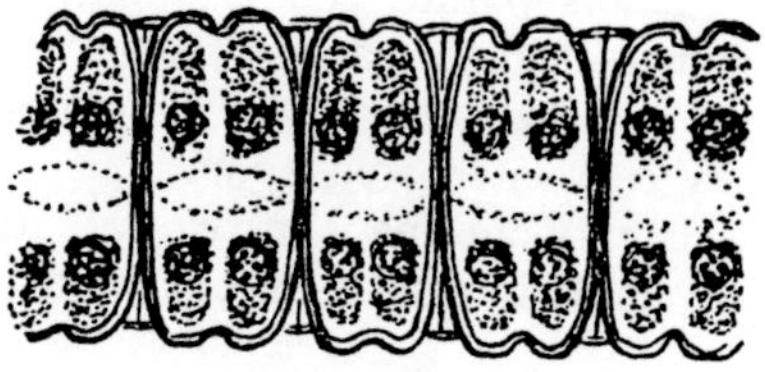

Fig. 139.—DESMIDIUM CYLINDRICUM.

The division between cells is very slight, and each semi-cell short, conical and somewhat blunt at the outer edge.

End-on view shows the cell outlines as elliptical, with a rounded boss-like protuberance at each extremity. Chromatophore longitudinally banded with two pyrenoids in each semi-cell.

Desmidium aptogonum (Fig. 140)

This is an example in which each cell has a cavity in the centre between each adjacent neighbour. It is a broadly oval thread or tube in which the cells lie side by side, taking a graduated twist as they extend along its length.

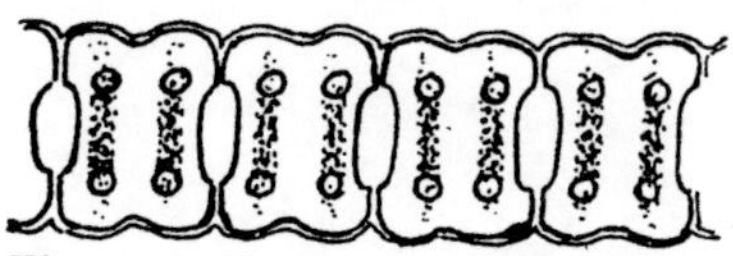

Fig. 140.—DESMIDIUM APTOGONUM.

The chromatophore is band-like, disposed lengthways, and the semi-cell have each two pyrenoids within. The layers are attached to the cell walls.

An end-on view shows each individual cell as triangular usually, sometimes quadrangular, and with the angles very broadly rounded. This desmid is relegated to a separate genus named "Aptogonum," by some authors. It is small, the width of the cell being but 1/1250th inch across.

Hyalotheca dissiliens (Fig. 141)

This takes the shape of a long cylindrical thread and forms a connecting link between the unicellular algæ and those consisting of cells united into filaments.

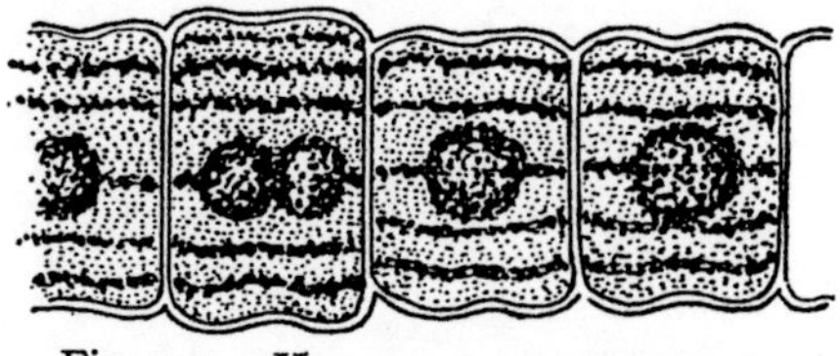

Fig. 141.—HYALOTHECA DISSILIENS.

The individual cells are disc-shaped, broader than long, lying side by side, and are encased in a very transparent, gelatinous sheath. This sheath is not readily discovered except by careful arrangement of the light or from small particles adhering upon it to disclose its outline.

Each separate cell is slightly constricted around the circumference of its disc at the centre, and at the junction to its neighbour is a shallow groove where cleavage takes place when division into new cells is entered upon.

Its endochrome is in two radiating plates containing one or more pyrenoids in each semi-cell. Breadth of cell about 1/1000th inch.

CHAPTER VIII

INFUSORIA

UNDER this heading is broadly classed all animal organisms having a single cell and found usually in standing waters. They may possess short hairs called "cilia" or have one or more whip-like processes known as "flagella." The former are ciliated infusoria and the latter flagellate infusoria.

The designation infusoria was originally given from their being found chiefly in the infusions, and again were styled "animalculæ," which means simply small animals. Under this name, however, were classed not only unicellular but multicellular animals, such as Entomostraca, the Water Fleas (Cladocera), Rotifers, Sponges (Porifera), etc. Even many minute plants, Protococcus, etc.

They form collectively Protozoa, whether flagellate or ciliate. The flagellated forms being distinguished by the title "Mastigophora" in distinction to the ciliates, whether they possess cilia throughout their whole life or only during the embryo stages, which retain the name Infusoria. Both are of world-wide distribution and found in every ditch and rivulet to the larger ponds and pools, etc., wherever situated. If a small quantity of dried hay be placed in water and left a week or so numbers of forms will have developed out, showing that decaying vegetation is their natural habitat, and in the similar natural haunts, among the last year's decomposing plant life, around the margins of lakes, etc., may be usually gathered a plenteous supply of their tiny forms.

Although living thus among material of decay, they are fond of fresh air, and if this is denied them they will quickly succumb and die. In gathering them it is necessary to remember this and not to keep the container corked up a longer time than is needed.

The cilia form the means of locomotion and also of directing to the mouth the various food substances found in the water. They arise principally from the outer coat or ectoplasm of the body, and may be restricted to certain areas or surrounding the animal on all sides. Tufts of them will be used to form temporary supports or to attach themselves to various objects with whilst foraging.

Occasionally they are blended together and form setæ or bristle-like stiff hairs, having the use of vibrating membranes frequently.

The bodies of the ciliates are generally of a very soft, almost gelatinous substance called the "sarcode." It has an inherent power of flexibility and elasticity which greatly adds to the agency by which their multiform movements and contortions are performed.

The protoplasm is cellular, but in other respects similar to the Amœba and the lower protozoans in its hyaline and colourless nature. Its refractive power is not much greater than water which renders some of the more delicate forms very difficult to decipher in their ultimate details.

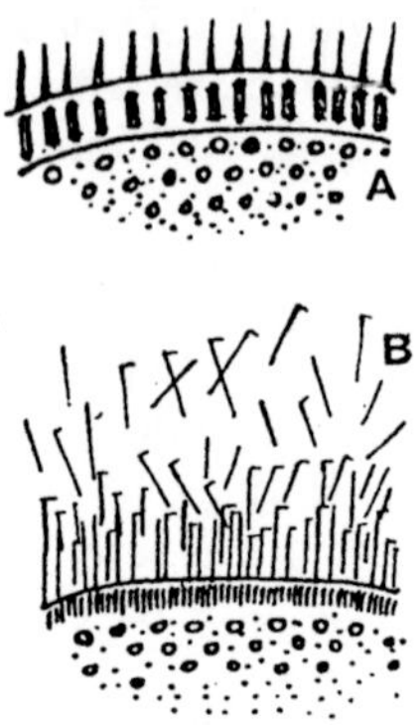

Fig. 142.—Trichocysts in Bursaria.

A. Dermal layer showing their position on margin.
B. As projected after acetic acid.
C. Detached spiral ones in the second stage evolution.

Many ciliates are attached to a support by a stalk or pedicel, sometimes rigid, at others flexible, as with Vorticellæ, sometimes short, as in Vaginicola, or as with Cothurnia the outer case is lengthened to produce a stem.

Upon some of the ciliates special threads often mistaken for cilia are found, they are the "trichocysts" or stinging threads and used as special organs of defence, as in Paramecium, Bursaria, and others.

In Bursaria they are about 1/2500th inch long, says Prof. Allman, and may be seen without any manipulation near the margin, arranged perpendicularly, and so long as the "animalcule" continues free from annoyance the trichocysts undergo no change, but when subjected to external irritation, as occurs during the drying up of the surrounding water, or upon the application of acetic acid or other chemical irritant, or pressure of the compressor or live box, they suddenly become transformed into long filaments, like a thin rubber ball when blown out to a tubular shape, which will then be seen projected from all parts of the "animalcules" surface (Fig. 142).

It is rather intricate to witness the procedure under the microscope, but if the Bursaria is rapidly pressed or crushed this will force out some of the trichocysts in an unchanged state. If one of these be now carefully watched, in a few seconds it will become, all at once, changed with a peculiar jerk into a little spherical body, as if by the sudden release of some previous state of tension. In this condition it will probably remain for two or three seconds longer, and then a spiral filament will become rapidly evolved from the sphere apparently by the rupture of some previously confining membrane, the filament unrolling itself so quickly that the eye can scarcely follow it, until it ultimately lies straight and rigid on the field of the microscope, looking like a very fine and long acicular crystal. This when completely evolved consists of two portions, the straight spiculum-like and longer portion and a second and excessively fine filiform appendage of less than half the length of the first, bent at an angle or frequently more or less curved at the free end. The form of these is best observed in such as have been thrown off and floated away to the margin of the drop of water and are there left dry by the evaporation. See Fig. 142C.

The resemblance to the thread cells of Hydra and other members of the animal kingdom is very striking, which affords some substantial analogy to their use as " poison organs " for defence. They are developed in the substance of the cell wall in distinction to special cells, however, as in the Hydrozoa.

Cilia are frequently present about the mouth, and lining the œsophagus and stomach in portions of their surfaces, facilitating the ingestion of foods and their aeration and the elimination of waste substances and gases. The action of cilia, and how the constant movement is set up and continued so persistently, has been the subject for much debate in the past, and even now is in a somewhat indefinite stage. The most that can be offered is that it is an inherent property of all living protoplasmic substances. The finer the filaments and the shorter their length the quicker are the movements as a general rule.

Returning to the Flagellates, these are either free-swimming or may be affixed to some support. Flagella generally arise from the deeper regions of the " sarcode " and are distinctly external appendages, commonly situated at one end or both, of the body, never from the sides. Their function is chiefly concerned with progression, by rapid spiral undulations twirling the body through the water in corkscrew fashion, or are used to withdraw food particles towards them in the vortex set up when the body is still. Some are used simply as trailers, dragging flaccidly along beside the body, as a long rope might hang from the bow of a boat and sometimes to well astern. These act as rudders and slow the pace considerably. There are many similarities or relationships between pseudopodia and flagella, and the interchange of one for the other has been noted in the case of the Rhizopod, " Vampyrella."

Among the various internal organs of the body is always present that most vital portion the nucleus, or in many organisms nuclei, there being frequently several. It can usually be seen as a more or less solid body, finely granulated with a yellowy or bluish tinge and generally opaque. In some flagellates it may consist of grouped particles or in others disseminated and scattered in various parts of the body.

In the infusoria it may take the form of a branched or crescent shape, and varies much in its position with the different species. Within the body of it will often be seen a much smaller spot, this is the "nucleolus," a truly vital spot to the life and future of the organism.

Curious to note about the nucleus in general is, that it is rarely a fixed object, but in the infusoria especially can be seen pushed backwards and forwards by the movements of the animal either in its progression or by the ingestion of its food. Infusoria also differ from other protozoans in possessing two sorts of nuclei within each cell, a larger one, the "macronucleus," and a smaller one, or "micronucleus."

The macronucleus is concerned chiefly with the growing life and subdivision in its multiplication by fission. The micronucleus is alone the body dealing with the purely sexual division in propagation. In some the shape of the nucleus is circular or oblong or may be kidney-shaped or sinuous and band-like, and will vary even among examples of the same species. It is a body that should be seen from all sides when studying, as often an oblong may turn out to be kidney or bean-shaped, as with the Paramecium and kindred families of Prorodon, and some Nassula. Again, in the elongated forms of nuclei they may frequently be seen coiled partially around other organs as the pharynx or œsophagi.

The nuclear body is ever in a transient state of change and elaboration within and will become now more transparent or again more opaque. Its walls are of a tough elastic texture, sometimes separated by a clear interspace called the "areola," and which will only present itself to view by the application of some chemical reagent, such as acetic acid or a solution of potash. If this be done with Vaginicola, for instance, the membrane separates and collapses into folds quite distinct from the inner body substance.

In the transverse division by fission of Chilodons, Coleps, Bursaria, etc., the nucleus always divides with the body as an essential nature of the act, and leaves half for either side to form and elaborate its own complements to. This may be taken as the rule, whether the division be longitudinal, transverse, or diagonal, the nuclei both "macro" and "micro" follow the process, such other structures as contractile vacuoles, mouth parts, etc., being grown as the division proceeds.

A means used habitually with the infusoria not only in association with the method of reproduction but also connected with the preservation of the individual is known as "encystment." This provides against natural adverse contingencies such as the drying up of pools and pastures which the creatures inhabit, and also enables them to resist climatic conditions unfavourable to their growth, and further provides, in the diffusion by winds and fowl, for their distribution to fresh habitats in the dried muds left open to such contingencies.

Encystment can take place at all ages, and when about to commence, the movement of the organism becomes slower, less active and ultimately ceases. While this is going on it folds in any prominent processes, cilia and others, in some cases casting them off altogether, as with the flagella, and contracts itself into a spherical mass. It next elaborates around it an excretion which gradually hardens, and the "animalcule" now completely encased begins a resting period. The outer wall of the case may become crumpled or reticular, or studded with little knobs or spines for the better withstanding of any hard knocks it may willynilly have to undergo during such times. When the season suitable for its growth arrives it will burst forth anew, maybe in quite a new locality, and so continue existence and disseminate its species abroad.

The process of conjugation occurs in both the flagellates and the ciliates and is often followed by "encystment" or the growth of swarm spores. These latter give rise to sexual reproduction, and gametes of varying size and opposite sex mating with others, a third and new body is formed by the two as a consequence.

In Volvox, a rather complicated flagellate, the sex cells are differentiated from the body cells and both male and female cells are developed, resulting in new colonies being produced by a union of the two, usually within the cover of the mother Cœnobium.

A protective sheath or "lorica" is secreted by the organisms in some cases, as a protective measure, into which they will rapidly retire to shelter when any untoward movement disturbs them. This action, moreover, assists in the compression and admixture of the soft foods absorbed, and is a means of security against currents too strong for their delicate structures. Most of these are examples of the fixed infusoria such as Cothurnia, Vaginicola, etc.

The "lorica" is generally transparent but may become with age a definite colour, reds, yellows, blues, greens and browns not infrequently being seen.

Upon various "lorica" sharp-pointed processes, setæ styles and other projections are formed, principally for protection and also as sense organs or antennæ in many species, notably Euplotes, Stylonychia and others.

In *Stentor mulleri* is an instance of an infusorian living either with or without a sheath and enjoying freedom of movement. Where present it will be seen carrying its house about with it, no small place comparatively, but a roomy ovate sheath capable apparently of accommodating three such animals as itself. The lorica is a soft, gelatinous structure open at one end with the trumpet-shaped stentor affixed securely to the bottom of the interior.

Within the sarcode or "endoplasm" of many of the flagellates (mastigophora) coloured corpuscles or "chromatophores" are enclosed. Within these are further pyrenoid bodies similar to the desmids, and have probably the same function, that of the elaboration and synthesis of starch foods. Many have oil globules and pigment spots, and noticeable with others as Euglena, etc., are the brilliant red "eyespots" usually placed at the front end near the base of the flagella. Whether these really operate with sight is extremely doubtful, they are usually nervous centres, however, and of a very delicate nature.

All infusorians are usually busy, lively creatures, and are seen at their best whilst living. This makes observation of their more minute details a matter of some difficulty; any simple method that will slow their activity down somewhat is often a considerable advantage. It must, however, not injure them, and to this end a little mucilage or a watery, thin solution of gelatin or egg albumen, preferably the latter, will be very useful; even the slow evaporation of the water from under the cover glass will effect this purpose, but not so well as the former. Sugary, syrupy solutions with water is sometimes equally effective.

In tracing the alimentary tracts in very hyaline species, a simple method used many years ago is still a valuable adjunct, that of rubbing a little indigo or carmine near the edge of the cover glass, on the slip, and allowing it to steal its way between. By this method many of the infusoria will partake of it, and it can be seen collecting within the food sacs, sometimes in a position little expected to be their actual location. The substances have no apparent injurious effect and are merely finely divided particles, which owing to their definite colour are readily seen and float about in the water without becoming in any way deleterious to the organism, at least for such time as it may be under observation at one sitting. Care should be exercised not to use too much, as the quantity of liquid under the cover is very small.

It will often be the means of discovering several vacuoles pulsating slowly within the body where none were visible before, and also of parts constantly hyaline which are therefore not connected directly in any way to the food organs, and, moreover, it does not stain the organism owing to its pigmentary character.

In the suctoria those infusorians possessing tentacular processes in lieu of cilia, or rather by age becoming possessed of tentacles, whereas in their embryonic stages they have only cilia, a distinct method of ingesting and procuring their food is observed. Here the tips are knobbed frequently with a tiny globule of sticky substance which readily attaches and holds small fry when passing near and happening to touch it. The tentacles have a very fine tube down the length, and by suction can withdraw the protoplasmic substance down it from the bodies of their capture. They can also absorb within their gelatinous substance, bodies suitable as food, whole, very much after the manner of the Rhizopods, if found particularly agreeable.

Nature seems to be noticeably apathetic among the infusoria, and without any nervous excitement tiny unicellular animals will almost play with, and deliberately loiter around, the long projecting arms of death until they are actually dismembered or unable to move, and are finally dispatched to that " bourne from which no traveller returns," as Catullus in "*Carmina*" records.

Euglena viridis (Fig. 143)

Where a stream is clear, shallow and slowly moving this infusorian is generally to be found in quantity upon the soft grey ooze at the bottom, especially in the spring and autumn. It is seen as bright green patches which at first sight may give one to think it an alga, so like in colour is it to the eye. These patches, however, can be noticed to move their positions if watched carefully. A favourable opportunity for this is when the sun is shining through a tree and casting bright loopholes through its branches upon the stream. As the shadows move with the sun so the Euglena will also move, keeping within the bright enclosures formed.

Fig. 143.—EUGLENA VIRIDIS.

On placing some of the green under the microscope it will at once be seen to consist of myriads of separate individuals staggering along, each with their narrow, flexible bodies squeezing and con torting themselves into all manner of shapes.

They are about 1/300th inch long, oval, and with a tail tapering to a short but fine point.

They are filled with many green chlorophyll cells, and have a bright red spot, consisting of a number of rounded granules near the head end constituting the " eyespot."

They possess one long flagellum in front about double the length of the normal body, which rises from a short notch-like depression, representing a mouth and by which they execute their movements of progression.

In the swirling of their lash small particles are drawn in the vortex towards them and have been thought to constitute food substances, but this is a mistake, there being no actual mouth opening. As the Euglena contains chlorophyll it is enabled to manufacture its own food by the aid of light, etc.

What it can do with its simple whip is astonishing, and would form an interesting exhibition for a lasso thrower to witness, not in the air, but in the denser medium, water. By its aid it can turn to right or left, drawing itself rapidly through the fluid medium, gyrating and touching with its quivering tip this little object or that, while at the same time it may be throwing coils with it in the middle, and still progressing in any desired direction, this without so much as a ripple or one semblance of turbulence or splash.

They are exceedingly fond of a bright light, and knowing this little weakness ? of theirs, one can obtain some amusing performances with them. For instance, with a flame edge focussed on the slide in a straight line they will " fall in " along its length; crossing this with another at right angles they will cluster along this and produce a brilliant green cross.

For the children's amusement the writer cut a stencil of his initials, and putting this near the flat side of the lamp flame, focussed these upon the slide so that the illumination there was limited by these outlines and in a very brief time the Euglena were grouped along them, rushing in from the " outer darkness " and shaping a print there with their dancing, quivering forms much to the delight of the youngsters when the stencil was momentarily removed, leaving them in the field of view alone.

Several varieties of Euglena have been named. Here are a few :

Fig. 144.—EUGLENA ACUS.

Euglena acus (Fig. 144)

With its long, thin, tapering body and tail, its eye nearly as wide as its narrow anterior end.

Euglena oxyuris (Fig. 145)

A flat and twisted body usually of three or four turns.

Fig. 145.—EUGLENA OXYURIS.

Euglena spirogyra (Fig. 146)

Rather plump, spirally beaded body and long curved tail. This one shortens itself at times into a sphere almost, and may be found without any flagella.

Fig. 146.—EUGLENA SPIROGYRA.

Fig. 147.—EUGLENA ZONALIS.

Euglena zonalis (Fig. 147)

Pointed at both ends, double convex in shape, with a band or zone partially around its width in the middle.

All have vacuoles, eyespots and nuclei as well as flagella, unless they are about to take on the sporing stage, when flagella are often absent.

Astasia tricophora (Fig. 148)

Literally meaning without a station, in contrast to its neighbour Euglena and similar organisms living in groups.

Its shape is very changeable, and from an elongated and cylindrical oval to an almost spherical one it may be found, compressing itself with all the curious contortions of which it is capable, into any outline between.

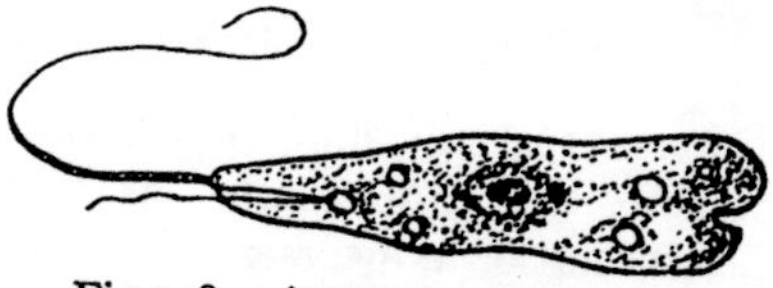

Fig 148.—ASTASIA TRICOPHORA.

It is about 1/350th inch long in normal position, colourless and flexible. Its cells are white, with one or two yellow, orange, or even red ones interspersed. It does not secrete chlorophyll, while Euglena, whose cells are green, naturally does so.

This is a recognized character which will readily distinguish it. It has one fairly thick flagellum, with which it swims along, and is used to produce a swirl of food particles towards its mouth end at the same time. It has a distinct indenture there, and was considered by the older naturalists to possess the rudiments of an alimentary canal. This is now considered untenable.

Its tail will at times be extended into a fine thread-like projection, but at others only a rounded end can be seen, this owing to its untiring changeableness. The body generally is wider posteriorly. It is common among the algæ or on the ooze in conjunction with Euglena, in muddy, shallow streams.

There is an elusive, shorter flagellum arising from the same base alongside the larger one, which it carries near to the body, directed backward usually, rather difficult to see unless specially searched for.

With Astasia there is no red "eyespot" as in Euglena. There is a small contractile vesicle situated near the anterior end; these organs perform a vital function in the circulation and metabolism of the individual compressing and relaxing the liquid contents within the body, eliminating waste matters and gases accumulating, and so provide an outlet necessary to the health and continuity of the species. There is a central nucleus. Length of body 1/400th inch.

Paramecium (Fig. 149)

This is one of the most prevalent and probably best known of all the infusorians. Its body is oblong, slightly flattened, and ciliated upon all sides. It is rounded at both ends.

The mouth is a deep and long fissure upon one side, extending almost to the centre. Owing to the shape of this, resembling the opening in a slipper, coupled to this its elongated body, it has earned the sobriquet of the "slipper animalcule."

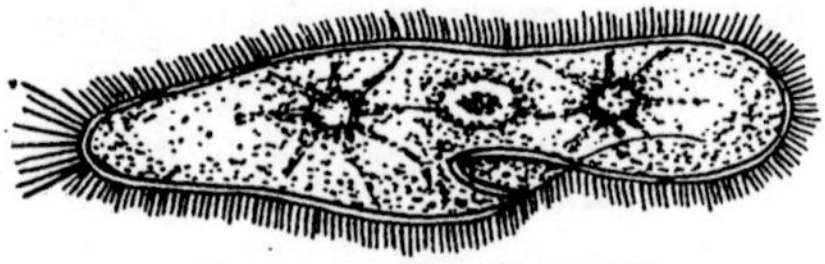

Fig. 149.—Paramecium.

At the caudal end is a tuft of cilia longer than the rest which when present forms a distinguishing mark of recognition. A separate species without this caudal tuft has been found, however, and so some discrimination is necessary.

Situated toward each end of the body is a star-like contractile vesicle regularly and constantly pulsating. They have been observed to contract and expand about eight times a minute. There are two nuclei, one large near the centre and another smaller a little farther towards the posterior end. The surface of the body is diagonally striated, and immediately beneath is a much granulated layer of small transparent or opal white bodies. Usually the interior is coloured, according to the food particles eaten. Generally it is of a yellowy green in appearance.

Its length is about 1/100th inch, and so forms quite a conspicuous object. It multiplies by division, either transversely or longitudinally. Two bodies, partly separated, may often be seen swimming side by side and obtaining their food quite agreeably together in that manner.

The body is profusely supplied with stinging cells called trichocysts similar to Bursaria, Hydra, etc. These are used in the procuring of food and also as protection for their soft bodies against enemy injuries. They are thrown off like darts, and may be seen at times floating near by, in the water surrounding.

Paramecium is fond of a bright light, especially while feeding. If one side of the field be darkened they will cross over to the lighter side and become stationary, and the procedure may be reversed to see them troop across to the other. This may not occur in all instances, but is observable where they are amongst a sufficiency of food in either direction. It is noticeable too the majority of the cilia around the body are all of one length, and excepting the special tufts mentioned and the "trichocysts" which have been more extendedly explained in the opening chapter, might be said to be uniform all over. This gives them a very neat appearance. The manner in which the organisms can reverse their direction and the means they use to accomplish this has formed much debate with the earlier writers and is a point well worthy of study when observing their actions, for the beginner to probe.

Bursaria truncatella (Fig. 150)

The body of this infusorian is white, turgid, and of an ovate or purse-like shape, broadly open on the ventral side by a furrow or "fossa" reaching as far down as the centre of the body in a somewhat oblique direction.

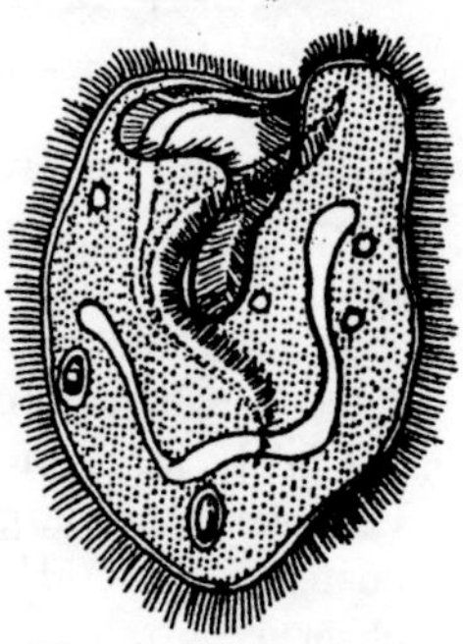

Fig. 150.—BURSARIA TRUNCATELLA.

The surface is profusely ciliated with even-length cilia disposed in close lines or rows. The mouth is fringed with longer and stouter cilia than the rest, which conveys the food particles to the narrow end of the deep groove for selection and absorption. The tremulous flap in the peristome is absent in Bursaria.

It is closely allied to Paramecium, but the general contour of the body and the larger mouth cilia readily distinguish it.

The nucleus, that is the macronucleus, as it has, like other infusorians, two kinds, previously alluded to in the opening remarks upon the infusoria, is waved and band-like, making a long sinuous detour around the lower half of the body and almost semicircular in its entirety.

Within the substance are also numerous contractile vacuoles, one of which seen just near the mouth entrance is large and active, and sometimes duplicated into two. The anal orifice is situated at the posterior extremity.

The infusorian multiplies by self-division either longitudinally or transversely.

Bursaria is interesting on account of the special cilia known as "trichocysts" which have the power of disarming smaller fry, and also as some means of defence from interference by its enemies, being shot out from the body like darts. See Fig. 142.

Bursaria leucas was minutely investigated respecting these bodies by Prof. Allman in 1854, and are akin to the "nematocysts" in Hydra in their action, but are not enclosed or developed from special cells of the Ectoderm as in that animal. Its habitat is ponds and ditches. The rotting leaves of Beech in quiet pools is a frequent source to find it in. It measures about 1/45th inch in length.

Loxodes (Fig. 151)

This organism is similar to Chilodon both in habits and outline. It is found generally in the older decaying vegetation. It has a flattened and flexible body often bending upon itself, and a sickle-shaped cleft for the mouth entrance on one side near the extremity of its elongated body. The upper surface is slightly concave and devoid of cilia, the under being covered with them and form its means of locomotion when creeping over the stems of conferva, etc., in searching for food.

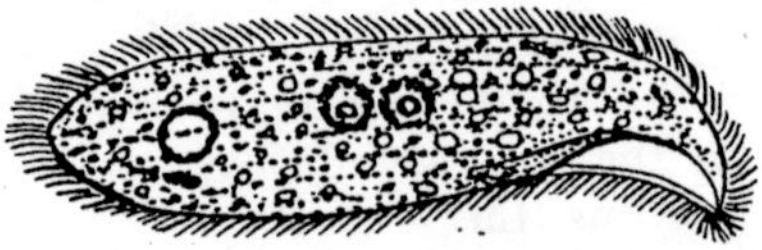

Fig. 151.—Loxodes.

Around the mouth is a fringe of longer cilia; these are used actively when swimming free. It has several nuclei, two of which are fairly large, near the centre of its body, and many contractile vacuoles, the largest of these generally appearing near to the posterior end. It is about 1/100th inch in length, and so readily observed.

Sometimes diatoms of quite large size may be seen within, eaten in its food. It has a striped dorsal surface.

Dinobryon (Fig. 152)

This small infusorian has a delicate sheath around it which is slightly expanded near the lip or open end and contracted about the base. It is a Euglena-like organism with two flagella, one of which is long and the other short. It has a small red eye spot near the anterior end. They form a colony of perhaps ten to twenty, more or less, their hyaline sheathes being attached to the lower ones at their sides, and so giving freedom for each occupant to ply its cilia at ease. They give a broken-branched and quaint appearance as they meander through the water by the aid of their cilia collectively in motion.

Evidently not designed for movement amongst many obstacles, the colony being ever halted against water plants, they are thus frequently found adhering along the surfaces of such specimens as algæ, Riccia, Lemna, etc.

Fig. 152.
DINOBRYON.
A. Colony of same.
B. Single organism (enlarged).

Usually they have two green stripes down the side of their bodies, a distinguishing mark when present, but these are often absent, and depend chiefly upon a plenteous supply of chlorophyll bodies around, suitable to be taken in as food. They increase by self-division near the centre of their bodies, either transversely or obliquely.

The longer flagellum is used in gathering in the food and is also mainly responsible for the movements of progression, the shorter one, lying near the mouth, being used to turnover and inspect the particles as they arrive before being consumed.

At the base of the flagellum there is a bright ring always seen by dark ground illumination, which is the best method of observation by the way, otherwise it is extremely difficult to make out their outlines. It is possible there may be a few cilia there, but this has not been determined satisfactorily.

The movement of the flagella and of the colony as a whole is slow and stately, and with regard to the green colour within the bodies a specimen with nine tubes had one absolutely clear body, two with a few granules within, and six with many sparkling white granules but no green or other colour.

The longer flagellum proceeds a good distance beyond the edge of the sheath, but the infusorian rarely, if ever, extends past it.

Dinobryon is usually a spring visitor preferring a temperate clime and seldom is obtained in the summer or autumn months.

The sheath is about 1/1250th inch long, which gives the indication of a very tiny individual and so not readily observed without much care.

Aspidisca (Fig. 153)

The body is rounded with a hyaline lorica, cut short across the posterior end, and having a sickle-shaped beak at its anterior.

The mouth peristome is situated at the left hand lower corner, as seen face downwards, and is furnished with many delicate cilia.

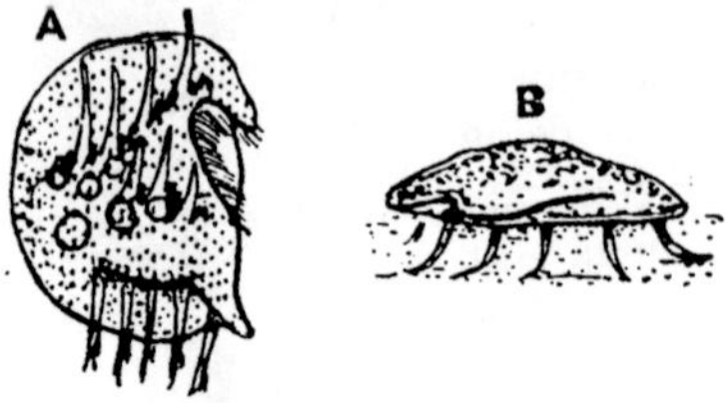

Fig. 153.—ASPIDISCA.
A. Ventral view. B. Side view.

There are no caudal setæ as in Euplotes, but usually five bristles fringed at their extremities extending out from the body. These are used in its progression as feet, along with three others frontally placed and four or five centrally.

There is a large contractile vacuole near the mouth and many smaller ones. As many as twenty have been counted. It has a habit of swimming on its back, and in this position will often be found also creeping along the under side of the cover glass as it is viewed in its element, rendering a very clear view of itself.

It is small, measuring about 1/500th inch in diameter. It is found frequently among conferva, and especially Lemna, as well as in decaying vegetable infusions generally.

The dorsal surface is arched slightly, and has five or six furrows upon it longitudinally. Its nucleus is extended in a band-like formation.

The method of multiplication is by self-division similar to many other infusorians, and two may occur swimming together, partially detached, both plying what cilia is free at the time to do so.

When in action it is constantly jerking backwards and forwards and round and round as though seizing a nibble and pulling away with it, and appears partial to the gelatinous external coatings upon algæ, etc., which it detaches with its busy cilia, perseveringly.

Pleuronema (Fig. 154)

This infusorian is remarkably and beautifully organized. In the method of netting its food especially so. One is reminded by it of Pope's famous lines :

> "Learn of the little Nautilus to sail,
> Spread the thin oar and catch the driving gale."

Only here we have the little organism catching particles of food by spreading its sail, at the same time driving its own gale and swirling them into its net by means of numerous cilia in rapid motion around its surface.

Pleuronema is of an egg-shaped oval, broader, that is, one end than the other, and finely striated longitudinally. It is about 1/300th inch in length. Upon the thickened striations of its outer membrane numerous fine cilia are situated, and amongst these are many much longer ones quite bristly looking.

Near the broad end these long setæ are in greatest numbers. They are flexible and occur among the shorter ones in fairly even distribution at all other parts of the body. Situated to one side is the mouth, at the end of a long depression, which extends the greater part of the length of the body, and from this is attached a net, a gossamer web-like curtain, which can be thrown out or withdrawn at will. It is attached on three of its sides and the fourth left free and open, making a loop as it is fluttered in the water. Into this is swept and caught food particles and conducted to the mouth at the farther end. The mouth, when the web is out, is at the end of a spiral vibratory vortex, which turns and selects the various particles it receives as they come, accepting or rejecting those considered suitable. If satisfactory they pass in, if not they don't and are rejected at once to make way for the steady oncoming tide of other such morsels.

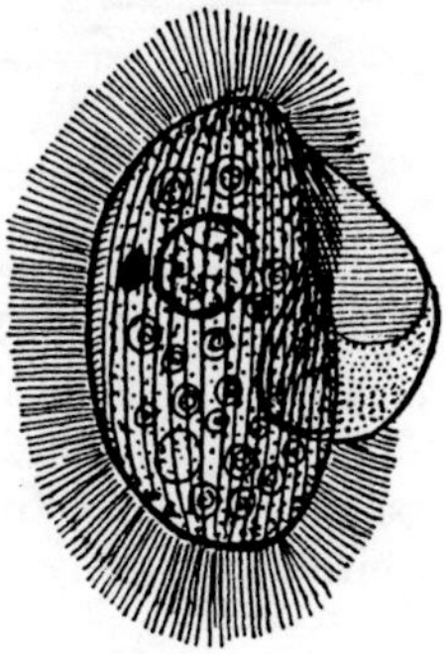

Fig. 154.
PLEURONEMA.

Pleuronema is to be found among decaying vegetation along with other infusorians, and is an object of real merit to witness in the act of feeding. It is hoped all may have the good fortune to capture it at some period in their lucky dips.

It is seen to greatest advantage under dark ground illumination, when it will be a miniature presentment of delight to watch. It is restless at times, flashing into view with a suddenness as rapid as its departure, but tracking it for a while it can eventually be held for a sufficient time to note the throwing of its net, looping it into a graceful fold, and the scintillating glint of this quivering drapery, waving into its open mesh the food particles as they are swept within its shade, is a sight worth all the trouble taken to obtain.

The organism has one vacuole at the anterior end, which in common to all its movements pulsates quickly, opening out and closing in with almost a run and a snap. Eleven pulsations were counted in one minute, the expanding taking much longer than the contraction. The body is slightly flattened upon the under side.

Cyclidium (Fig. 155)

As this is often present in gatherings taken from stagnant waters it might be mistaken for Pleuronema, at first sight. Its characteristic differences will soon obviate this. It is small, about one quarter the length of Pleuronema or 1/1250th inch on its longer axis.

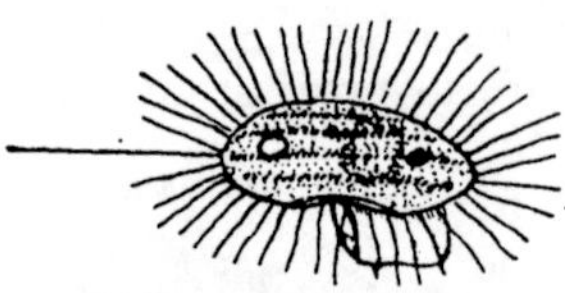

Fig. 155.—Cyclidium.

It is an elongated and narrower oval in outline, and its cilia are fewer and wider apart comparatively. It has a long straight bristle carried in line with its body at the posterior end, which is absent in Pleuronema, that stands off quite stiffly when at rest. It has a little collecting net of less pretentions, and its rapid changes of position are somewhat similar.

Darting here and there, stopping instantaneously and then away, before you have time almost to be aware of its presence, makes it quite an amusing " here we are again."

Uronema (Fig. 156)

Is another infusorian intermediate in size to the previous two, about the 1/800th inch in length, fewer in numbers, but sometimes found with it. This has a long bristle at its tail end.

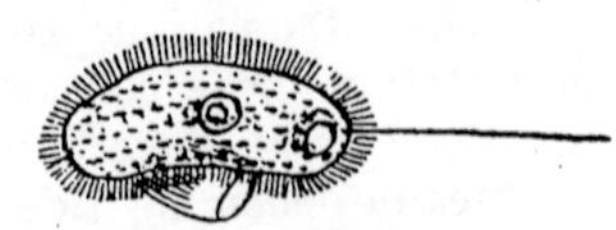

Fig. 156.—Uronema.

The contractile vacuole is posterior and the nucleus central. Its cilia are very short and exceedingly vibratile; they are irregular and independent. It has its little net, but the depression it arises from is curved concavely while its back is convex. This gives the organism a kidney-shaped appearance. Around its mouth the cilia are densely arranged in a furrow.

Ctedoctema (Fig. 157)

Another similar infusorian about the 1/1000th inch in length but without a net. It is coarser rather than the others, the bristles are few, stouter and longer comparatively. The bristle at its posterior end is carried hooked at the tip; this is its chief distinguishing feature. It is often very abundant among the fresh-water algæ.

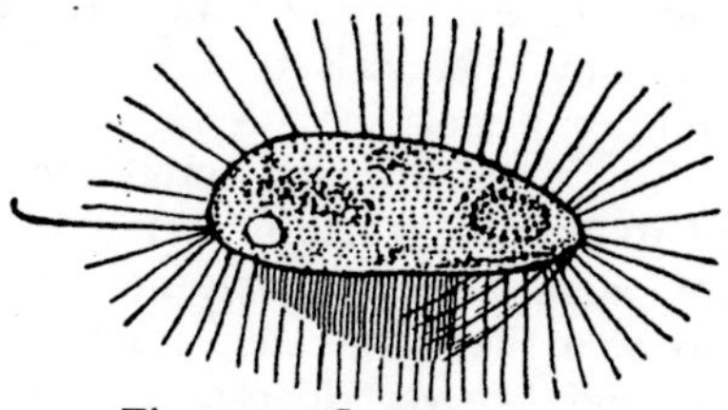

Fig. 157.—CTEDOCTEMA.

The mouth vent is situated at the lower end of a groove, and has a border bearing a row of rather large curved cilia, which become shorter towards the entrance. These help in differentiating it, and in place of the net these border cilia act in guiding the food to its mouth.

Lembadion (Fig. 158)

There are many infusorians obtained which are often passed over for want of being able to name them. This is a common one of such. During the spring and summer months, and whenever there is a portion of Sphagnum or Moss with the gathering, one is frequently able to find some specimens of this organism.

It is well to familiarize oneself with its form. It averages the 1/400th inch in length, is of an oval shape, flattened on the one side, and usually this is uppermost, upon which is situated its voracious mouth. It is a hyaline shell, and when little food is present makes its contour difficult to grasp entirely whilst moving. It is striped longitudinally, and upon these are set the numerous cilia by which it effects its rolling, spiral movements. At times it will swim rapidly backwards.

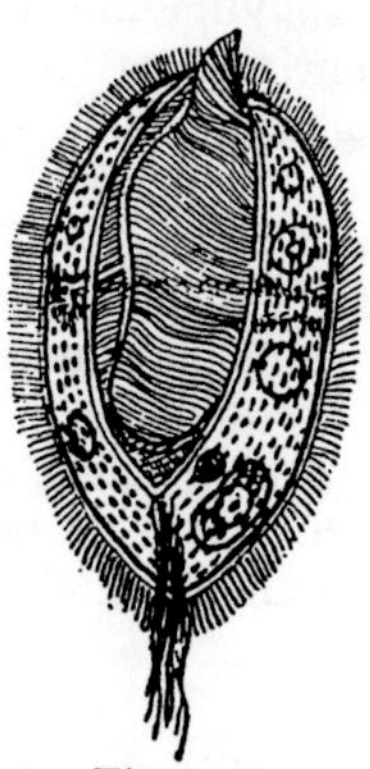

Fig. 158.
LEMBADION.

At the posterior end are several long lashes, ten to twenty, depending upon the species, which are in constant play and which are used to help it crawl and push itself when in contact with any objects. These may be seen curved upwards when touching the cover glass, presenting their glistening and reflective tips. They act very powerfully in swimming by a wave-like rhythmic motion very elegant to watch, but at such moments their rapidity leaves only a " flying view " before they are out of the field again.

It requires a good dark background and as wide a field as possible. It is very active and requires some little patience in following round to a more quiescent state, to see its details. I prefer not to add any chemicals to the water, but occasionally use a watery solution of gelatine.

The anterior end is scroll-like, and near where the normal part of its mouth in use is situated. Should, however, an extra large and dainty morsel come near, it can present an alarming cavern, opening the whole side practically, which may then be perceived to consist of very long cilia closely set across a wide gap extending nearly the length of the side.

The writer saw a specimen make a deliberate dash from a distance equal to its own width and snap up with sudden vigour, quite startling to witness, a double flagellated monad, with its one trailing lash peacefully steering along, the other browsing around in front, and with a few deft turns over and over of it lengthways, just to tuck in the straggling ends, apparently, open the wide peristome the full length of the breach and drop the whole bodily into its abyss. Here it could be seen for a time settling into a cosy corner and finally becoming a granular mass along with the rest.

At another time Lembadion seemed to show a selection in its choice, and though a Euglena persistently would get in its way it would only pick it up, turn it over on its cilia, and then reject it at once. Possibly his taste does not aspire to chlorophyll, and certainly they are not seen with much green among their food particles as a usual occurrence, its other food of size being the brown or golden circular cells of the Trachelomonas.

When feeding it has a nibbling, snappy action with constant short retreats and an incessant revolving habit. At ordinary feeding times the anterior end and about one-third of the side peristome is in use only.

The cilia upon the scroll end meet from either side, forming a definite projection there. It multiplies by division, and may be seen frequently in pairs, partly separated, swimming about, similar to *Euplotes charon.*

Halteria (Fig. 159)

A very pretty, transparent and lively little creature this, darting spasmodically in and out of view. It will often be encountered among water from decaying vegetation in still ponds along with other infusorians. It is not unlike a small pear in shape, flat at one end with a rounded point the other.

It has several vacuoles, one or two large proportionately and a circular nucleus. On its flat end is a circlet of long fine cilia, constantly in movement. By these it can bring in small particles of food and also effect the rapid locomotion of itself through the water. In addition to these it has about a dozen long stiff bristles twice the length of its body's diameter by which it takes flying leaps well out of the field of view at times. In relation to these there is a little point to record hitherto apparently overlooked. In previous descriptions and figures they are given as long straight bristles. They are, however, divided at the extremities and open. Those pointing foremost usually carried so, while the others to the rear are kept quite straight. Further, they are not in a line or row around the centre of the body but disposed unevenly about the upper half. The mouth is close to the ciliated crown, situated at one portion of this margin.

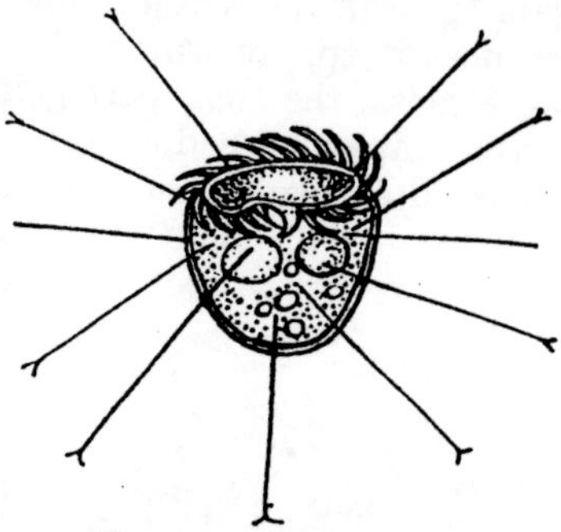

Fig. 159.—HALTERIA.

A very much larger specimen the writer has seen of this species with larger crown cilia and longer bristles, about 1/400th inch diameter and rather more pointed at the lower end. It had several circular, brown and yellowy food particles it had eaten, within, among numerous vacuoles and a distinct nucleus. The larger ciliated wreath was especially fine to witness, and having more scintillating colour was altogether a superb object.

Coleps (Fig. 160)

This infusorian makes its home among decaying vegetation, the sedges and rush stems and aquatic debris on the margin of quiet ponds and similar places. It is often present in large numbers, multiplying rapidly.

Its lorica is in the shape of a cask with longitudinal stripes, and in the commonest species is cylindrical with bluntly rounded ends. It has long had the sobriquet of "the barrel animalcule" from its apt appearance to that object.

It is about 1/500th inch in length, and in the specimens here described the 1/1100th inch wide. There is a much larger species occasionally presenting itself with a more tumid centre.

The lorica is usually drawn and described as consisting of square armoured plates. This is a natural view in some species, but will not fit correctly all specimens alike.

A figure here given represents the markings most commonly seen upon those in the writer's collections. The plates are not squares, nor does it appear armoured. Their shape is not unlike a champagne glass, the stem part going obliquely deeper. The upper is of a D-shape attached closely to a longitudinal rib.

The outer coat is smooth and shiny and of a chitinous nature save for the ridges from which the cilia issues. Beneath this coat is a finely granulated layer fairly evenly spaced. From whichever angle oblique light is brought the granules present their front. If side lighted they are longitudinal, if endwise, transverse, etc. It is through these two layers the larger markings are observed. They are therefore never sharply focussed with wide angled lenses. It is including too much in one plane that gives rise to the armoured appearance.

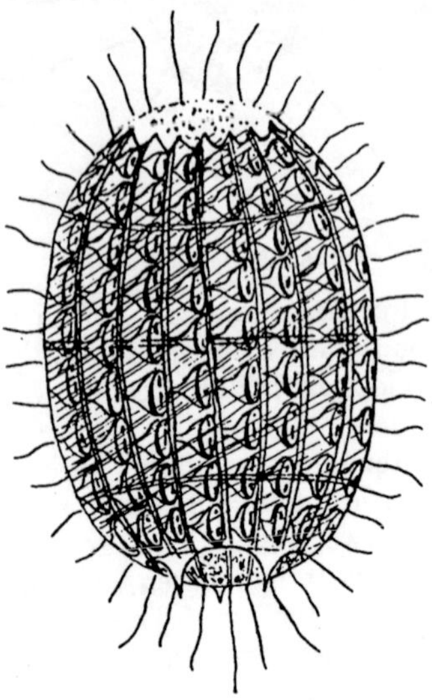

Fig. 160.—COLEPS.

The mouth is situated at one end, and around it is a saw-tooth ring, upon which one sharp point arises from each longitudinal rib end. At the distal extremity are three spines, set a short distance from the apex. The cilia there are longer, with one central, of twice the length of the others. At the mouth end the cilia are also long.

It is a rapid swimmer, revolving as it goes along. It is also a voracious feeder, and with its sharp teeth and its rapid revolutions will bore big pieces out of suitable morsels. It is a useful scavenger, devouring any plasmic bodies like Chilodons, and even its own kind, which may have come to grief either by injury or other cause and that might be left to decay and otherwise help in fouling the waters.

Coleps in this respect seems especially solicitous for its own species, and if a member becomes quiet and at rest for any abnormal length of time others in the vicinity will not cease paying it their attentions, prodding it with their spiky mouths or jostling it in one way or another with their bodies as an inducement to bestir itself and "move on." Life to them is no place for the sluggard or the dullard, either get on or get out seeming to be the motto.

It multiplies by self-division, and may often be seen with a fully formed half lorica and a thinner and more transparent half attached. It has several vacuoles pulsating within and a nucleus.

Its colour is usually opal white, but it may be green or brown according to the food recently eaten and which may be seen through its semi-transparent case. It can be induced to ingest coloured substances added to the water, but at heart Coleps is a cannibal. His food is ejected posteriorly and not from the anterior end as was formerly conjectured.

Between the central division line and extremities either side is another transverse band which becomes the next separating line as the upper half is detached on propagation of a new individual.

To illustrate the flexibility of the outer case Coleps has been seen with its mouth dilated half around an alga resting spore feeding upon the gelatinous outer coat and giving it quite a flask-neck appearance.

Chilodon (Figs. 161 and 162)

The body is irregularly oval, convex dorsally and flat or slightly concave on the under side. It is a rather thin object and can be seen through readily.

It is ciliated upon the under surface and but sparsely upon the upper. It is longitudinally ribbed, and at its anterior end is projected a lip-like extension to one side. From this "beak" a line of cilia extends to the mouth at its base.

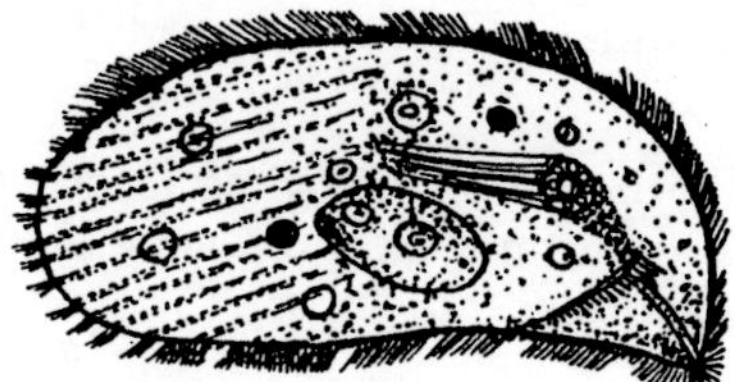

Fig. 161.—CHILODON.

The mouth opens into the pharynx, which consists of a bundle of rods fascicle-wise arranged into a tube. The rods do not touch closely, and the pharynx can be expanded when large objects of food are undertaken.

Chilodon is very variable in size, and may be less than 1/1000th inch long to as much as 1/200th inch. Its numerous cilia twinkle prettily as it glides along touching this and that object. It is often studded with coloured alga and other spores of red and green and golden hues.

It rubs its mouth over algæ, and on one occasion was seen to do this to a piece of narrow Oscillaria, which by a suction action it bent in the middle and drew it, doubled so, into its pharynx, but finding it too long and difficult it ejected it (Fig. 162).

Another time it endeavoured to engulf a Synedra diatom, a very long, slender and needle-like object, but by the time it was well inside and at the limit of its capacity, fully three times more remained outside, and in this ludicrous fashion it was swimming or crawling about for a time, finally having to expel it.

So long as the diameter is within compass it evidently has no discrimination regarding length of objects. It multiplies by self-division, and may often be seen partly disjoined foraging about together.

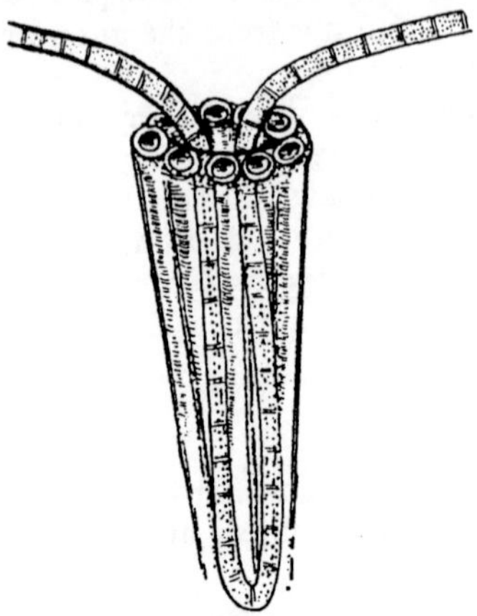

Fig. 162.—Mouth of Chilodon.
With a portion of Oscillatoria doubled in.

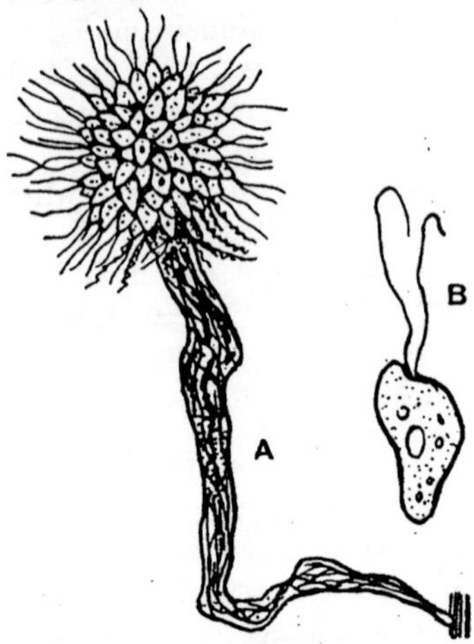

Fig. 163.—Anthophysa.
A. Colony on pedicel.
B. Single zooid.

Anthophysa (Fig. 163)

This forms a cluster of zooids similar in appearance to Synura when swimming. There may be as many as forty in a colony or only a few. One has been seen left by itself.

They are without colour and are very small, distinguishing them from Synura. The whole bunch may only measure the 1/1000th inch across, whilst the individual zooid is but the 1/5000th inch in diameter.

They are either free or fixed to a pedicel upon plant stems, moss fronds, etc. This pedicel is a stout but irregular bundle of threads, joined together as though each had contributed one, and are not parallel in position but bent in awkward angles and give a thickened end upon which the zooids are grouped, tapering to a thin, slender attachment at the root.

The zooid is pear-shaped, with an oblique hollow or flattened portion at the mouth end.

They possess two flagella of equal lengths, but very often only one can be seen. Occasionally they withdraw one, leaving but a shortened tip projecting close to the mouth, and with the other make their customary movements for procuring food. The short one appears useful in holding and turning over any tiny morsel submitted before being retained or rejected.

Size is of no inconvenience to them, they deal with proportionate matters and are as capable as the larger company to extract their needs and fulfil their mission in life with equal efficiency.

When they wish to become free from their pedicel a rapid rotary motion is continued and the cluster eventually twist themselves away, leaving the old support behind.

They have a nucleus and several pulsating vacuoles. The stem becomes brown with age and particles often adhere to it.

Stylonychia (Fig. 164)

Is a flat, elongate-oval body ciliated along the edge and constricted slightly in the middle. The anterior end is truncated somewhat and is the greatest in width. It measures 1/250th inch in length. It is soft and flexible, clambering over and around any obstacles.

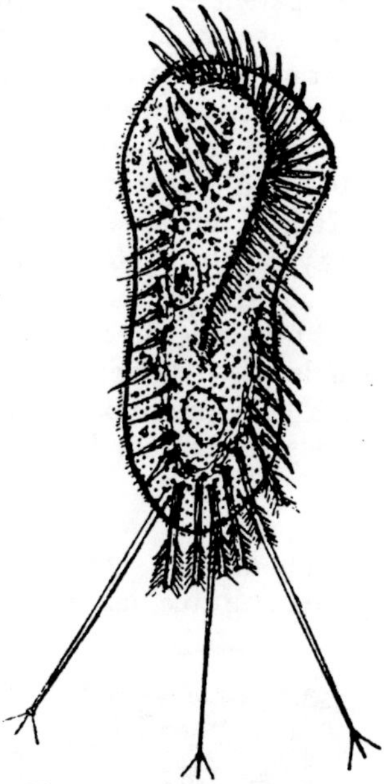

Fig. 164.—STYLONYCHIA.

The mouth is upon the under side, at the end of a long ciliated fissure or depression. The front end has bundles of stout cilia extending from the right margin across to the left and around the mouth. The posterior end is transparent and has a number of setæ upon it. Five are stout and fluted along their length with feathered sides near the extremities. Two others there are, striped, wavy feelers. Along the mouth half of the body are nine more of these, and all are used in crawling, as feet, when out foraging. They are set in oblique rows of threes.

A distinguishing feature apart from these are three long needle-like styles even distances apart at the posterior end, twice as long as the setæ. Usually they are figured as plain points, but the writer has always resolved them into three exceedingly fine projecting "toes." They are not carried open, however, at all times, and are apt to be overlooked unless the specimen is in a clear position and quiet. They are sensitive, quivering feelers, and have a knob-like portion at their point of origin.

The animal as a whole is often a pretty object to see owing to the varied coloured spores and food particles it has eaten, red, yellow, white and green often being intermingled within its glistening, transparent envelope, giving quite a studded gem appearance to it.

It has a peculiar forward and backward movement in its habit, starting as if in a hurry to obtain some morsel it is aware of and then as suddenly withdrawing and reversing the move, imitating a skater very faithfully executing the outside edge forward and backward consecutively.

It is a sensitive creature with no war-like tendencies, hurrying away at the slightest touch from others. It is found among decaying aquatic plants and especially infusions containing Oscillaria.

It multiplies by self-division, and may often be seen two together, partly severed, swimming united like "Siamese twins," both plying their cilia and apparently digesting and taking in food as if single. It has one large pulsating vacuole near the mouth and two nuclei.

Urocentrum turbo (Fig. 165)

This at first sight might be taken for the head of a Vorticella swimming free, as they often will, minus its stem. It may be distinguished, however, by its more consistent globular outline and in having a short tail.

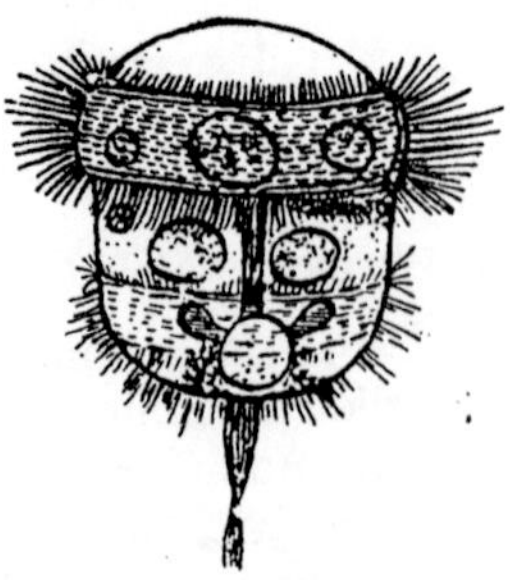

Fig. 165.
UROCENTRUM TURBO.

Its body is hyaline and gelatinous, of a somewhat bluish tint. Around its crown is a zone or band of cilia arising from a number of fine ridges, ten or twelve, set horizontally. These form its principal means of locomotion and procure its food.

The central half is but sparsely ciliated and the lower portion completely so, but not in so distinct a band form as the upper. It has a short tail at the centre of the posterior end consisting of a bundle of longer and stouter cilia, knit together for the most part and free to open or close at the tip. Thus at one time it will be seen as a pointed end, at another tassel-like. It literally wags its tail, and is evidently an organ of much sensitiveness. It helps in steering its course, and can be bent in a hook shape. It swims in a rapid spiral manner.

The body has one pulsating vacuole and many cells containing granules. Its nucleus is band-like. The outer surface under high power is very rugged and pitted. It measures from 1/280th to 1/350th inch in the widest part. The mouth is a ciliated sac lying at one side close under the upper ciliated zone.

It is a busy creature and seldom at rest, gyrating along with tail a-wag without a moment to spare. It has a method of checking this headlong career at times, very useful and ingenious, by throwing out a string of a sticky and extensile substance from its tail which attaches to some object or other and so anchors itself while still revolving. The substance bears the torsion strain well.

Urocentrum was timed to make one hundred and six revolutions in one minute whilst thus held to one spot, its travelling speed working out at about 125 yards an hour.

It will revolve in either direction, rapidly reversing itself occasionally. It has also another peculiar movement, darting swiftly from side to side with a jump.

Its food is discharged from the caudal end, this part of the body being hollowed triangularly. It multiplies by self-division, and is often to be seen in twain, head to foot, swimming along.

Euplotes (Figs. 166 and 167)

Body broadly oval, convex dorsally, and fairly flat ventrally. The anterior end slightly truncated or cone-shaped. It is a creeping infusorian similar to Stylonychia, Stichotricha and others.

The mouth is at the extremity of a wide funnel-shaped opening, strongly ciliated all round. The cilia at the edge and front end are plied in bundles, giving a toothed appearance. The upper surface is ribbed longitudinally. Near the centre is the nucleus, and beside the mouth a large pulsating vacuole.

On the under side are nine stout but flexible cilia, and set around the nucleus portion five others, these are used when crawling along seeking its food. They take the place of feet, and when seen pattering upon the under side of the cover glass, with the animal upside down, quite a normal position, they are bent and inclined at any irregular angle, and are gelatinous enough to stick securely and hold it in that posture.

At the posterior end are two extra stout spurred bristles, forked at the ends, the spurs able to feel and vibrate about as they are dragged behind, the bristles are fluted lengthwise.

If the animal is viewed from a side position, which may be obtained as it crawls around algæ and the stems of aquatic plants, etc., the next style would be seen carried stiffly, like a tail, standing up from the edge, giving notice of anything about it at a good distance from its body in that direction (see Fig. 167). It is fluted for two-thirds its distance down, and ends in an exquisitely fine point, the remaining third. It is very sensitive to touch and immediately darts away if any moving object brushes past it.

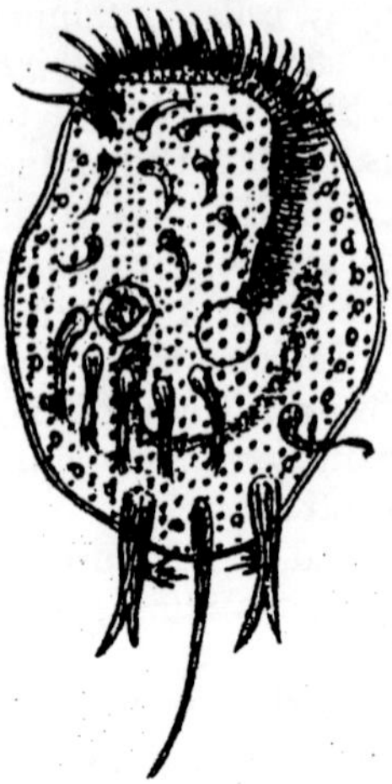

Fig. 166.—EUPLOTES. (Vertical view.)

The last cilia, next to it, is a flexible one, kept constantly waving around in a spiral way, seldom seen straight. The body near its circumference is thin and transparent. Within this are a number of granulated, hyaline cells set along the border in single file. In the focus of a good light these shine out brilliantly and appear as a row of pearls. Owing to this "Euplotes" has had the sobriquet of "The little coster" applied to it.

The thicker, central portion of the body is usually coloured with various spores, unicellular alga, etc., taken in its food, giving a bejewelled presentment of brilliant reds, greens, yellows and whites, quite beautiful to behold.

Euplotes measures 1/280th inch in diameter. It multiplies by self-division, two together, head to foot, and may occasionally be seen swimming along, both plying what cilia there happens to be free; no waste of time, no delay manifested, until completion is effected.

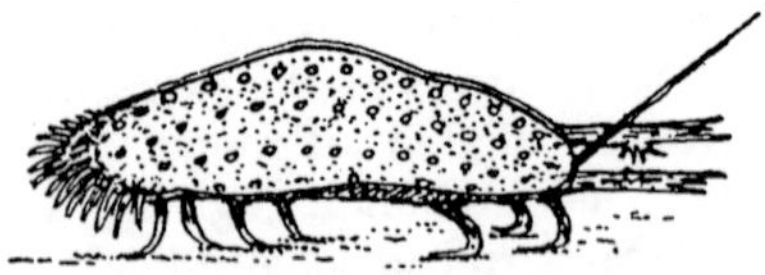

Fig. 167.—EUPLOTES. (Side view.)

A busy life these infusorians lead. Euplotes has been under survey six hours at a time, and never once was it observed to have its cilia still. It is somewhat variable in its outline. The writer has noted over a dozen slight variations. Some of the general markings, as the pearly "buttons" around the edge, may not be visible at times. The number of seta and foot cilia do not vary, however, nor its usual movement habits.

It is seen at its best by dark ground illumination, and may be considered a highly organized and sensitive infusorian, offering an excellent example of untiring energy and scrupulous care in the selection and quality of the food it consumes.

Upon one occasion only was it observed to have a second biparted and shorter style pointed upwards beside its usual one carried to the rear.

Stentors

The body of these may be described as ovate or conical in outline, varying with considerable irregularities owing to the plastic nature of their substance and habits of contortion. They are found free swimming or attached by their narrow extremities to aquatic plants, etc., and even to one another at times.

At the wider anterior end is a spiral wreath of long cilia surrounding the crown, ending in a small coil at the mouth entrance situated there. The crown has small elevated rings or spots upon it. The body is striped longitudinally and covered over all with short cilia, constantly in motion. With the aid of the crown cilia and these the animal progresses as well as procures its food.

Among the short body cilia are many longer and stiffer ones which, though flexible, are moved but slowly, forming, probably, organs of touch to a greater range around it. At times a fringe of long cilia extends from the crown down the side in a sigmoid wavy line, ending in a small coil near to the middle of the body. When this is seen multiplication is in progress.

Increase is made by subdivision, either obliquely or transversely. It is also accomplished by a budding process from any eccentric part of the body. It has been observed propagating a new stentor from the crown, within the ciliary wreath, finally breaking off from an attenuated point completely ciliated and in "full working order."

There are several vacuoles within, and one large one, situated near to the entrance of the mouth or œsophagus. The orifice used in ejection of the waste material is situated at the side beneath the ciliary disc.

Stentors are among the largest of the infusorians, some being quite visible when separated from the plants, etc., surrounding, to the unaided sight. They are found in most stagnant or standing waters, and are of several colours. Red, green, brown, blue and black being some of their hues, depending upon the strength of light sent through in viewing them ; opal white is a usual appearance also.

Lake red seems to run fairly frequently in their tissues when seen under good conditions, and even the green ones will have splashes here and there of this tint. The black will tone down to a browny colour if strong enough light is used, and the browns will reduce to a reddy, ruby tint once more under similar circumstances. Blues and greens differ chiefly in the amount of yellow in each.

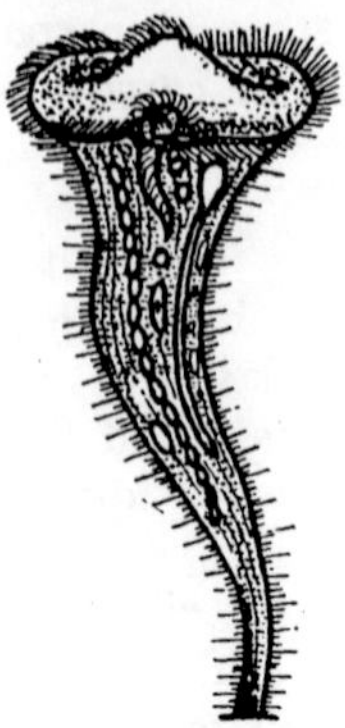

Fig. 168.
STENTOR POLYMORPHUS.

The colours are actual pigments in the cuticle and not due to coloured foods which may have been eaten.

Stentor polymorphus (Fig. 168)

This is a green, trumpet-shaped species fairly common, found either solitary or in groups upon decaying sticks, leaves, aquatic plants, submerged stones, etc.

It may be fixed or free-swimming. If fixed, as is most usual, it is attached at its narrow extremity by a sucker-like disc. This disc may be seen upon detachment to be composed of a bundle of threads, which may close together as soon as freed or remain as cilia-like, flexible projections from the end for a time.

The body is entirely ciliated, and has the longer pliable ones previously mentioned interspersed among them. When grouping themselves, their habit is to throw around them a mucous-like sheath, dividing it into separate compartments and allotting one infusorian to each.

Upon the under side of a frond of Lemna ten of these were observed so encased, all hanging head downwards, plainly visible to the unaided eye. Their length, fully extended, is about 1/20th inch.

The nucleus is moniliform, i.e. like a string of beads, and runs longitudinally downwards at one side. The outer cortical layer contains many green chlorophyll granules.

Fig. 169.
STENTOR IGNEUS.

Stentor igneus (Fig. 169)

This is smaller than Polymorphus, being 1/70th inch in length. It is often in a contracted state, becoming an oval at such times with a short, pointed tail. It is also frequently found free-swimming, when its cilia disc performs the principal propelling action, and also its means of capturing food.

Its nucleus is spherical, a distinguishing feature, and its ruby, lake red colour is distinctly evident. Its frontal wreath is not completely surrounding the crown.

In the spring they are often plentiful.

Stentor cœruleus

Resembles very much Polymorphus, only that its colour is blue. Its nucleus is bead-like, it is trumpet-shaped, and its frontal cilia complete round the crown.

Vaucheria is a frequent habitat of theirs. They are small, measuring but 1/450th inch in length.

Stentor niger

Intermediate in size, and of a dark brown to black in colour. Its nucleus is spherical, like *S. Igneus*, and its frontal cilia is also spherical, which distinguishes it, apart from its colour.

It is about 1/100th inch in length, and favours an oval shape rather than a trumpet form.

Dendromonas virgaria (Fig. 170)

These infusorians form a tree-like colony, having rigid branches and a single rigid pedicel. The plasmic little zooids are affixed upon the extremity of each branch, and may number more than a hundred in all.

The main pedicel quickly branches after leaving its base of attachment, and these form almost the bulk of the object. There appears quite as much branch as zooid in the aggregate.

The body of the infusorians is somewhat oval or pyriform in shape and possesses two very fine flagella, one shorter than the other, very faint and difficult to see unless conditions are particularly favourable. They are situated a little way down, at the side from the apex or tip of the zooid, which gives a lip-like extension beyond their point of attachment.

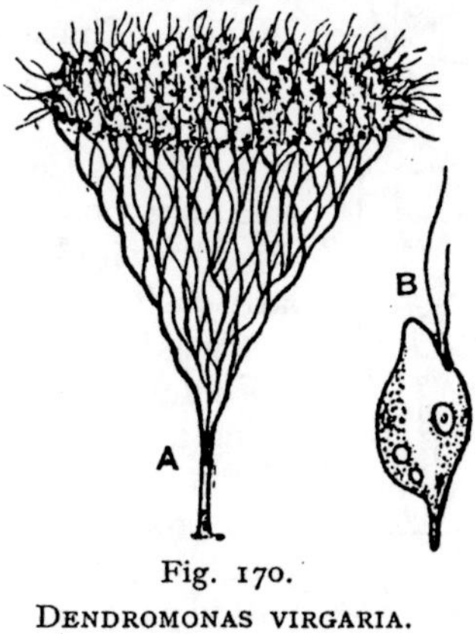

Fig. 170.
DENDROMONAS VIRGARIA.
A. Colony of same.
B. Single zooid.

There is a single and fairly large-sized nucleus near to the middle of the body, and at the posterior end are two small pulsating vacuoles. There is no definite mouth organ, and there appears to be no general covering to the body. A tiny particle of food whipped by the flagella upon any portion of the surface seems to sink in and become digested.

They are extremely small. The length of each zooid is about the 1/2500th of an inch only, and the colony in height altogether little more than the 1/1250th of an inch, so their details are principally for high-power, accurate work. The colonies are frequent upon Myriophyllum and various alga plants at times, in ponds and still waters.

Sometimes the branch will have two of the zooids upon the tip, and the scintillating little flashes given to their outline, under dark ground illumination, will often reveal their presence when only the one-inch objective is in use ; but further than that a one-fourth or one-sixth will be needed to study them fully with. The zooids seem rather crowded and are somewhat compressed in shape.

Vaginicola, Thuricola valvata and *Pyxicola carteri* (Figs. 171, 172 and 173)

These are soft-bodied, cylindrically shaped infusorians. The upper end is flattened and lid-like, usually set at an angle and with a wreath of cilia surrounding. To one side is situated the mouth orifice into which food is swept by their action.

Fig. 171. VAGINICOLA.

Taking Thuricola as typical of the class, this is encased in a flask-shaped lorica, perfectly transparent, and into which it retracts every now and again or upon any disturbance, emerging but slowly and expanding its cilia once more. It is attached to the bottom of its flask directly, without any short stalk, and the lorica is similarly fastened to aquatic plants or other water objects (see Fig. 172).

Frequently two bodies will be encased in one sheath, living amicably together, extending and contracting in unison.

It propagates by self-division and also by a budding process—gemmation. Both processes always take place within the sheath. In gemmation the buds are formed about the base of the animal. The young soon acquire a ciliated wreath, and, escaping from the sheath, are free to wander away and provide a location and form a lorica upon their own and continue life in the same way as their parents.

The sheath is fairly stiff in texture and measures about 1/200th inch in length. At a short distance on the inside, from the open end, is a valve or lid which automatically closes across as the animal retracts, covering it in. It is transparent, and needs careful looking for, to see. As the Thuricola slowly glides along, in extending itself, rubbing close to the sheath, when it reaches the lid, its slightly pointed body will be seen to shunt on to the main line, as it were, and become centrally placed, and the lid will as slowly bend back and remain at the side while the animal is outside foraging.

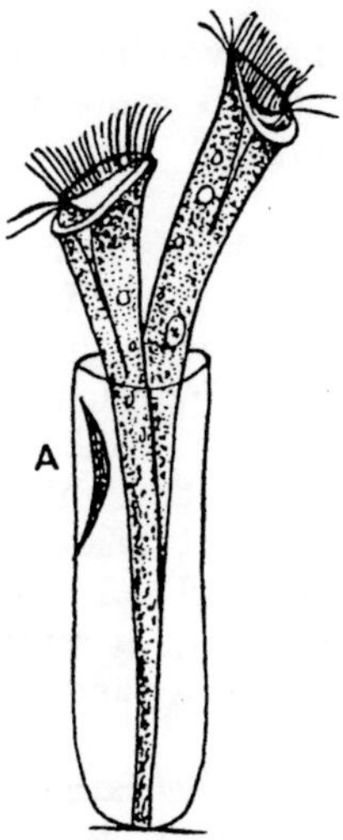

Fig. 172.
THURICOLA VALVATA.
A. Valve attached to inner side of sheath.

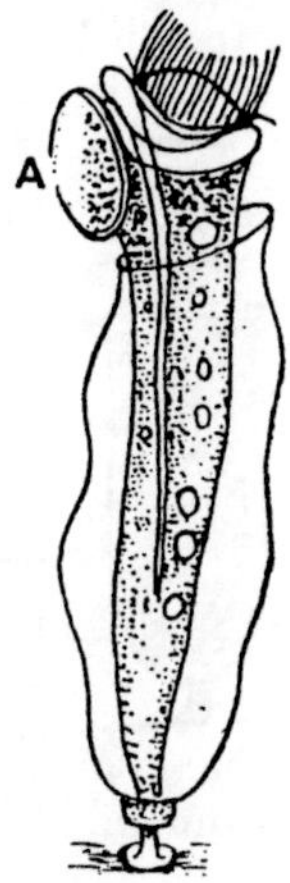

Fig. 173.
PYXICOLA CARTERI.
A. Lid of chitin attached to zooid.

The body is finely and transversely striated. In self-division this occurs across the centre of the infusorian, and new cilia forming a fringe may be seen there sometimes in motion before the bodies are definitely separated.

There are several contractile vacuoles, one fairly large, usually at the anterior end beneath the ciliary wreath. While feeding, it projects a good distance beyond the upper edge of its lorica, perhaps a third of its length, and is perfectly free to sway from side to side within its sheath, having no attachments other than its base.

The body will occasionally present a pale greenish colour, and is usually very finely granulated.

Vaginicola is the former name for these ciliated infusorians, but the genera has been subdivided. Vaginicola is now only retained for those that are devoid of a valve. There is, moreover, a short foot stalk attaching its base to the sheath, which distinguishes it from Thuricola (see Fig. 171).

In the other genus, *Pyxicola carteri* (Fig. 173), the chitinous lid is attached to the body of the zooid itself, immediately beneath its peristome. This functions in closing the opening when it retracts within.

Cothurnia (Fig. 174)

This is a more slender, cylindrical infusorian, tapering to quite a pointed posterior apex and encased within a flask-shaped sheath, similar to Vaginicola. The margin of its sheath is not everted or bent over like that organism's, and is often straight and tubular. It is stiffly formed and transparent. With age it becomes a yellowy brown, and is finely striped longitudinally. At its lower part it is attached by a short foot or pedicel; this forms a distinguishing feature when comparing it with similar ciliates.

It will be found upon aquatic plants, etc., and a favourite place is upon various entomostracans, especially Cyclops and Water Fleas. In this manner it obtains a variety of foods.

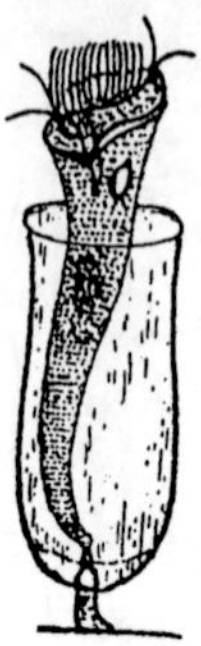

Fig. 174. COTHURNIA.

Its body is attached within the sheath by its pointed end and free to move sideways. At the floor of the case is a mucous substance for a short distance up. In this, gemmation, one of its processes of propagation takes place. At its anterior extremity is a ciliated wreath around its flattened crown, and with eight or ten short, curved bristles standing out at distances around. These are probably organs of touch. To one side of this disc is the mouth, and near by a circular contractile vesicle. On the opposite side is a band-like nucleus.

Its body is finely granulated. The anal orifice is beneath the ciliated wreath and near to the mouth. Its body contracts now and again. It is a timid creature, gradually expanding as it feels confidence restored.

It never extends far beyond the level of the sheath, like other and similar infusorians, when its cilia are closed and it is within the case, little more than a rounded end or pointed oval may be discerned.

Its length of lorica is about 1/500th inch, Vaginicola's being twice this size. Its mode of multiplying is by transverse division, and will sometimes occur also by longitudinal fission.

Platycola (Fig. 175)

This infusorian is very similar to a Thuricola in construction, were a single specimen of that species adhered by its lorica at the side, instead of at the end. The lorica is oval, seen from above, flattened beneath, and convex upon its upper surface. Upon the lower side it attaches itself to Myriophyllum and other water plants.

Within, its tapering, cylindrical body is fastened to the farther end, and through an opening, closely fitting it at the other, its head is protruded. The head has a flat crown, fringed with cilia, as in Thuricola, and the mouth placed similarly at one side. Near by is a fairly large circular contractile vesicle.

Owing to the position of its body being parallel to the stem or other surface it is adhered to, the upper portion is bent outwards, at right angles, to ensure the free working of its cilia. Upon its crown are a few short bristles, seen standing out to a distance, and used as feelers or organs of touch

Fig. 175.—PLATYCOLA.

It is a shy creature, and darts back instantly at any slight disturbance. Its body is finely granulated, with a nucleus near the centre, and in appearance is almost colourless.

The lorica at first is transparent when young, but later with age becomes yellowy, and finally of a dark brown. It is then almost hidden within. It usually extends in action considerably beyond the aperture of its sheath, equal almost to the length. The lorica measures about 1/275th inch in length, and is of a softer nature than the vase-shaped sheaths of Vaginicola or Cothurnia, which are placed upright. It increases by self-division, the central portion producing a new wreath of cilia and finally separating into two.

Peridinium (Figs. 176 and 177)

These are ovate bodies, having a convex dorsal and concave ventral surface. They possess a lorica composed of about twenty polygonal plates, the principal characteristic of the organism being a ciliary furrow around their circumference. From this central furrow runs another short one at right angles to the apex of the anterior end. At the junction of the two arises one long flagellum, and near to this the mouth entrance is also situated.

Fig. 176. PERIDINIUM CRASSIPES.

They are commonly of a brown colour, and sometimes, though rarely, green. There are no horny processes upon them as in Ceratium, but several short spines are prominent upon the outer lorica, seen to advantage when an empty shell is viewed. It has also a reticulated surface, and is non-contractile, with a space separating an inner and softer layer beneath. This has been called the primordial utricle, and forms the cover around the principal substances within, and to which they are directly or indirectly attached.

The general appearance of the Peridinia resembles very much some forms of pollen grains outwardly. The flagellum is the chief organ of movement, while the ciliated wreath gives the peculiar rolling and oscillatory motion to its progression.

Occasionally a red speck or specks can be seen within, which, however, becomes absent, or obliterated, with age. The organism swims with considerable activity. It has a central, irregularly oval nucleus and a finely granulated interior. No pulsating vacuoles are observable. It multiplies by longitudinal division, and sometimes by spores, with the usual resting stage, as in Euglena, etc.

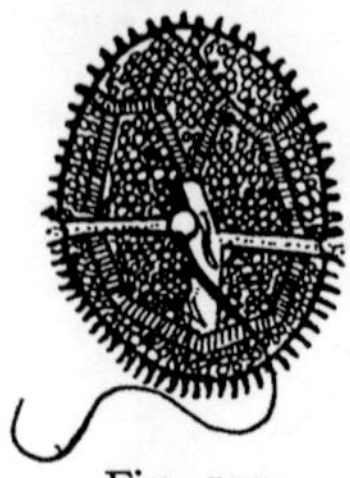

Fig. 177.
PERIDINIUM TABULATUM.

Peridinium is found among aquatic plants in still fresh water, and seldom, if ever, in infusions. It also inhabits salt waters, and most of them are marine in the fullest and finest forms; yet so plentifully have they been found in fresh waters, notably in a pond in Phœnix Park, Dublin, that it is recorded a white disc held but five or six inches beneath the surface was almost invisible through the immense number of their brown bodies.

In the marine species they are of a yellower colour and phosphorescent.

Ceratium hirundinella and *Ceratium tripos* (Figs. 178 and 179)

This organism forms one of the Dinoflagellates and is not unlike Peridinium; only in the latter genus there are no horny processes,

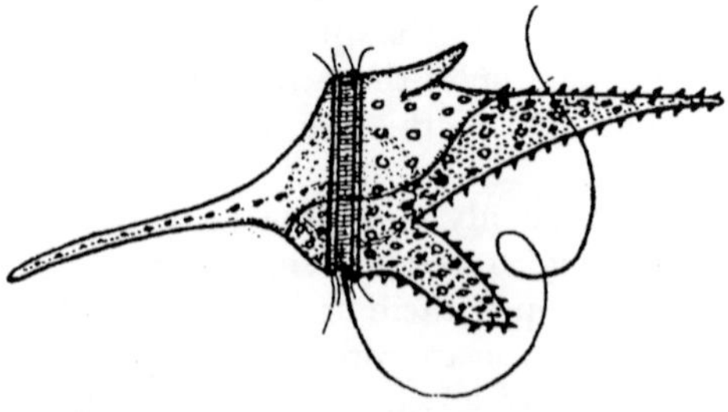

Fig. 178.—CERATIUM HIRUNDINELLA.

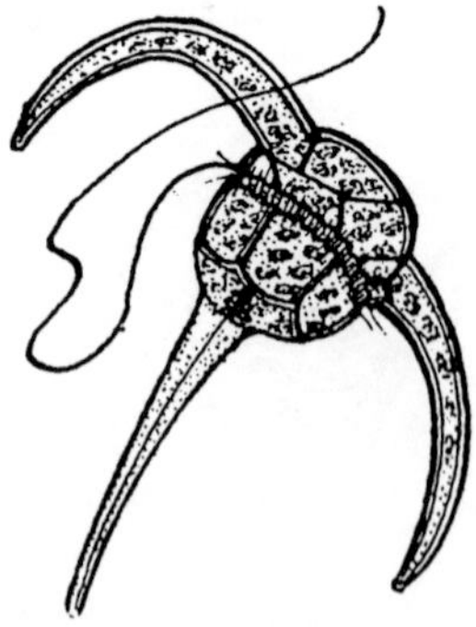

Fig. 179.—CERATIUM TRIPOS.

while in the Ceratium there are usually four. The sub-class to which it belongs is the Dinoflagellidæ.

The exterior coat is of a chitinous, horny nature, with a long process, to the rear of an ovate, spherical body. Upon the anterior extremity is a further long horn, and upon either side of its base two shorter ones. All are knobbed and ribbed profusely, giving it a rugged appearance.

Around the centre of the body is a groove, from which at one point arises a long, waving flagellum. It swims with a swaying, rolling motion upon its longitudinal axis. Along the groove are many fine cilia in rapid motion. There is a central nucleus, rather difficult to locate sometimes, near to the centre of the body.

The general colour of the organism is greenish, or sometimes a yellowy brown. It is a species usually found in the open and more central portions of lakes and reservoirs, etc., and is at times prevalent in the plankton or surface skimmings.

The flagellates are said to be distinctly phosphorescent when in considerable numbers, emitting a dull glow at dusk as they are turned over by the oar or splash from the boat.

Between the anterior end and the middle groove several tiny red flecks may be observed occasionally. The whole body is more or less granular in structure and often reticulated exteriorly.

The length is from 1/275th of an inch to the 1/150th, tip to tip, the horns varying considerably in different specimens, according to age and growth.

Although *Ceratium hirundella* is an inhabitant of fresh waters, there are other species found as marine in bays, harbours, etc., around the coasts (see also *C. longicorne*, Fig. 179A, facing p. 114).

Vorticella

These lovely ciliated infusorians are generally found in clusters together, sometimes occurring in such profusion as to thickly cover the stems of tender aquatic plants for an inch or more, and are then visible as cloudy grey patches along the surface. Normally they are quite invisible to the unaided eye; but should such appearances be noticed, a touch will send them in and the patch vanish.

They will attach themselves to all conceivable kind of objects, every class of plants, floating logs and branches, even to shells of snails and entomostracans. Quite a novel sight, but not at all unusual, is for a cluster to be seen careering along affixed to the back of a Cyclops, transported "wherever he may roam."

Like the hired hunter the novice failed to control, which jumped him over the hedge, took him across the well-kept flower and kitchen gardens, through the squire's orchard, resting him in the greenhouse. "Just like one who knows each tit-bit to visit," so Vorticella takes his free, unceremonious rides at intervals, hoping for the best, and a halt sooner or later for refreshments to ply his cilia in momentary peace once again.

Vorticella is shaped much like a bell, campanulate, as it is called, and has the appellative of the "Bell animalcule." Around the rim of the bell a wreath of cilia is situated. Covering the opening is a domed lid, ciliated also upon its edge excepting where the hinge might be. The lid usually is kept closed, or nearly so, and a passage is thus formed between the two rows to a point on the circumference where the mouth is situated. The mouth is a rounded and spirally ciliated depression.

The crown of cilia, lashing in a wave-like motion, creates a whirlpool in the water, and particles of food are drawn near and dashed with some force into the food channel.

The nucleus is a long and very finely granulated horseshoe-shaped body, placed across one side. The colour of the body is either a semi-transparent watery white or a bluey slaty-grey tint and opaque. In one species the body is green throughout, save the stalk.

Vorticella has several pulsating vacuoles. It can break away from the stem and continue unimpeded. Its ciliated crown then becomes its means of locomotion, giving a spiral course to it through the water. In this free state it will sometimes contract into a ball, draw in its cilia, and cover itself with a strong gelatinous overcoat, and thus stay in a resting or "encysted" form until suitable weather conditions allow it to open out again and become active once more. It has been said, after being in this condition, it will throw out monad forms fully flagellated to become adult Vorticella by and by.

The outer coat is usually spirally or transversely finely striated, which is not easy to observe at all times unless very good objectives can be used, but out of seventy species known only about twenty are recorded as actually smooth, so the probabilities are, even if they are not seen in the commoner species, they are there awaiting better opportunities of research.

Vorticella is very sensitive to shocks, and, when fully extended, if the table or stand is given a rap, they will instantly coil their stems up tightly and pause, before unfolding once more. They increase by longitudinal division. The nucleus divides, and one-half goes either side, each side becoming eventually a complete and separate individual.

At certain times a second wreath of cilia is produced upon the lower half of the body and suggests the possibility of their increasing by transverse division as well.

The stems of Vorticella are transparent in all species, and have the means of contraction and expansion. This has usually been attributed to a thread running spirally down the inside. It is a point upon which much more observation and experiment is needed. The writer has seen a stem without a thread, or without a substantially filled tube within, that has continued to contract and expand. The stem may contract, or start to, from any part along its length.

In a specimen deviating from a straight line into a U-form in the middle, the expansion would only extend to this shape each time, never fully to the limit. The stem contracts quite independently of any effort from the head end, and often whilst watching would seem to form quite a nuisance to the Vorticella, just as it has put out its cilia for action to be relentlessly jerked into a coil again.

In Carchesium there is the same thread running down the stem spirally; but it never contracts its stem. The stem is more complex than might appear at first sight. The so-called spiral muscular thread lays upon a flat, transparent, spiral band. The whole stem is said to be a hollow tube; yet it has never been seen flattened, as one might expect to find it occasionally, or even when tightly folded up in coils. It appears to keep its rounded shape as an elastic solid at all times.

The spiral muscle contains numerous nodules, having the appearance of short columnar bodies, highly refractive either end, embedded in the substance irregularly and along its entire length. These are not favourable to elasticity. They are probably spores. The muscle having pulled the body, supposedly, into a tight coil, there is then left expansion again to accomplish.

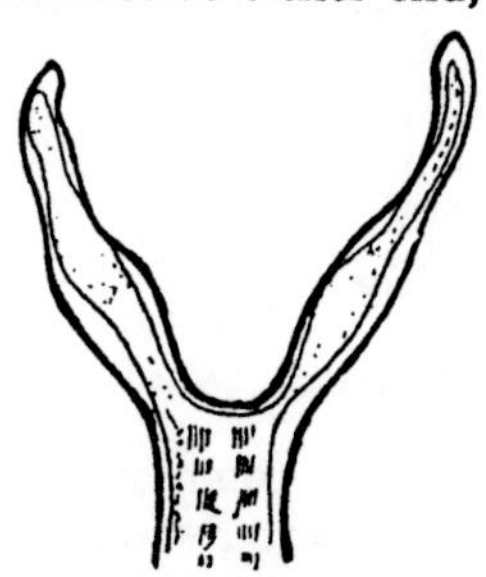

Fig. 180.—VORTICELLA. Stem, × 1800 (after Everts).

Organized bodies possessing muscles need abductor as well as adductor operatives. The abductors are not apparent in Vorticella.

Further, the thread is always longer than the stem it is in, whether coiled or uncoiled. This is a provision its spiral form permits, and preserves it from strain. It becomes then a matter of some difficulty to account for the spasmodic jerkings common to all the Vorticella. The writer considers the final solution to lie, in all probability, in the peculiar substance of the outer stem itself and akin to the same that causes the movements of diatoms.

A sarcode, having the natural property, whilst living, of contraction and expansion under constant stress and strain, that is colourless, and of a plasmic and distinctly tough texture, light and temperature may have some influence upon its action. It awaits a further systematic investigation for its ultimate explanation.

The stem is striated longitudinally, and is a very sharp test to see when in its *aqua media*. This striation is often seen broken into short threads by transparent bands transversely occurring across at intervals of about three in the width of the stem. The threads are said to be muscular fibres (see Fig. 180).

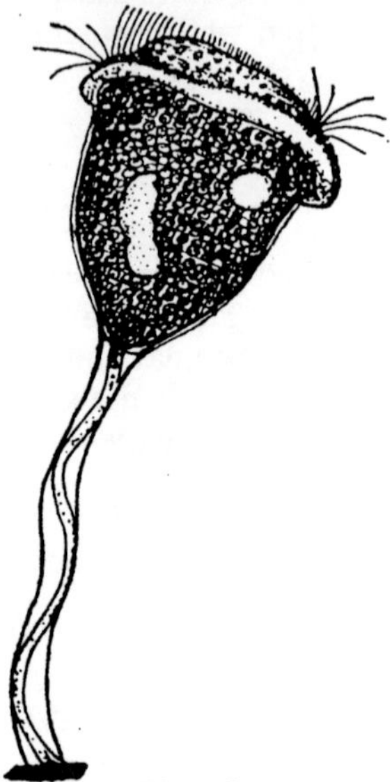

Fig. 181.
VORTICELLA MONILATUM.

Vorticella monilatum (Tatem) (Fig. 181)

The body is widely bell-like, and the ciliary disc-edge bent backward into quite a thick everted lip, and when fully expanded will exceed in width the entire length of the body. When in contraction it is almost spherical, and the throat cleft is deeply prolonged.

The outer coat is hyaline, and can be easily distinguished from other species by the entire surface being ornamented with hemispherical bead-like elevations closely set in transverse annular rows, one above the other, those nearest the ciliary wreath being the largest.

The nucleus is short, compact and band-like. The stalk is about four times the length of the body and fairly stout. The body length varies from 1/400th inch to 1/1400th inch.

It is common on Myriophyllum, forming social colonies and in fresh water generally.

Vorticella microstoma (Ehr.) (Fig. 182)

The body of this is ovate or pear-shaped, contracting at times to globular. Its outer covering is finely striped transversely. The ciliary disc is narrower than the central circumference of the body, and is not everted or bent over like many other species are.

The nucleus is band-like and horseshoe-shape, lying about the middle. The stalk is from five or six times the length of the body.

It is a very common object in infusions and stagnant waters generally, and is found either singly or more usually in social clusters.

Its length averages about 1/500th inch. In multiplication, which is by longitudinal division, a girdle of cilia is produced around the posterior end previous to separation.

There are several pulsating vacuoles, and the throat cleft is unusually prolonged.

Vorticella chlorostigma (Fig. 183)

The body of this is bell-shaped and much longer than broad, giving it a narrow, attenuated appearance. Its outer coat, as well as the inner substance or parenchyma, is of a decidedly bright

Fig. 182.
VORTICELLA MICROSTOMA.

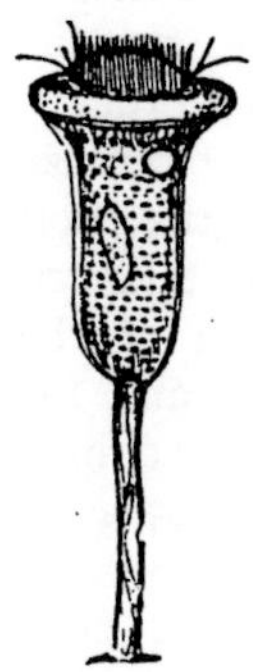

Fig. 183.
VORTICELLA CHLOROSTIGMA.

Fig. 184.
VORTICELLA GLOBULARIA.

green colour. It is very granulated, and along the cuticular surface are many transverse rings, or striæ, evenly spaced.

When found it is usually in profusion, covering aquatic plant stems and other submerged objects so thickly as to present to the unaided vision patches of brilliant green mucus, not unlike that of the ciliated protozoa *Stentor viridis*.

Its length is 1/240th inch without stalk. The stalk is from four to five times longer and fairly stout. It is found as a social colony, and rarely single.

Vorticella globularia (Mul.) (Fig. 184)

The body of this is, as its name implies, globular or spherical, having but a narrow ciliary wreath, much contracted in proportion to the central circumference. Its stalk is very slender and long, from six to seven times that of the body. The diameter of the body is about 1/150th inch.

It is often found attached to a species of Cyclops, one of the most active and abundant of fresh water Copepods. The ciliary wreath is not everted, and stands up at right angles when in action.

It is not found particularly abundant, but once seen is easily remembered on account of its bulging, turnip-like shape and the delicate long pedicel, which marks it as being almost top-heavy and weakly constituted. It contracts its stem spirally, expanding again but slowly.

Vorticella nebulifera (Ehr.) (Fig. 185)

This has a conical body seen extended, contracting almost to a spherical outline. The ciliary wreath lies obliquely upon a crown, but moderately dilated. It is common at all times of the year, and especially found upon the roots of Lemna among other aquatic plants.

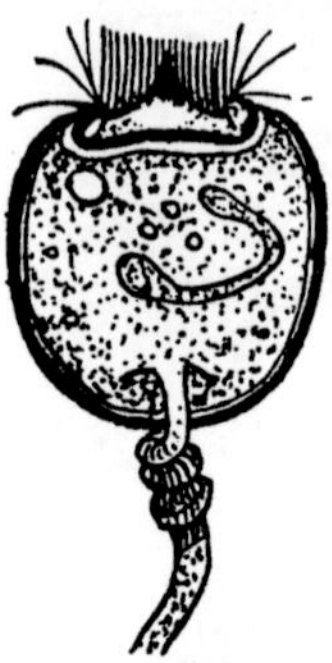

Fig. 185. VORTICELLA NEBULIFERA

Its body is colourless, and the outer surface smooth and plastic. Sometimes it is seen in folds and puckered. It has a slender stalk about five times the length of the body. The nucleus is horseshoe-shape, and many transparent granules are to be observed in the central mass of the surrounding endoplast. Its length averages, without stem, about 1/350th inch.

Multiplication is similar to other Vorticellæ by division. Several pulsating vacuoles may be seen. When coiled upon its stem the inner tube affords a good view of ridges along its course.

Nassula ornata (Fig. 186)

The body is entirely plastic in any direction, ovate, globular or cylindrical in outline, and completely ciliated along closely placed rows. The visible side may count twenty such rows of even-length cilia in rapid motion.

It has a twisting, gliding movement, and can creep through openings much smaller than its normal size by extending a portion of itself and thinning out into a narrow cylinder, passing its substance forward until its bulk is upon the one side (Fig. 186A).

The mouth is about a third of the length down from the anterior end, and has a throat or pharynx surrounded by a collar of rods, much in the manner of Chilodon.

When it wishes to eat, or abstract the gelatinous outer coats of Oscillaria or other alga, it protrudes this quite an appreciable distance out from its surface and will draw objects in such as diatoms, etc., or bend the tender filament of an alga at whatever point it meets it, and may often be seen with a bent strand of such drawn half-way down the pharynx with the two extremities projecting outside. Here it will rest for a time, either until further withdrawn or else ejected altogether, minus some of its outer mucous coat, which has been absorbed as food.

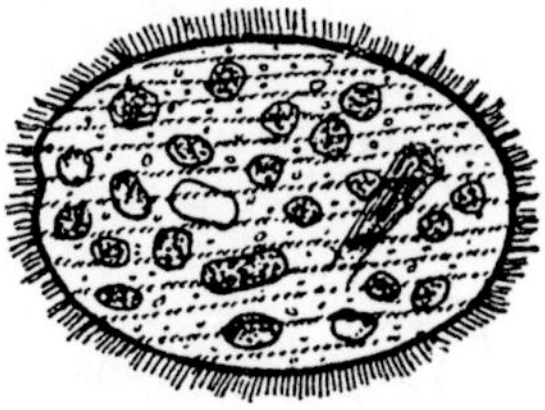

The body generally is of a pinkish tint or inclined to brown, and is variegated with numerous violet, oval or circular patches upon its surface. These are very characteristic of the genus, and give it a dappled appearance. They are of fairly large size. The macro nucleus is situated posteriorly, and is a large spherical body.

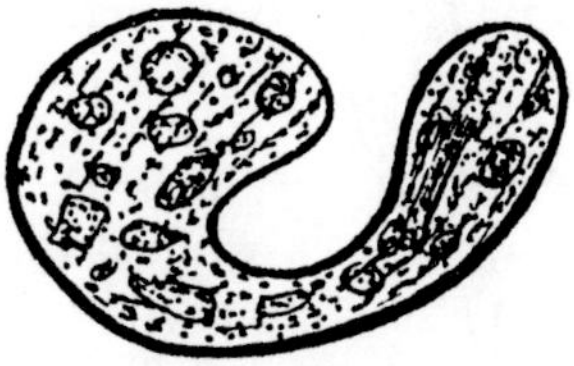

Fig. 186.
NASSULA ORNATA.
A. A position assumed among its many contortions.

There is a conspicuous contractile vacuole pulsating, situated near the base of the throat. Numerous small granules are seen within the body and appear to flow with the general substance as it squirms and pushes its way about. Oscillatoria is a frequent alga amongst which the organism is found.

Its length is about the 1/100th of an inch, normally.

Stichotricha secunda (Fig. 187)

Body lancet-shaped, elongated at the front end into a flat neck. Posterior end rounded. The peristome is narrow and the cilia long and slender. There are three oblique rows of setæ upon the lower side and no frontal or anal styles. The nuclei are two, situated near to one another about the centre of the body.

Fig. 187.—STICHOTRICHA SECUNDA.

The animal sometimes secretes a gelatinous sheath, from which it protrudes the anterior half of its body, or will vacate this and swim freely without it.

The body is of a hyaline substance and very contractile, and will assume many wave-like contortions in its habits while in search of food. It swims actively in a spiral manner, and occasionally will be seen creeping over aquatic debris. It is frequent among Sphagnum, and measures about 1/125th of an inch long.

Phacus longicaudatus (Fig. 188)

The body is oval, flattened, and with a twisted portion about the middle, continuing into a conspicuous pointed tail. It is leaf-like, and its surface is longitudinally striated. The stripes follow the curve and become closer together as the tail point is approached.

Fig. 188.
PHACUS LONGICAUDATUS.

It is generally of a green colour, and has a red eye speck at its anterior end, close to a long and very active flagellum difficult to observe at times. It is found in ponds, often along with Euglena.

The outer coat of the body is of a substance admitting little change in its coiled outline, and without contractility. It also resists the action of reagents, which, while removing the green colouring matter from within, leaves it intact and quite transparent.

Near the base of the flagellum is a large contractile vesicle. The nucleus is situated at the centre of the body. Where the flagellum projects is a fairly deep cleft.

In another species, *P. pleuronectes*, the body is not twisted and the tail quite short and curved. It is a smaller specimen, measuring about 1/500th of an inch in length. *P. longicaudatus* is 1/250th inch long.

Chilomonas paramecium (Fig. 189)

This is a very common infusorian in stagnant waters. It is generally colourless, but has been found having a decidedly pink tinge. It is ovate, wider anteriorly than posteriorly. There is a small notch at the front end which, being set obliquely, presents a lip-like projection above the "mouth" as the notch is called, which forms a characteristic feature of the organism. The "lip" is furnished with two very fine flagella, which gives a rapid movement to the busy little creature as it wends its way among and around decaying algæ and aquatic plants in quest of food.

Fig. 189.
CHILOMONAS PARAMECIUM.

It has a method of suddenly halting in its hurried progression and of affixing itself by one of its flagella to objects in the water, or to the cover glass when under observation, while the free flagellum continues spirally in motion.

The endoplasm has many dark-coloured corpuscles, regularly placed, along the inner wall, which turn blue on application of iodine, showing their substance to be allied to starch.

At the posterior end is a comparatively large nucleus of a faint pinkish tint, and at the anterior end, near the base of the flagella, is a circular transparent contractile vacuole.

It will form quite a grey cloud occasionally when in considerable numbers, and to see a cluster of such in the water is not infrequent, with infusions from decaying pond vegetation. It is said that soft water, in which wholemeal bread has been steeped, will produce quantities after standing a few days out of doors.

Chilomonas is a small object and measures from 1/1000th inch to 1/600th inch in length. Its mode of propagation is by self-division, longitudinally.

Lacrymaria olor (Figs. 190 and 191)

This infusorian has a cylindrical body with an extremely long and highly contractile neck. At the extremity of the neck is a conical projection surrounded by a crown of long cilia, and at the tip of this the mouth is situated.

The body is somewhat spindle-shaped and is short in comparison to the neck, ending in a pointed extremity; it is, moreover, obliquely striated. Occasionally the striæ are transverse. The whole surface is profusely covered with short even-length cilia, kept in constant movement.

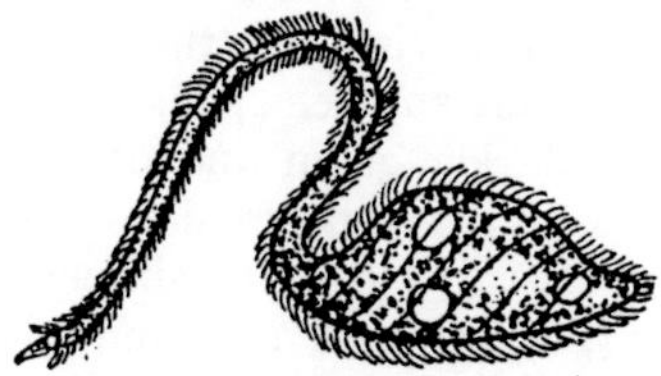

Fig. 190.—Lacrymaria olor.

There are two, if not three, contractile vesicles usually present, one posteriorly always. The nucleus is situated on the ventral side near the centre. The infusorian was formerly known as the "Swan animalcule" on account of the graceful movements of its neck, often assuming a similarity to that of its namesake, the aquatic fowl and "royal" bird, so well known.

Fig. 191.
Lacrymaria olor.
(Contracted.)

One of the habits it delights in is to secrete itself beneath a mass of decaying vegetation, there to stretch and contract its wonderful, lithe and extraordinary neck in and around, with intricate and most difficult turns, the various loopholes within the substance, and as rapidly extricating and commencing again in another direction.

When swimming, like the swan in flight, its neck is fully extended before it, probing to right and left as it goes along. When the body is completely contracted it presents quite a different appearance, and is then a uniform cylinder-shaped body with the mouth tip and its cilia alone projecting from the anterior end, and without vestige of spindle-shape extremity or tail. The oblique striation in this condition is then very apparent throughout (see Fig. 191).

When thus contracted the infusorian measures about 1/400th an inch in length. With neck fully extended it was found to be 1/40th inch on an average.

Volvox (Figs. 192 and 193)

is one of the best-known and probably most admired of all the algæ. It is a spherical body of a gelatinous and hyaline material, studded with small green cells evenly spaced over the whole of its surface, and each bearing two tiny flagella, by which it effects its rolling and progressive movements.

At times it will perform its onward way without turning upon its axis, keeping the same front first, at another it may spin upon its axis like a top, without progressing. It is seen occasionally with fine threads connecting each tiny cell, but not always, and the microscopist must not be alarmed and denounce his apparatus if they be not seen at all times. Its diameter is about 1/50th inch, and is just visible to the eye when held to the light.

Several smaller spherical bodies may be seen within, generally of a darker green and of varying sizes. These are the daughter cells, and in one mode of propagating will divide their contents into four, then eight, sixteen, and so on, always some multiple of four, growing in size and increasing more rapidly at the centre than the sides, in a similar way to Protococcus. Finally, the parent colony bursts, and they are then free to follow existence upon their own.

Volvox prefers a warm shallow pool where green vegetation abounds, and any sudden change in temperature is usually fatal to them. They have, however, been found in winter beneath the ice enduring great cold, and it seems that, provided the change is gradual and not sudden, they can withstand considerable variations. Within the hollow spheres may sometimes be seen a small parasitic Rotifer freely swimming about contentedly. It is called *Hertwigia parasita*. Volvocinæ at present are included under both Protophyta, i.e. a group of the simplest plants, as the unicellular alga, and the flagellated Protozoa or small animal zooid forms having contractile vesicles, and are therefore upon the dividing line between animal and vegetable organisms.

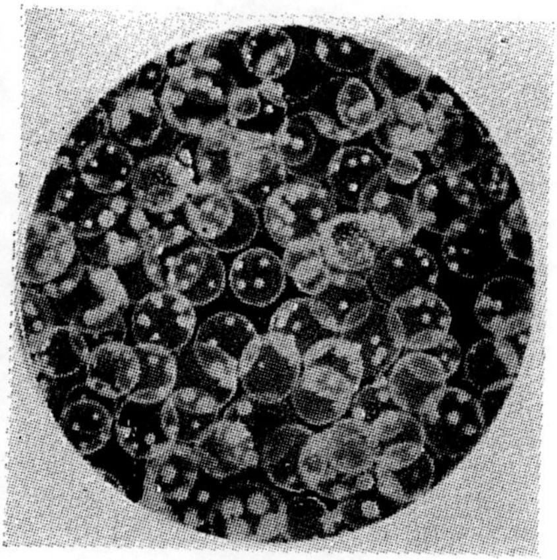

Fig. 192.—VOLVOX GLOBATOR.

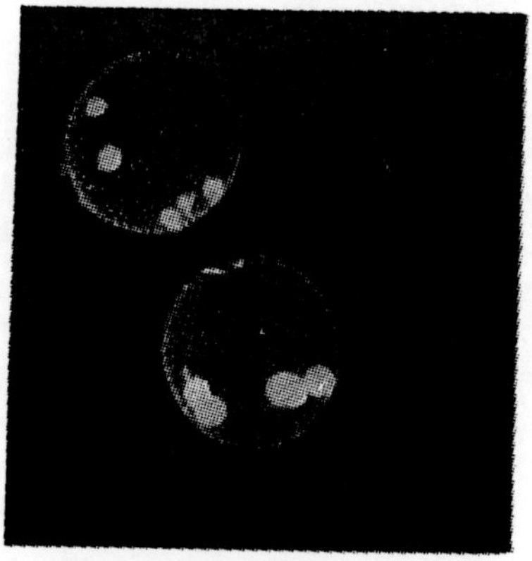

Fig. 193.—VOLVOX.
(Enlarged view.)

Fig. 224.
BRACHIONUS MILITARIS.

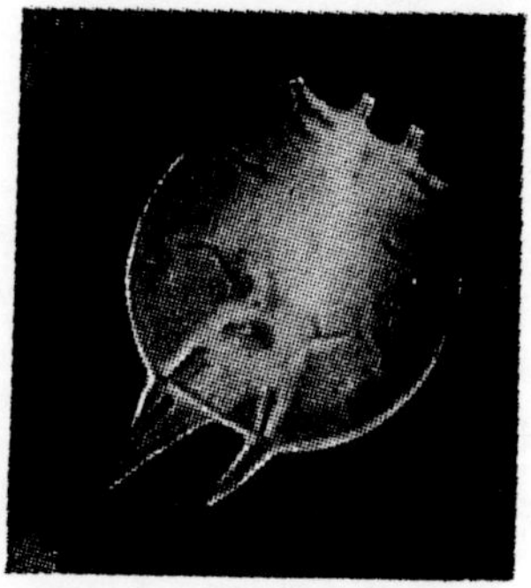

Fig. 227.
NOTEUS QUADRICORNIS.

[To face page 168

If the daughter cells divide as described above, and continue secreting chlorophyll, they are considered as on the plant side. If, on the other hand, it takes place from the interior plasm by gonidial development, contractile vacuoles appear, the cells become flask-shaped, then spherical, yellow antherozoids penetrate, blending with the green ones, selective male and female units operate, changing all to bright orange, the outside becomes studded with spikes and a third separate body found, the line thus followed most closely conforms to the Protozoa, and they are placed upon the animal side.

As Dr. A. C. Stokes in *Aquatic Microscopy* tersely puts it, "If those scientists are right who say the presence of contractile vesicles, always, indicates an animal, they do exist, and Volvox is therefore not a plant. If they are wrong, as they probably are, Volvox manifests no anxiety, but floats along in graceful curves, and remains as delicately beautiful, and as carelessly free, as if it had never excited a moment's interest or a moment's discussion."

Pandorina morum (Fig. 194)

This is a distinctly beautiful colony of green plants growing in a very hyaline spherical or spheroidal envelope, with a rolling movement not unlike Volvox, but rather slower.

The cells enclosed are in eights or multiples thereof, sixteen and thirty-two occurring almost as frequently. They are heart-shaped, and closely packed at the centre, the wider ends being outermost. Each cell has two long flagella, separating widely apart as they leave the outer surface.

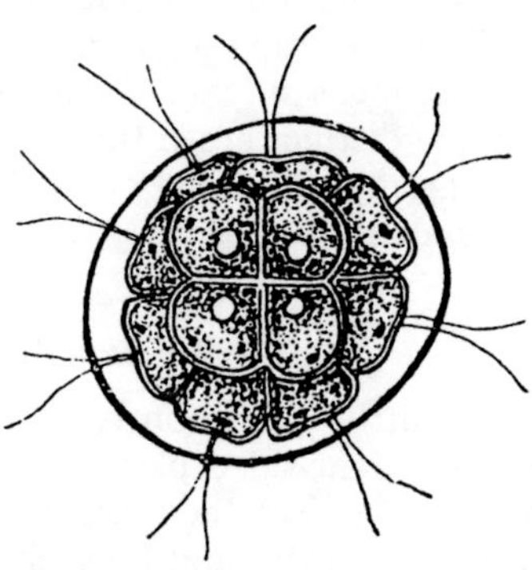

Fig. 194.—Pandorina morum.

A red pigment spot occurs near the base or largest upper portion, and a single pyrenoid at the inner and narrower apex. Each cell is complete in itself to produce its kind, and eventually, a new colony, and will divide into groups of smaller cells, which escape, as young colonies, from the older one, by the bursting of its gelatinous covering.

At the point where the cilia arise is a short process with a colourless space, close to which is the red so-called "eye-spot."

There is no pulsating vacuole observable. Light appears to form a stimulus in the conversion of the asexual spores, for if seen by night many will be found motionless at the bottom of the tube, while during the day they rise and assemble upon the sides. The spores may then be seen to have clustered together into separate colonies and to have left the usual formation, and are irregularly disposed within the parent envelope, devoid of any cilia, the cilia being discarded as soon as division is entered upon.

When the colonies are sufficiently advanced the outer envelope is broken, and each separate individual of the group is now seen to possess two long fine cilia of its own which it immediately knows how to use, and swims away freely. After a short time of this it will join another and group into pairs by a fusing together, first into a rough figure of eight form, and finally into a single globular cell. This becomes the zygospore and is bright red in colour, and in this state they remain at the bottom of the water until the following spring, when a rejuvenescence into a number of cells takes place and they enter life again as perfect plants.

In this reproduction by successive division in each cell there are as many new colonies formed as there are individual cells, and so it will at once be recognized how rapidly they can accumulate when each may produce sixteen and even thirty-two, as previously stated, from an original one. It is remarkable that where the clusters are without cilia within their mother envelope often only fifteen can be counted, not sixteen as might be expected.

When collecting algæ in the spring one may be sure to find a few specimens at least of Pandorina enclosed among their filaments. It is the distinct green tint and the crystalline envelope that under dark ground illumination makes it such a beautiful object as it sways along so gracefully in the water.

Gonium pectorale (Fig. 195)

This little plant consists of a colony of sixteen cells arranged around a four-sided plate, and form a flat single layer. Each side has three cells at the margins, and the remaining four spring from within the square and fill each corner. All of them have two fairly long fine lashes to each cell, and as these are moved, first the edge and then the front of the plate may alternately come into view.

It is rare to see them swim or push themselves backwards, always pursuing a direction forward, and following their quivering flagella in the progression. This distinguishes them from all other flagellates.

The cells are oval, with an almost invisible coat of gelatinous substance surrounding them. There is a red pigment spot in each cell. The chlorophyll is attached to the inner walls and contains one pyrenoid. The cells may revolve upon their edge, wheel fashion, as they go forward, but the plate invariably keeps its edge at right angles to the direction in all its progressive movements.

The colonies are exceedingly regular and very beautiful, the clear bright green of the cells, with the scintillating tips of the flageila as they reflect the illuminant, adds a charming, elegant appearance to the whole organism as it moves along.

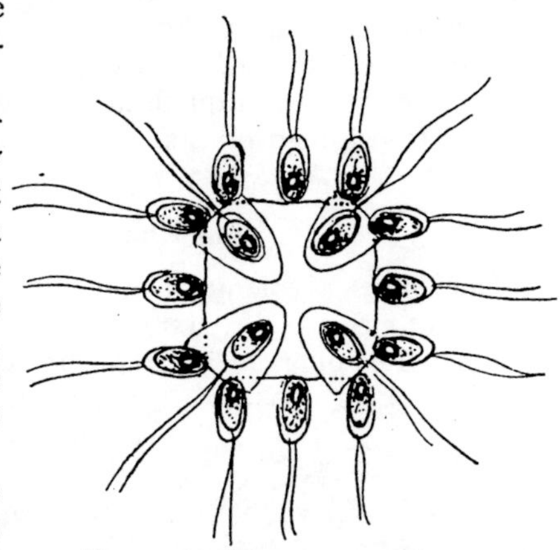

Fig. 195.—Gonium pectorale.

Gonium forms one of the commonest genera of the Volvocaceæ. It multiplies by division, each colony of sixteen single cells producing sixteen daughter colonies while still within the mother cell wall. Neither this general coat or cœnobium nor the individual cell walls themselves are concerned in division, but only the contents of each cell.

The production of the sixteen colonies does not take place all at once but in four successive stages, and naturally distends to a considerable size the parent cell wall to suit the extra increase of the daughter colonies as they continue to grow. As the colonies mature and the temperature suits, some will burst through the wall, leaving a vacant space in the group until all are eventually emerged. These are then free to break up and develop on their own, producing two tiny cilia to each little cell, swimming about as swarm spores for a time before coming to rest. Among the number, some will approach, like Pandorina, in pairs, their points coming together and the bodies gradually coalescing. A single zygospore results, which germinates independently after a short resting period.

The appearance of the free spores is very similar to Chlamydomonas, the cells having a neck-like portion, clear and transparent, from which the two cilia arise.

In the species *Gonium sociale* only four cells are united together instead of the sixteen in *G. pectorale*, and are somewhat larger in size than these.

The habitat of Gonium shows a preference for stagnant waters, and is often found among the decaying vegetation on the muddy stratum at the bottom of ponds and quiet pools. The width of the plate averages 1/200th inch across and the size of the individual cells from 1/600th to 1/1200th inch. The four central cells are usually larger than the rest.

The specific name "Pectorale" is taken from the likeness to the sixteen jewels, in quadrangular form, on the breast-plate worn by the Jewish high priests.

Eudorina elegans (Fig. 196)

This forms a colony in a transparent, spherical, gelatinous envelope of either eight, sixteen, thirty-two, or sometimes sixty-four cells, oval or globular in shape, attached to the inner surface, and each have two cilia projecting at right angles to it. The cilia are invariably found widely divergent after emerging from the outer coat. The cells have each a red pigment spot, and the chlorophyll is attached to their inner walls.

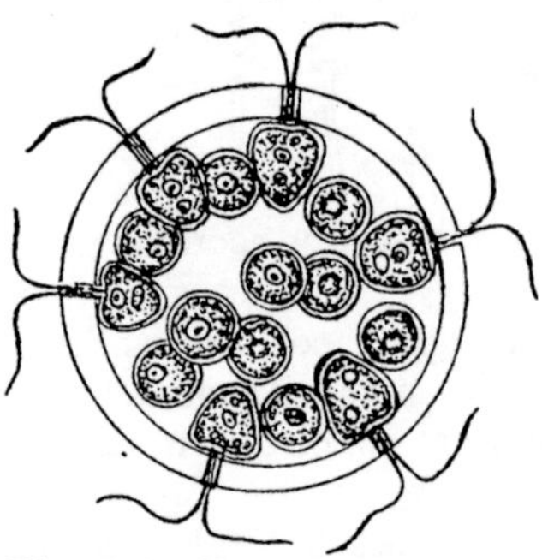

Fig. 196.—EUDORINA ELEGANS.

Their method of multiplication is by division of the cells, forming at first a plate-like colony, which eventually becomes rounded and spherical like their kind. All the vegetative cells may become either male or female, and in each male cell sixty-four fertilizing bodies or antherozoids are formed.

The union of the antheridia with the oogonia is to produce a separate body, the oospore. This is of a brownish colour, spherical, and with a smooth external membrane.

Eudorina is found in similar situations to its class in lakes and ponds among the green vegetation generally. Eudorina revolves in its progression similar to Volvox, and at times may be in amazing numbers, imparting a decided green colour to the water. They are very fragile and delicate owing to the nature of their envelope, and do not keep active in aquaria for any length of time, probably not more than a day or two at most.

It is better to retain only a few in one vessel and to frequently replenish the water supply if it is desired to preserve them. They prefer an abundance of water and plenty of chlorophyll plants or green algæ near. This affords a shade and imparts a nutriment to the water apparently congenial to their tastes.

Synura uvella (Fig. 197)

Formerly this protozoan was separated into two genera, Synura forming one genus and Uvella the other. Synura is a spherical colony of about fifty tiny zooids with pear-shaped bodies attached by a thread-like tail either to the base of its own production or to the centre of the cluster among which it exists. The outer membrane is very delicate and often spinous.

There are two narrow yellowish-green bands extending along the sides of the body. At the anterior end are two unequal-length flagella by which the colony rotate as a whole and also use to procure food with. They are rapid swimmers.

Two contractile vesicles are situated near the posterior end of the body not readily seen but always present. The zooids are capable of stretching themselves a considerable distance by means of their extensible filiform tail, and also by the gelatinous nature of their bodies.

No eye-speck is present. Length of body about 1/800th inch. Common in pond and shallow pools among algæ in early spring.

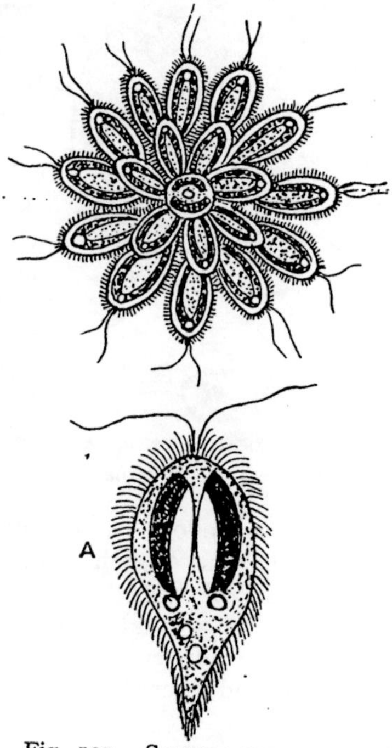

Fig. 197.—Synura uvella.
A. Single zooid of same.

Synura is an organism like Uroglena, noted for the peculiar pungency of an odour produced during its growth to the waters of reservoirs, among which it is frequently found.

Prof. G. C. Whipple says: "It is an oily substance with an odour of ripe cucumbers and a bitter and spicy taste which may be recognized when the water is diluted to the extent of one part in twenty-five millions."

"Some punge" for such a miniature creature, as the Americans might say.

Uroglena anglicana (Fig. 198)

Small free-swimming colonies of zooids adhering upon a spherical gelatinous substance. The individuals are very numerous, and are separately arranged around the circumference of the globular mass. Each individual has usually two yellow chromatophores and a red "eye-spot."

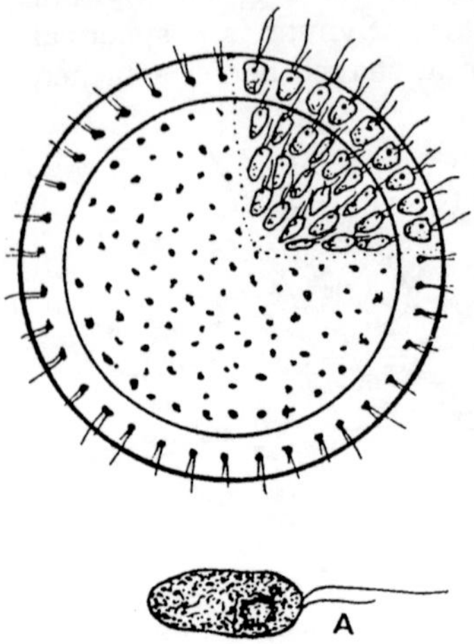

Fig. 198.—UROGLENA ANGLICANA. A. Single zooid of same.

There is a short tail, distinguishing it from *U. americana*, embedded in the "lacerna" or base, but no visible connection one with another.

The mode of reproduction is by self-division, but when the zooids divide the "lacerna" enlarges, without undergoing any fission itself in the process. The zooids are thus provided with the necessary additional space, but are still kept quite separate one from the other as before. At the outer anterior end are two unequal-length flagella.

Uroglena is often present in the waters of reservoirs, giving it a distinctly fishy odour caused by an aromatic oily substance produced during the growth of the organism, and of all the fishy odours, says Prof. G. C. Whipple, Uroglena is considered the worst, resembling that of cod liver oil.

Diameter of cluster about 1/90th of an inch. Found in turfy, peaty pools and reservoirs.

Carchesium polypinum (Figs. 199 and 200)

This is a most magnificent object to see under dark ground illumination, very similar to a spray of lilies-of-the-valley with a hundred or more flower heads each branching from a single stem, every flower as clear as crystal and each crowned with a gossamer-like halo of cilia in rapid movement.

Unlike Vorticella, the stalk never contracts, only the "heads" or zooids individually. They may do so one at a time quite independently, or may all contract at the same moment. Each has the same delicacy and fear of any untoward disturbance, and in contraction appears to find some safety.

The bodies are bell-shaped, and each is placed upon the end of a separate branch. The main stem is always the stoutest, and is affixed at its base by a cement, generally to the stalk of some aquatic plant or the under side of lily leaves, and frequently to stones and submerged twigs. Here they extend into the water, playing their double row of cilia around the crown, producing a vortex, and drawing in food particles which are carefully touched and scrutinized before being allowed to pass into the mouth.

Fig. 199.—CARCHESIUM POLYPINUM.

The mouth is situated upon one side near the rim, and is spirally fashioned with cilia of its own in constant movement. Near to this is a fairly large vacuole, pulsating, opening and closing in steady, regular motions. This assists in the breathing and elimination of waste fluids and gases in the metabolism of the animal.

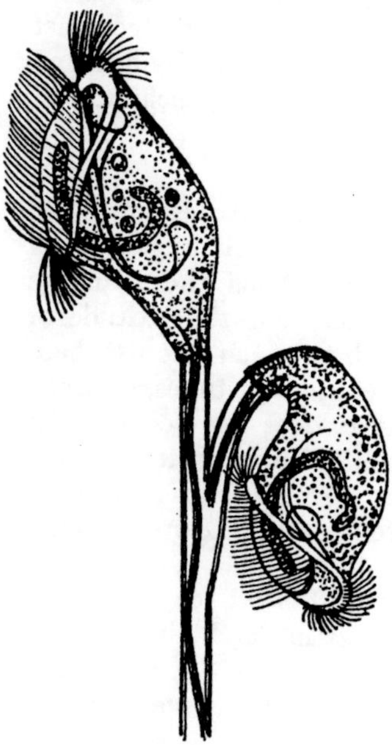

Fig. 200.
CARCHESIUM POLYPINUM.
Zooids in detail.

The height of the colony may be as much as $\frac{1}{8}$ inch, each separate zooid about 1/500th inch in length.

The nucleus is an elongated body curved into a horseshoe shape, somewhat lying to one side, and with one end near the ciliated margin. The outer coat of the body is spirally striated and the interior much granulated. The colour is a watery blue tint and almost transparent.

The stem is a more detailed structure than at first appears. Like Vorticella, it has a spiral band down its centre, which has been considered to be the means whereby they all contract their stalks, but this stands out in contradiction, because it never does so. The band has, moreover, many granular and reflective bodies studding its length, probably of the nature of spores, while the rest of its substance is very finely granular or transparent.

Upon the outer surface of the stalk are many longitudinal lines or flutings; about eighty have been counted all round. The stem itself is but the 1/1600th inch across, so that the difficulty of their observation may be understood. Transversely also are plain bands at rather irregular intervals between, and number about four in the space of the stem's width.

To see the ciliated crown in action at its best, select one, if possible, facing head on towards you, then the intensely fine silken threads as they swirl around the periphery give an appearance not unlike the undulations a long rope or lasso will make as it is waved through the air, each cilia rhythmically following its neighbour in such perfect relation, the whole movement becoming a wave succeeding waves, coursing around the crest of the zooid—truly a marvellous sight.

Always use dark ground for the ciliated protozoa, it is unsurpassed; the numerical aperture it gives is ample and the contrast most effective, allowing a clear view quite restful to the eye and full of the most beautiful silvery detail which is quite enchanting.

Epistylis flavicans (Fig. 201)

A very similar ciliated protozoan is this to Carchesium. The zooids forming the colony have been styled the "little gentlemen who nod, but never bow." Instead of the stems and branches contracting as in Vorticella, Epistylis merely has the latitude of bending its head upon the tip of the branch to which it is attached. This it does very suddenly and with a severe jerk at times, so much so one wonders when witnessing it that it does not come off or impale itself upon the end of its branch. Often as it has just extended its ciliary disc in full spectacular working order this horrid spasm will take place and put a stop temporarily to its foraging labours.

The bell-shaped body has a rather broad head, or disc, held diagonally. The stem is affixed to a base, as in Carchesium, and is beautifully ribbed and twisted lengthwise. These flutings count about forty all round in a stalk whose diameter was measured to be 1/2500th inch.

The branches do not join directly with the main stem, but finish as somewhat solid curved ends at their juncture, and of a "U" shape more or less, rather than straight attachments. This feature is characteristic and distinguishes Epistylis from Carchesium.

The Coronal disc has five or six circles of cilia around it, and when a zooid can be found and seen active, end on to view, a very fine display, interesting to watch, may be obtained. After a sudden contraction it will slowly extend again, and first one row and then another of the cilia will be set going. The cilia forming them are rather sparse in each circle in comparison to Vorticella and other similar protozoans.

The body of *E. flavicans* is of a pale yellow tint which makes a decided contrast to Carchesium when seen together. The main stem is forked into many branches at its higher end, and to the extremities of each, one individual zooid only is attached. In contracting, their shape assumes a globular form when completed, the cilia being closely drawn in. The macro-nucleus is curved and band-like, placed near to one side of the body. Numerous micro-nuclei are seen in constant movement as the body writhes about.

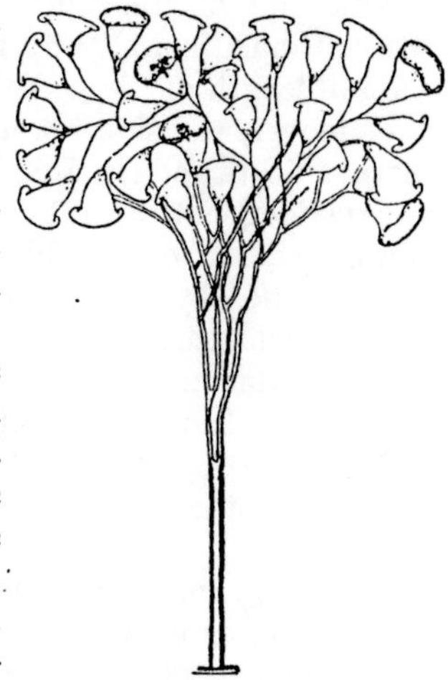

Fig. 201.
EPISTYLIS FLAVICANS.

As the colony grows old the stalk loses its strength and ability to support the full-sized "polypidom," as the tree is called, and will often be seen collapsed into a tangled mass, with the busy zooids brought down with it, continuing away, bobbing up and down, until their term of life ceases or, as frequently happens, are able to detach themselves.

It is quite normal for the heads to detach even in full vigour and to find the zooids freely swimming without their stalks, and in the existence which follows they are capable of producing new individuals by the conjugation of separate sexual gametes and so to increase and multiply their species. A further method of reproduction is by self-division. In this, one of the zooids will divide itself longitudinally, separating its macro-nucleus in half in the process, and allotting a portion to each division.

The mouth is rather difficult to make out while the organism is active; it opens into a slightly coiled and tapering tube situated near to the centre of the frontal region. Close beside it is a conspicuous contractile vesicle. The outer surface of the body is finely striated either obliquely or sometimes transversely.

The main stem has usually been considered hollow, but the

writer has been able to resolve the outer case into a bundle of long spiral threads intertwined. Further, the very fine strands appear to take a right-hand turn in their course. If the first surface view is taken, the stem being upright and the tree also, then the threads, commencing on the right side, dip under obliquely to the left and are continued spirally over to the right in a long helical manner the full length of the pedicel.

Moreover, the side branches are the same, and as they become more delicate and situated farther out, the strands become less and less distinct, but in so far as any structure from the outer surface can be seen, it follows this spiral collection of threads. The hollow is occasioned by seeing only one side of the central thread, which follows right through the U-shape junctions from the zooids to the base of attachment. Where the branches fork, the centre strand divides and continues to either side.

A full " polypidum " will have from fifty to sixty zooids all busily occupied at the branch tips. Each branch is secreted by the zooid it supports.

Epistylis is found attached to Riccia, Lemna, Myriophyllum and aquatic plants generally. Its average height as a colony is about 1/9th inch, and a single zooid, extended, will measure about 1/190th inch.

Acineta fluviatilis (Fig. 202)

Is a species of Suctoria, commonly met with upon the stems of aquatic plants, Lemna, Myriophyllum, Ranunculus, etc. It has a triangular-shaped body set upon a short stalk. At each upper corner are placed several retractile tentacles. These form its means of capturing food. Near the centre of the anterior margin is a single contractile vesicle, which may be watched, slowly contracting and dilating.

It is a small object, the full length of its lorica being little more than 1/500th inch and its stem only a quarter that in length. Where the tentacles protrude an opening in the lorica is made, so that they do not actually stand upon the surface. They are hollow, and capture monads and other soft-bodied infusoria, which stick upon their gelatinous tips.

The body is compressed and very delicate in structure, and does not always fill completely its covering. Usually near the upper and broadest margin of the lorica will be noticed a clear space, but this is not always present.

The tentacles are knobbed with the snare product of the organism, and when an object is affixed the Acineta absorbs some of the protoplasmic contents of its captive, which as often as not, in the writer's experience, makes off, somewhat lamely, perhaps after affording some sustenance to its captor, and fills out again in time with the appearance of being quite well and contented.

At other times, however, the tentacle will withdraw it closer to the body, other radiating arms bend to the needs, more gelatinous substance will be sent out along the tentacles, sliding over the outer surface, which makes a fairly certain end to the possibility of escape, and the whole will be withdrawn into the inside. The movements are very steadily executed, and when completed the tentacles are extended again ready for more.

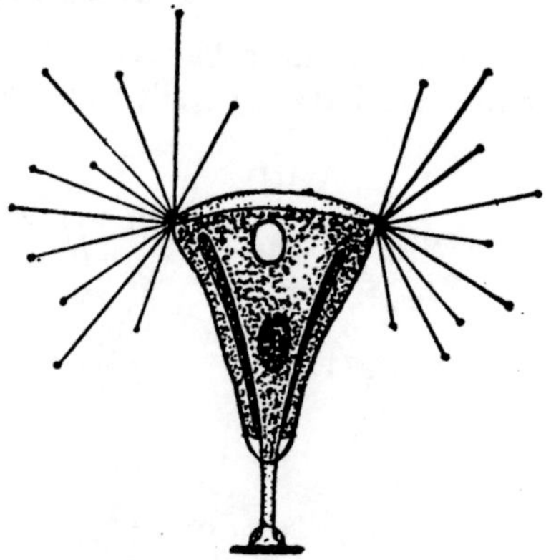

Fig. 202.—ACINETA FLUVIATILIS.

In some situations the Acineta may have to wait many hours before any suitable fry comes along, it appears in no great hurry, or to shift its position, but sits tight on its pedicel and just trusts. Probably an imbibition of water takes place all the time and ultra microscopic substances contained in it passed within, so that it may not entirely depend upon the comparatively larger fry one witnesses it absorb under the microscope.

The size of the body rarely increases once it has become an adult. Its method of multiplication is by division. There is a single nucleus near the centre of the body.

Some of the Acineta, for there are several species, have their tentacles tapering, while most have them of nearly the same width throughout. They never vibrate, but are pushed out in straight lines of gelatinous threads, and when extended appear as if quite stiff and inflexible, swaying gently, or again slowly retracting. The motion of the plasm along their length may be seen flowing in irregular small patches towards the tip or again in the opposite direction.

The whole of the tentacles may be retracted together, leaving the outline almost smooth, and at such times it is difficult to decipher what the object can be. This is rather a rare occasion, however, and the Acineta is usually busy with its arms stretched when seen in its natural conditions.

The stalk is so short in some that it seems absent altogether, but if you try to dislodge it you may find it adheres very strongly at one point which gives the negative information, at least, that in all probability it is there, but unseen.

Again, in

Dendrosoma radians (Fig. 203)

the animal forms colonies of its plasmic self, several fusing together, and these have no stalk whatever, but are attached to an object by a repent stolon, which may gradually creep along bearing upright and rather thick branches of its substance rising from it. At the summit of these will be situated many tentacles, which are used in the same manner as in the Acineta generally.

The body portions contain many vacuoles pulsating and ribbon-like nuclei which ramify into the various branchings. The tentacles are knobbed at the tips, and the colony is of large size. Its height being about 1/25th inch and even larger.

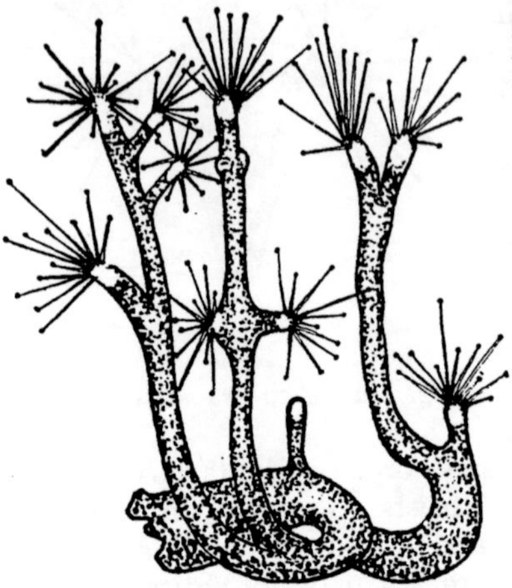

Fig. 203.—DENDROSOMA RADIANS.

Another form,

Rhyncheta

is parasitic on the Entomostracan Cyclops and consists of only one tentacle, formed of a single, simple, movable process arising from a small body mass, with which to fulfil life and carry out its mission successfully and satisfactorily.

Suctoria (*? Podophrya*) (Fig. 204)

The organism about to be described does not appear to have been figured in any work that can be traced. It is small and broadly oval in shape, with even rounded ends, measuring 1/500th inch on its longer axis.

It is armed with two distinct sets of tentacles of about thirty in each, placed near to either end. These are three times the length of the body when fully extended and are tipped at the ends with a sticky fluid, not in the shape of knobs exactly, but more of a thickening for a short way along their length. They are suctorial tubes, hence the class of infusoria they are placed in.

There is one large vacuole and several smaller ones and a distinct nucleus. Its body otherwise is of a clouded grey, much granulated, without green or any other distinctive colour. Across its waist line can be seen a ciliated scroll depression which is in constant slow wave-like movement from end to end.

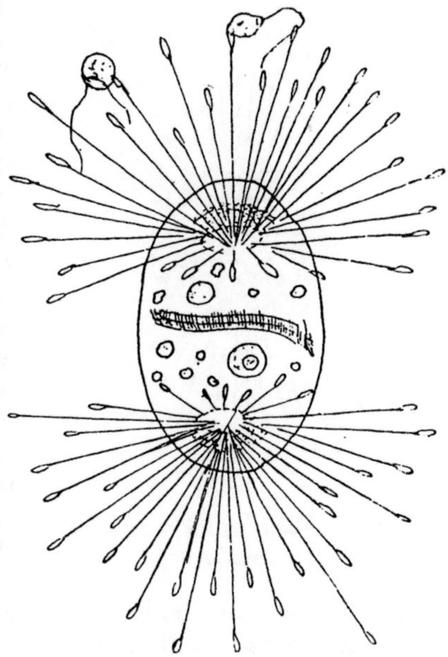

Fig. 204.—SUCTORIA (? PODOPHRYA).

It lies almost motionless while in wait for its food until some unfortunate monad or other soft gelatinous morsel adheres to its tentacles' tips. It then commences to shorten this, drawing the particle nearer in consequence, until touching others they fall to and assist in holding it. One of these then pierces the substance, which in the instance noted was at the root of the flagellum of a Heteromita. The sucking action takes place. Some of the contents are then dissolved and absorbed by the tentacle into the body.

The flagellum was seen to float away finally and the rounded cell at the same time cast adrift in another direction. Most probably it was a species of Podophrya, but these usually have a short stalk attaching them to the stems and fronds of various water plants, and such was not apparent. It gives, however, a fair outline of the Class Suctoria and so is here included.

Actinophrys sol. The Mimic (Figs. 205 and 206)

From almost all collections of water weeds and pond surfaces, or from among moss at the edges of pools and streams, this may be found. It is only a small object about 1/600th inch in diameter.

It has a spherical body so full of vacuoles as to appear a tiny, reflective, bubbly mass. It stands upon its tentacles. It has one or more cilia not always visible. Its tentacles have numerous sticky particles to be seen shining along their length. When a monad or other soft-bodied small creature touches these they become numbed and slow in movement and are drawn closer to the body until a drop of protoplasm is advanced and the captive surrounded and ingested. This may take place at any part of the body.

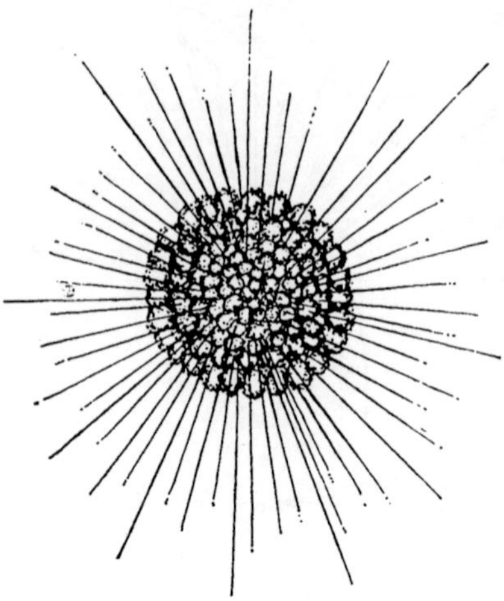

Fig. 205.—Actinophrys sol. (Normal view.)

There is no set ingress or egress, no mouth nor anal canal. It has one or more cilia, and these aid it in its locomotion, which is usually very slow.

An exceptional and interesting instance of mimicry was exhibited before the writer for fully an hour by one of these recently, an incident worthy of record. Beside a fragment of moss on a life slide under the microscope was a still and stately specimen of *Actinophrys sol*, stretched forth, clear and crisp to view. Suddenly it gave a bounding jump nearly out of the field, and as it came to rest it withdrew most of its tentacles, leaving a half dozen or so upon the under side, extended but a short length. At the same time rapidly pushing forward a small portion of its body into a head-like piece, it produced in straight line ahead two or three flagella and several short tentacles.

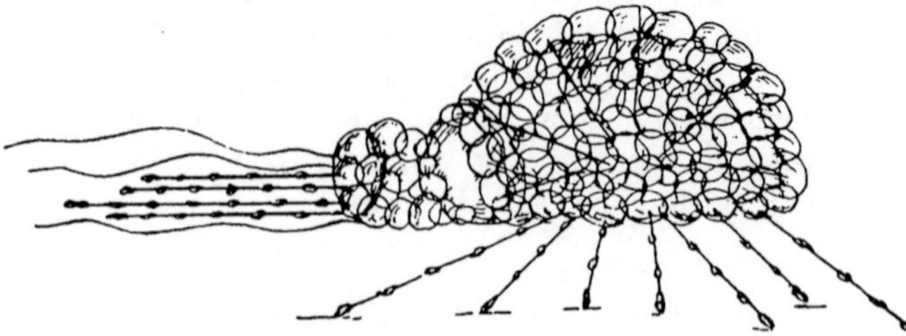

Fig. 206.—The Mimic.

The flagella began quivering up and down, and with the shortened tentacles under its body and larger part, was crawling along at a brisk pace as if anxious to find something lost, head part close to the ground. Between the body and the head was a transparent, slightly fore-shortened vacuole, and at the head was another. The effect as it strutted busily along was to reproduce very realistically an insect in outline, a beetle or ant in mimicry, save that it was opally white in colour.

After fifteen or twenty seconds of this it suddenly snapped into a sphere again, produced its tentacles normally once more, and so remained for perhaps a minute. Then the same sudden "fly away Jack," well out of the field this time, needing quick adjustment of the slide, and the same process was in operation as before.

After following him quite an hour the writer finally gave in and left Actinophrys to his own ingenious frolics.

It was not observed to eat or to interfere with other organisms while in its mimic form. It possessed some points of resemblance to *Actinomonas*.

Actinosphærium eichornii (Fig. 207)

This Rhizopod is not unlike an *Actinophrys sol*, but whereas that object is vacuolated evenly all around its globular mass, "Actinosphærium" has a margin of distinct and larger vesicles, as an "annulus," in which are two vacuoles situated, which are extensile and assume a much more prominent size to the others before contracting. This "annulus" constitutes a fairly even depth all round the spherical shape of the body. The outer surface is also reticulated with hexagonal divisions between the rounded "bubbles."

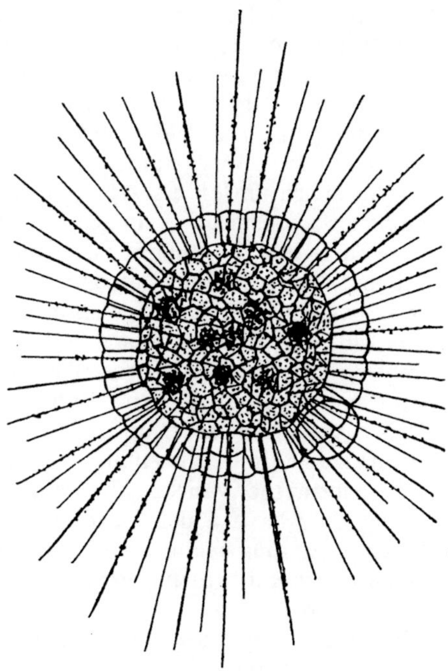

Fig. 207.—Actinosphærium eichornii.

At each centre of these geometric divisions (which sometimes are five-sided) the pseudopodia are extended, dotted along their length with a flowing and sticky substance used in the capture of their food. Within each "bubble" may be seen many small granules and bacteria-like bodies constantly jigging away in a Brownian-like movement within the apparently fluid contents. Sometimes there are large clumps of these also dancing away in a similar manner.

The outer layer of cells are of a clear nature with very few granules within, and as the object lies before you under the microscope, appears as in the illustration to form quite a conspicuous clear ring around the periphery.

The two larger contractile vacuoles seen in this edge view form important organs in ridding the Rhizopod of waste fluids and gases and of substances taken in the food, forming residues to be eliminated. They also act the part of lungs, performing the office of breathing apparata in the interchange of oxygen and carbonic gases necessary to their needs.

One of the peculiarities connected with the pseudopodia is that the animal can throw these off and disband them if roughly pressed or becomes inactive for any untoward reason. Many nuclear bodies may be seen as rounded, dark masses scattered throughout the inner protoplasm.

Each pseudopodia may be seen with care to have a fine thread-like filament running centrally down its middle, appearing of a different colour and texture to the outer and sticky part of the ray. It has its origin deeper down than the outer protoplasmic layer. The pseudopodia are seldom withdrawn entirely into the body, but extend from all parts of it.

"Actinosphærium" is a sluggish object, and where seen, generally remains for a considerable time in that spot. Like "Actinophrys," it likes sunlight. Its habitat is among aquatic plants. Lemna is a frequent source to find it among.

It is not common, and yet seems to be fairly prevalent in all ponds and quiet waters at some period during the summer months. Rarely do you go a season through without finding it in England where the net and tube are used for collecting.

It is a voracious feeder and will ingest animals about one-third its own size at times, distending its area ludicrously.

The contractile vesicles always reappear in the same place, showing they have a definite positional function, and after contraction expanding again as before and continuously so. They will frequently extend into finger-like processes of unequal lengths in the operation.

The vacuole at greatest expansion is thin, smooth and balloon-like, and it is upon the return journey or contraction that the little pustules project and are forced out of the vesicle with considerable vigour. The performance, as Dr. A. C. Stokes has likened it to, is that of a person " blowing into a kid glove to see the inflated fingers leap out."

As a further distinguishing feature between Actinosphærium and Actinophrys, in the latter is always present a lighter and more transparent circular nuclear spot, which is never visible in the former animal.

The average diameter of Actinosphærium is 1/100th inch, though some have been recorded as large as 1/25th inch.

CHAPTER IX

ROTIFERS

ARE probably the most attractive and beautiful group of micro-organisms coming under the amateur's microscope. Their fantastic, diversified shapes and brilliant colours occur so frequently in all his fresh-water gatherings, and these, added to their lively and sometimes amusing mannerisms, have installed them among the first favourites almost as long as they have been observed and examined.

The most noticeable feature on first seeing them is a ciliary wreath or wreaths whirling around like wheels about the crown, either as a single disc in the centre or upon stalks to either side. This prompted their name to be known as Rotifers or rotating, wheel animalcules.

As we now know, this is simply an ocular deception, and is but the rhythmic play in regular sequence, back and forth of single, fine projections or cilia. Glistening in the brilliance of the illuminant, they cast a silvery sheen of great purity. The ciliary disc around its edge may be convoluted, waved, cleft, or simple. It may have one, two, or even three rows of cilia around it. The purpose of the cilia is to bring food into the mouth and also to assist in locomotion.

The body is conical, ovate, or cylindrical, usually terminating in a more or less pointed foot, having three toes, one of which is used sucker-like to attach itself where required. Their whole habit seems quite human in a way; they busy themselves to obtain proper food to eat and oxygen to imbibe, to find suitable accommodation and surroundings in which to dwell, arrange for the careful growth and production of their offsprings, and instal around themselves coats of armour, studded with knobs and bosses, spines and points, to protect them from enemies and the unknown forces of nature, and so to adapt themselves to withstand adversities and be amongst those fittest to survive. They have found and proved their place " in the sun," and have excelled us humans in that they have been dried and resuscitated again in water, and as the story book says, lived " happily ever after."

The body takes a great variety of shapes. In Trochosphæra it is spherical; in *Rotifer neptunius*, long and exceedingly attenuated; in *Polychætus colinsii*, it is turtle-like; in Stephanoceros, beautifully flower-like with its long ciliated and radiating arms.

Surveying a typical common Rotifer, and probably the most abundant, if not the most distinguished, *Proales werneckii* (Fig. 208), will give perhaps a clearer general outline of the structures and aid in understanding the main points common to most species. Coming first to the cilia, their office is manifold. These insure supplies of fresh oxygen, throwing currents of water constantly backwards over their bodies, at the same time bringing food to the mouth. The body has no special organs for respiration, and the interchange of the carbon dioxide, CO_2, from the body and the surrounding oxygen in the water is accomplished by its entire surface.

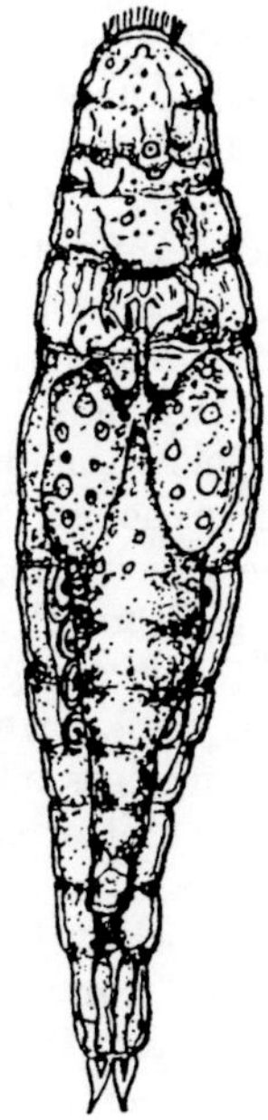

Fig. 208.
PROALES WERNECKII.

The cilia also carries the animal forward into new pastures and supplies, and in the free-swimming forms is the principal means of progression. There is a twofold salutary action in this. As the food is drawn backward the Rotifer moves forward and the two meet, giving a quicker and more abundant selection. Again, the products of excretion and respiration constantly accumulating, are more rapidly and readily removed by the swifter water currents. The cilia in some species also aids in the construction of a tube or nest, as in Melicerta and the Floscules.

The mouth, situated at the back of the disc, opens into a cavity with thick muscular walls, constituting the mastax. The mastax contains a pair of rather complicated jaws known as the "trophi." The trophi consists of several members, but two principal pieces can be distinguished, a middle, fork-shaped piece known as the "incus," and the two side parts called the "mallei." In the incus portion at the middle is a projecting piece like a handle to a two-tined fork which is known as the "fulcrum," and the two tines are the rami. The rami are so joined to the handle they are free to move from side to side in a scissor-like action. These often have short indents upon their edges called teeth.

In the mallei or side parts these may be divided into two, a basal piece called the "manubrium," and joined to the top of this almost at right angles a piece called the "uncus." The duplicate of these parts forms the opposite side of the jaw. Their combined action is to open and close and so cut and masticate any foods passing between. They are unlike any other animal jaws. Muscles are attached to the various parts, and the whole form the main portion of the mastax.

The trophi vary a good deal in different species both in form and development. Some become thick and stout, the unci flattened into broad plates having many ridges across, and form the malleate type. These do not cut the food so much as pulp or grind it, and with soft gelatinous single-celled alga and the like are quite suitable and fully answer their purpose. Some are long and slender, the unci ending in a single sharp point. These are the forcipate type, and may be thrust out of the mouth, seizing upon living prey in a direct manner. Both these types and many intermediate between the two may be found among the common Rotifers.

Below the mastax is a small pair of glands called salivary glands. After the food passes the mastax it enters into a wavy, slender tube, the æsophagus, and leaves this on its upper side. It then enters the large stomach and is there digested. At the front or anterior end of this two large glands are attached ; these contain the gastric juices and are known as the gastric glands. After leaving the stomach the undigested portions pass into the long slender intestine to the outer exit or cloaca. The cloaca lies on the dorsal side near to the foot.

Assisting to eliminate waste nitrogenous matters is a special set of fine tubules running alongside the alimentary canal, consisting usually of two either side. One of each set has thick walls and the other very thin ones. Along their course are minute club-shaped organs, closed at their free ends, having a membrane, in constant vibration, within. From the appearance of these, giving a flame semblance in miniature, they have been called "flame cells." They are well observed in the Floscularia and are much convoluted, branching variously around the inner margins of the corona and down the sides within.

Prof. H. S. Jennings says, "There is reason to believe that the walls of these tubes absorb the nitrogenous waste matters from the fluid of the body cavity." The waste matter passes backward, driven by the flame cells to the region of the cloaca. Here is to be found "a small sac into which the tubes from both sides enter, along with the intestine. This sac or contractile vesicle as it is called, contracts at intervals, expelling to the outside the fluid with which the tubes have filled it. The contractions take place frequently, so that a large amount of fluid is expelled."

The sexes in Rotifera are separate, the females predominating. The male is usually a minute degenerate creature, lacking an alimentary canal, and where the above descriptions are observed it will at once distinguish the animal as a female. In place of the ovary, in the male can be seen the spermarium, in which many spermatozoa freely swim about. It is a sac-like structure with a large ciliated, tubular opening, from which they are discharged.

The foot is an ingenious telescopic extension of the body, and can be used in different ways (see Fig. 209). If the surface is a flat one it will extend the three toes after the manner of a tripod, and one of them will become a flattened disc adhering by suction through a folding inwards of its tip as if pulled from within, while the remaining two rest strident upon the outside. Should it wish to hold to a twig or narrow object it will bring two of its toes upon one side and one upon the other, and by its telescopic withdrawal from within will grip it tightly as a lathe might hold a piece of metal with a three-jawed cone chuck.

Fig. 209. Foot of a Common Rotifer.

In the control of the various organs of the body is a number of sense parts and a nervous system. This resides in a large ganglion known as the brain, situated at the front portion on the dorsal side of the mastax. This radiates numerous nerve fibres to the different organs used in digestion and for its progressive movements. In Floscularia, Synchæta, Hydatina, etc., various kinds of sense organs are found about the corona, consisting often of little tufts of short cilia, doubtless for touch or as guards to the various ducts leading outwards.

Rotifers when free swimming, apart from the propeller action of their cilia, make a spiral rotary movement in their forward progression, and the position of their pointed toes acts much as a rudder in the guiding and direction of their course. The toes also have glands, usually two, near their base which can exude a sticky mucus to assist their attachment upon hard surfaces. Sometimes this is spun out to great lengths, anchoring the Rotifer at long distances, which can only be seen with difficulty at times, save by the small particles adhering to its surface, so tenuous is its substance and fineness. Often a Rotifer will be seen swaying from side to side, apparently unattached, its cilia plying rapidly in quite an impossible position but for this gelatinous thread holding it securely. It is evidently of great strength, withstanding a great amount of torsion, as well as of direct pull, during the many revolutions and gyratory movements of its captive.

Occasionally it will be seen dragging a diatom or a desmid in its mesh like the appendage of a kite. Twisting appears to break it after a time. Upon the head is situated two red, sometimes black, spots at a distance apart, supposed to be organs of perception and known as "eye-spots." In age they are often absent or unobservable.

The coat of mail, or lorica, in which many are covered, is a firm cuticle, chitinous in texture and often embellished with knobs, spikes and processes, just as the higher vertebrates possess hairs, feathers, claws, horns, etc., and in the same manner have their rise and point of origin in the epidermis.

The stomach is capable of great expansion. Beside the large amount of food, at times it has enormous eggs, bulging it outwards, in the females, almost making the body as wide as normally it is long. The cloaca, or passage by which both the eggs and the contents of the stomach are discharged, can dilate enormously and at such times may be everted and thrust through its external orifice.

The food of Rotifers is omnivorous, consisting of the lower algæ, diatoms, desmids, Protozoa, Entomostracans and other Rotifers, even the smaller members of its own species come as material to its voracious appetite. One of them was observed with a long flat ribbon of a Fragilarian diatom, distending its stomach ludicrously. Another, an Asplanchna, who find Anurea rotifers especially tempting, had partaken of one which it shortly ejected again, and in about half an hour the captive was swimming about apparently as lively as ever.

Rotifers are subject to a parasitic disease occasionally, known as Ring disease, particularly noticeable upon the Hydatina, producing circular vesicles all over the body, which generally ends in the death of the victim.

It is rather strange, but none the less true, that the majority of rotifers one sees are females, the males being scarce and in some instances are rarely if ever seen.

The reproductive parts consist in a rather large ovary or germarium lying upon the under side of the intestine and near the root of the foot, divided into two different portions. The larger part is called the vitellarium, and prepares the yolk for the developing egg. It contains from six to eight prominent nuclei. Adjoining is a smaller section known as the germarium, with many small nuclei and in which the germ receives one of these in the production of each egg. There is a passage or oviduct leading from the ovary backward into the cloaca by which the eggs are discharged. The eggs are laid in diverse places.

On Vaucheria the common Proales makes a nest, bulging out into a sack a portion of the outer sheath, laying an enormous number of them within, almost to bursting point, a hundred or more being a common occurrence, with the female rotifer literally covered, lying hidden within the centre of the mass.

In the Floscularia the eggs are deposited in the hyaline sheath about the foot stalk and may be seen as oval bodies, sometimes discharging their embryo with all the characteristics of its class, completely equipped at once to swim off and fend for itself. They may also be attached to varied aquatic plants *en masse*, or even within the frond cells of mosses, etc., or again left freely swimming about in the water. Oftentimes their colour is of a pinky nature.

Megalotrocha alboflavicans (Fig. 210)

This rotifer forms itself into colonies attached by the foot end to aquatic plants. It has no sheath as in the Floscules and similar Rotifera. The body as a whole is elongated and trumpet-shape, and the coronal disc somewhat kidney-shaped, being flexed or incised about its margin and appears to be set obliquely upon its body.

Within the disc seen expanded are four opaque "warts," so called. They are roughened elevations and form a distinguishing feature of this genus. The body is free and transparent when first hatched, with two red eye-spots, but as age and adult form is attained the organism becomes yellowish. It attaches itself in radiating clusters to Myriophyllum, frequently, or other water plants.

Occasionally the whole colony may be seen free, using their ciliary discs to provide locomotion, and in this way seem perfectly comfortable, obtaining change of environment, fresh food localities, and capable of again fixing themselves in pastures new, more congenial to their nature.

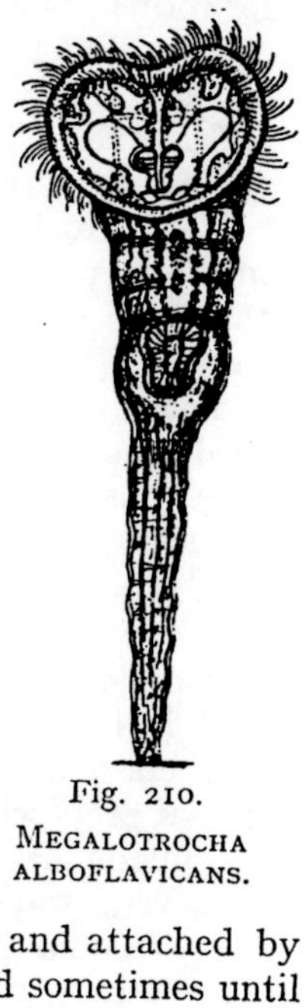

Fig. 210.
MEGALOTROCHA ALBOFLAVICANS.

They usually increase themselves by producing one egg at a time, very rarely two, which is finally expelled from the lower part of the body and attached by a single transparent thread, with which it is held sometimes until five or six eggs may have gathered before hatching commences.

During this time, and while floating in this manner, like a number of toy balloons in the water, the embryo may be distinctly seen moving and turning within the shell, the two tiny eye-spots also visible and the jaws moving in the act of chewing as if free. What there can be to eat is a mystery, and is due probably to a reflex action of the muscles while in their semi-protoplasmic state.

In some of the few species of Megalotrocha the corona is squarish, *M. semibullata* (Hudson), for instance. The length of the single Rotifera is about 1/35th inch.

The colonies when separated may be seen with the unaided eye. The rotary organ is so definitely incised sometimes as to form two " lappets " or flap-like portions to it. The digestive organs consist of the usual maxillary jaws, a pouch or stomach, two glandular swellings, and two openings or cæca passages.

The water vascular canals, outlined in the " Proales " as a type, with their tremulous tags, are present in Megalotrocha. The eye-spots become obscure or obliterated in the adult rotifers, and when seen are usually situated upon the upper edge of the ciliary wreath, between the two rows of cilia. This is a rather uncommon position for rotifers as a general rule.

Several muscular bands will be seen operating the various parts of the disc, foot, etc., giving a great variety of free movements, and as the individuals of the colony are watched bobbing up and down irregularly they are an engaging spectacle.

Megalotrocha is often infested among its older colonies with an infusorial parasite, which apparently feeds upon the mucus around the outer surface, and may be seen running about over this. It is called *Chilodon megalotrochæ*, and is very similar in outline to the ordinary Chilodon. It is necessarily a very small creature, and its food supply is therefore not great, offering little or no inconvenience to its host. Occasionally it seems to get a nip within the folds as the rotifer snaps back in its up-and-down movements.

Philodinidæ

This forms a distinct family of rotifers of the order Bdelloida, and includes the few genera of Callidina, Philodina, Microdina, Adineta and Rotifera. The latter genera has the two eyes in the dorsal proboscis.

Philodinidæ differ considerably from the general types of rotifers in several ways. Prominent among these, the family is devoid of a lorica, and the corona is structurally distinct, consisting of two simple rotary organs borne on stalks, almost circular, resembling wheels when the cilia about their margins are in motion.

The body is fusiform or spindle-shaped and worm-like in its movements. The skin is flexible and leathery and the base of the disc stalks often clothed with short cilia. On the back of the corona there is a long protuberance or " tentacle." The foot ends in four short unequal toes, which are seen best when the animal is creeping along. At such times it protrudes them, and almost as quickly contracts them, which gives sufficient opportunity, however, to observe their outlines. These form its points of attachments. It also bears two " spurs " on the dorsal (upper) side a short distance from the end. They are long and tapering to a point. The jaws are prominently seen within, opening and closing amongst their embedded muscular substance, and the ridges fit closely in a manner to form grinding teeth, as they are actuated.

The various portions of the body can be thrust in and out like the tubes of a telescope, and by this means can take long caterpillar-like stretches to almost a thin line before detaching their foot or in the search for food.

Fig. 211.
PHILODINA ROSEOLA.

The Bdelloid rotifers differ from most others in the fact that they all have two ovaries in place of the usual one, and on this account have been classed apart under the name Digononta as distinguished from Monogononta, and, strange to say, male Bdelloida rotifers are not known.

As the difference in the species, however, is but slight, and the greater part of them belong to the Callidina, their study has been confined mainly to specialists in this group.

Philodina roseola (Fig. 211)

In this the body is smooth and the red eye-spots oval in shape. The processes of the foot are horn-like and short. It has a distinct power of withstanding drying. It may be found as little pink spherical masses in the dry deposits of eave troughs and tanks, and when placed in soft water the little balls quickly expand and swell out, taking up their rotifer form and commence activities uninterruptedly as though continuing where they had left off and without hurt or hindrance. Many such have been obtained from dried sphagnum and mosses that have been laid by a considerable time in this dry condition.

P. roseola is about 1/50th inch in length normally, but is capable of stretching itself to quite three times this length at will. The body is furrowed longitudinally all round. The eye-spots are placed cervical, i.e. some distance down from the front end and not upon the proboscis or frontal column, and this forms the particular distinguishing feature from the various species of the genus "rotifer" mentioned as belonging to the Philodinidæ.

The amateur may find the genus "rotifer" apt to lead him astray on account of speaking of individual specimens generally as rotifers, so it is here mentioned to remind him that such name has been given to a distinct section of the Philodinidæ or common creeping rotifers. These worm-like Bdelloids were the first rotifers to attract the attention of microscopists, and it is to these that the names wheel-animalcules and the Latin rotifer and rotator are due.

The ovaries usually extrude the developed eggs before the embryos are hatched, completing this portion always in the water. The tube at the neck has a few short cilia upon its extremity, which act as antennæ. They are usually the first portion to be protruded, as if feeling its way gently, when the rotifer is extending.

Philodina aculeata (Fig. 212)

The chief peculiarity of this species distinguishing it from its neighbours is the body has about ten or twelve short, curved, soft spines upon it. Sometimes there are as many as twenty. They are situated upon the back and not readily seen unless carefully looked for, being transparent like the surface of the skin and lying flat. When the animal is stretched out they are closest, but when rounded in a heap by contraction they may be seen standing off slightly, and this forms the best time to view them. When the animal telescopes his body and produces folds in the surface the front spines will sometimes be seen pointed straight to the front anteriorly in an opposite direction to the normal.

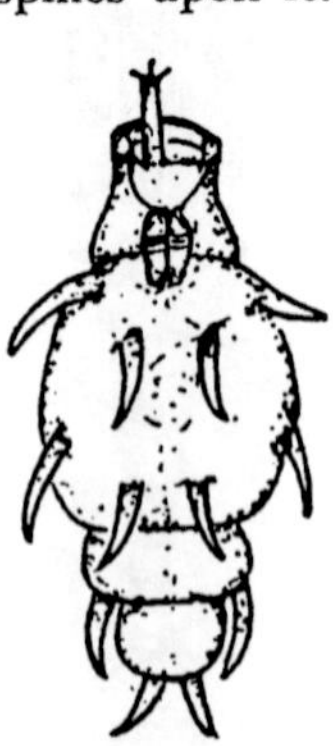

Fig. 212. PHILODINA ACULEATA. (Contracted.)

The eyes are round and placed cervical as in other of the species. The projecting antennæ or tube in front comes to a somewhat rounded end, and is divided into three small sections or points. The foot bears the usual pair of "spurs" and also the four toes as with other Philodinidæ.

It is a fairly large animal, slow in its movements, and is rarely seen swimming, but keeps to its natural creeping habit of locomotion. Its length averages 1/70th inch when in normal attitude.

There are three stout teeth, slightly divergent, in each jaw, like strong cross ribs. The jaws themselves are coarsely striated, opening and closing like the two knuckles of the hands held together.

It is a fairly common species in some localities all the year round and is known to withstand the rigors of winter quite contentedly. Altogether it is a robust specimen of its genus.

Taphrocampa annulosa (Fig. 213)

The body of this Rotifer is an elongated oval. There is no lorica, and the distinguishing feature about it is its cuticula. This has many transverse ridges and folds, about twelve in all, which give it a decided insect appearance. The tail has two short conical points or toes.

It has been said the rotary organs are wanting, but the specimen before the writer has two ciliary wheels which are protruded upon the head, one either side like the horns of an ox, and both are busy with their cilia in motion. There is a single cervical eye.

The folds of the cuticula are flexed inwards and outwards as the Rotifer moves and turns about and would seem to be of some assistance to its motion. It is not a very speedy traveller, reminding one very much of the movements of a Chætonotus grubbing around, and is found in very similar habitats among Riccia, Mosses and Ceratophyllum.

It is evidently a primitive organism and of the family Notommatidæ. The length of body is about 1/100th inch.

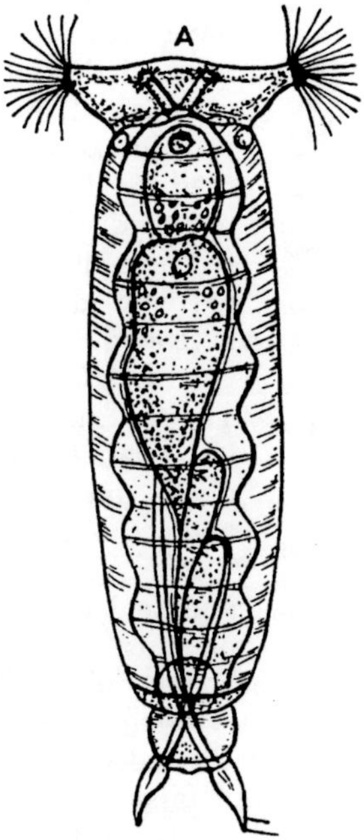

Fig. 213.
TAPHROCAMPA ANNULOSA.
A. Shows position of rotary organs when not in use.

Rotifer vulgaris (Fig. 214)

It has often been asked what is the most common rotifer, and the answer comes readily, *Rotifer vulgaris* in all probability. Yet this is open to much doubt. Again, what is it like? and as unconsciously is given an illustration of a Philodina with its two eyespots placed cervically.

R. vulgaris belongs to the family Philodinidæ, whilst the most abundant rotifers are the Notommatids, which has Proales as a type.

Upon Vaucheria thousands of these *Proales* eggs will be laid along its length in pouches (see Fig. 65), hatching out in swarms at suitable seasons. Nothing like this is ever forthcoming with the genus Rotifer. Nevertheless it is an abundant species and was one of the first worm-like creatures attracting microscopic observation by that famous discoverer Leeuwenhoek of Amsterdam and Delft, *circa* 1660, and for this fact microscopists generally have grown to recognize it as the most common. The work of that great naturalist with his home-made tiny globules of molten glass with which his studies were mainly executed is a marvel of history and should be read by all lovers of the microscope.

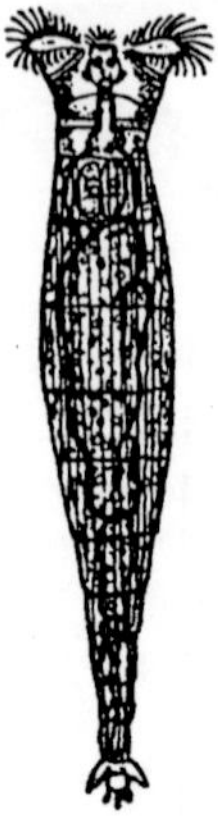

Fig. 214. ROTIFER VULGARIS.

The distinguishing feature of *R. vulgaris* is its two red eye-spots, which are always situated upon the extreme edge of the proboscis or snout, best seen when extended. The body is without a lorica, almost cylindrical, and tapering to a long extensile "foot." The foot has three toes upon the end with which it can either adhere by a telescopic suction action or placed above and below a narrow filament of conferva, etc., and so hold fast (Fig. 209).

It is a creeping animal like the other Philodinidæ and seldom takes long to swimming, though at times or when necessity arises it will use its ciliated crown for this purpose and so change its locality.

The rotary organs are double in all the species and can be contracted or expanded at will by muscles attached to their bases. Between this bifid corona is the ciliated tip of the cylindrical projection which bears the two "red eyes" situated near to one another.

R. vulgaris invariably produces her young fully "fledged," so to speak, and in this respect differs from other Philodinidæ. They leave the lower intestinal region by the cloaca passage into the surrounding water, but the problem for the specialists up to the present has been to account for how they manage to pass from their free position in the body cavity to this orifice. A collapsible oviduct has been assumed.

Another point the microscopist may find interest in may be the statement that of all the many, many specimens of *R. vulgaris* observed the male of the species has yet to be discovered. That is to say, so far as at present known, or ascertained, all are females that have come under observation.

The jaws or mastax is very similar to the Philodina, and the passage leading into it is ciliated the whole length. This is an effective sight when seen under dark ground illumination, and in action appears like flowing water, trickling along, as the light scintillates about the busy lashes in motion.

As the rotary organs close in, during contraction quite a lid-like cover is formed at the proboscis extremity, and as it protrudes again first the little tufts of cilia are pushed gently forward and when satisfied everything is safe it literally turns back, inside out, the whole upper end and displays its wheels ready for instant action. The trunk or proboscis tip therefore forms a kind of "fez" over the cilia wheels. A little below the head portion a hollow tube or "antennæ" is placed, generally at about a right angle to the body, at the end of which a few delicate cilia occur. The throat passage is slightly constricted before entering into the mastax. Upon the back flutings of the surface, skin-like longitudinal and semi-transparent plates are situated down the length of the body.

The normal size of the animal is about 1/50th inch.

Rotifer neptunis (Fig. 215)

An exceptionally long, spindle-shaped body, terminating in a still longer foot, having three horn-like toes at the extremity and a pair of spurs some distance up at about the third joint.

It is a curious creature, almost uncomfortably ungainly as it stalks about upon the foot end, bending this side and that as if with difficulty it reaches any spot it desires. How this form has evolved is a mystery, unless it once was a tube dweller.

In the build of its body it resembles the ordinary species of the genus. The two eyes are situated on the edge of the proboscis, the little jaws and the ciliary wheels are similar. There is also a short antennæ, ciliated at the tip, near the proboscis and situated at one side.

When the animal is in the act of crawling the rotary wheels are generally withdrawn and the antennæ is then seen. The full length of the animal will measure as much as 1/20th inch.

Fig. 215. Rotifer neptunis.

Callidina (Fig. 216)

Similar to Philodina, but distinguished by the absence of eyes. There is the usual proboscis and the spindle-shaped body, capable of contracting into a globular mass as it hunches its way about worm-like or leech-like.

The rotary organ is double, but not affixed upon a pedicel, and has a thick ciliary margin surrounding. There are two jaws set with numerous delicate ribs or "teeth." The alimentary canal, narrow and thread-like for the most part, terminates in a bladder-like expansion posteriorly. Generally the ovarium is well seen and often a single large ova within. There is a short spur-like projection on the neck. The normal length of the animal is about 1/60th inch. The genus is under revision at the present time and will in all probability be broken up into several genera. There are many species, which differ but slightly from one another. It is of the order Bdelloida=creeping Rotifera.

Fig. 216. CALLIDINA.

Stephanops (Fig. 217)

Is one of the Coluridæ group of rotifers resembling Euchlanis, the hardened cuticle forming a solid lorica all round with an opening at each end for the head and foot. A portion of it extends beyond the head into a sort of hood, and somewhat depressed at its upper edge.

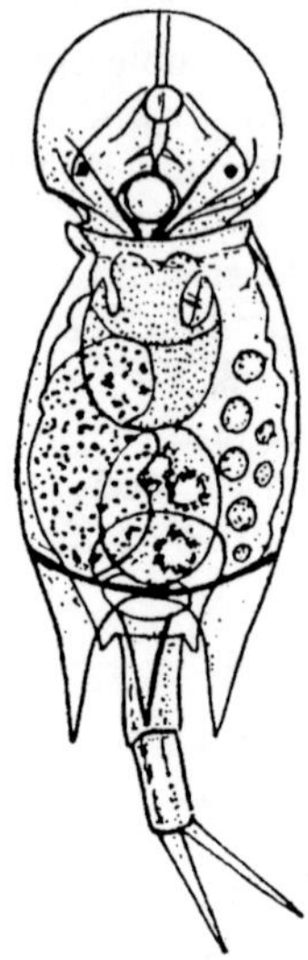

Fig. 217. STEPHANOPS LAMELLARIS. (Front view.)

In two species the lorica has thorn-like processes at the posterior end. The species generally are all small and belong to the "nibblers," so named from the action they have in procuring their food, and are commonly to be found creeping about aquatic plants, confervæ and debris, always appearing very intent and busy in their actions. The foot is furcate, that is with two horny processes projecting, which open and close like a pair of scissors as it steers and touches with them in its course.

The animal possesses two small red eyes situated at the front of the head, one either side. When the head is withdrawn into the lorica the hood still remains extended.

Stephanops is a very transparent organism, and on this account its internal anatomy is somewhat difficult to follow in its entirety, especially as it is very quick in its movements. The length of the lorica is about 1/300th inch. The hood attains varied shapes and sizes in different specimens, often appearing like a halo around the head. In some it is almost circular, in others flat towards the hinder portion and semi-circular, in others a wavy fluting will occur along its outline.

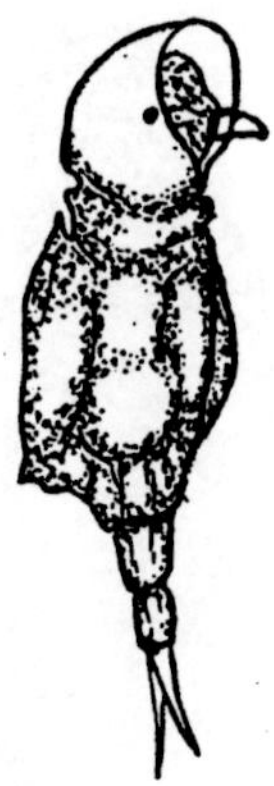

Fig. 218. STEPHANOPS MUTICUS. (Side view.)

In some Stephanops there is a long pointed bristle or spine upon the back, extending towards the rear, attached at its base to the lorica, which is movable and flexible, and as the lorica itself possesses a fair amount of flexibility it can obtain a considerable latitude in its movements. Its use is problematical, possibly a relic of defence, or of use in affixing itself in certain situations as a holdfast while it bites and tears its food, as is its habit in eating or rather "nibbling," its jaws or mastax being protruded from the preoral aperture (mouth) near the front and lower surface of the head at such times.

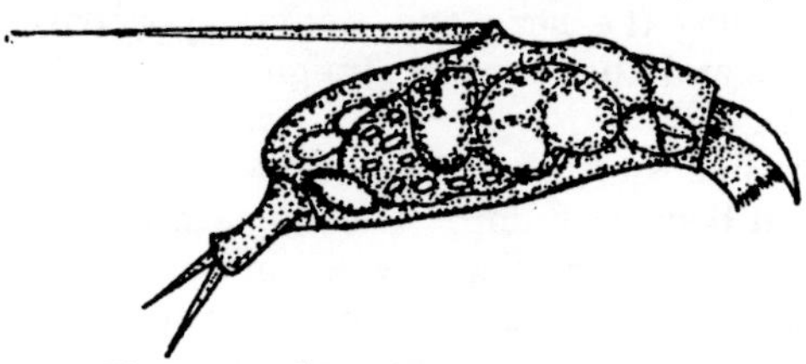

Fig. 219.—STEPHANOPS UNISETATUS.

To distinguish Stephanops in four of its species :—

S. lamellaris, the lorica has three spines posteriorly (Fig. 217). *S. cirratus*, the lorica has two spines only. *S. muticus*, the shield is without any spines or points and is cylindrical (Fig. 218). *S. unisetatus*, has the long spine upon its back as described and figured (Fig. 219).

Pterodina (Fig. 220)

The genus Pterodinadæ is distinguished by a soft, smooth, and flattened lorica, shaped something like a tortoise shell with rounded edges. It is perfectly transparent.

The ciliary disc is in two rows, fairly widely separated, but running parallel around, similar to Melicerta, and in fact it has several points in common with that genus. There is the same remarkable type of jaws, the malleo-rammate type as described; they all possess two eyes, while most other rotifers have only one; even the foot ends in a tuft of cilia which is characteristic of the young Melicerta. It is probable therefore that it is closely related, although a free-swimming rotifer, in contradistinction to the fixed habit of that genus.

Fig. 220.
PTERODINA PATINA.

Upon the front border of "*Pterodina patina*," figured, there is a scalloped portion over which a rounded central lip is projected.

The foot is simple and tail-like, with a suction disc at the extremity by which it can attach itself. It is fringed with a ring of cilia, and is exceedingly flexible and wrinkled, almost band-like, and regularly circular. It has the habit of frequently withdrawing this entirely beneath the lorica.

There are two long conspicuous muscles running from the head to the lower portion of the lorica which are coarsely striated and in fairly constant action. Withdrawing the head and frontal appendages, they are swung to the sides and then tensely straightened out in a manner interesting to observe.

The two eye-specks are opposite the scallop of the corona, and between these is a rounded body or hump. The stomach and portions of the intestines are lined with cilia in vibrating movements assisting the food passage. The ovary is horseshoe-shape. A peculiarity of the rotifer is its habit of lying quiescent for long stretches of time, half an hour to an hour sometimes.

There is a rather small contractile vesicle near the centre of the body not readily seen always. The length of the lorica is about 1/120th inch. The rotifer is found among Lemna, Myriophillum and Conferva generally.

In *P. elliptica* the lorica is narrow and the front border is not scalloped. Its outline is elliptical.

Scaridium longicaudum (Fig. 221)

This rotifer belongs to the family Dinocharidæ. It stands out prominently by having two long toes. There is a single red lenticular eye, placed cervically, i.e. upon the neck portion of the head.

The lorica is very transparent and thin. It is hardly noticeable.

The body is vase-shaped, the ventral side flat, and the upper convex. There is no marking or sculpturing upon it. The foot is twice as long as the body and formed of two joints slightly enlarged at the extremities. The toes about half as long as the foot. The foot has a noticeable bend inwards ventrally and its movements are often of a springing or leaping character, hence its name.

Behind the eye there is a fold in the neck transversely where the head is drawn into the body, and similarly the foot shows a fold when it bends. The skin is rather stiff.

Its habit of springing about wildly is very similar to the Furcularia, and possibly has developed a long foot and toes along this line of descent.

The rotary disc is compound, ciliated, with a bristly bunch of cilia projecting centrally, almost like a hook, and used doubtless as a feeler or antennæ in its food collecting. It frequently is found among the patches of Oscillatoriæ in quiet waters. The foot has two thickened club-shaped muscles, the movements of which are very remarkable to observe.

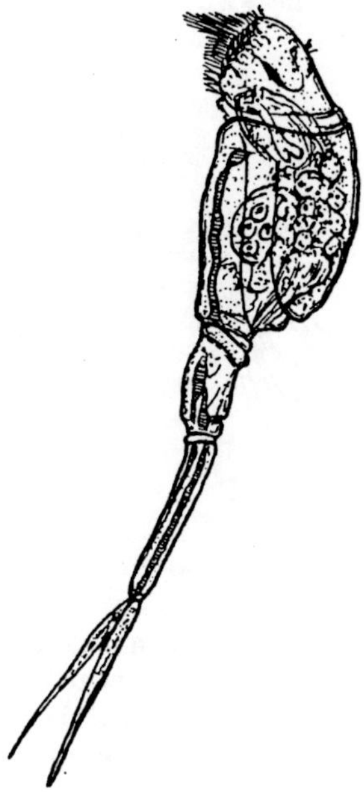

Fig. 221.
SCARIDIUM LONGICAUDUM.

The shell of the winter egg is usually clothed at either end with many scattered hairs. The length of the body is about 1/210th inch, and with the foot included it is as much as 1/75th inch.

The jaws are somewhat unequal in size and possess double-pointed teeth by which it cuts and masticates its food.

Polyarthra platyptera (Fig. 222)

This rotifer belongs to the Synchætidæ family, and the first noticeable feature about it is the long serrated oar-like appendages which are attached to its sides. There are twelve in all, set in groups of three, dorsal left three, dorsal right three, ventral left three, ventral right three, all of which project backward. Their bases are on the breast part of the body.

The animal is a rapid swimmer, and by the aid of these "pinnæ" will often leap about like the Water Flea, *Daphnia pulex*.

The rotary organ consists in bundles of cilia, each set in muscular sheaths, about six in all. When in movement they give the coronal edge the appearance of being double, similar to "*Brachionus*."

The swimming motions, as well as the procuring of food, are effected by these. While they are in use the "pinnæ" are depressed, but in leaping, which is done entirely by the "fins," the pinnæ flash out and the body instantly darts forward, sometimes out of the field of view, in quite a sudden and startling way. It will somersault over in its flight occasionally.

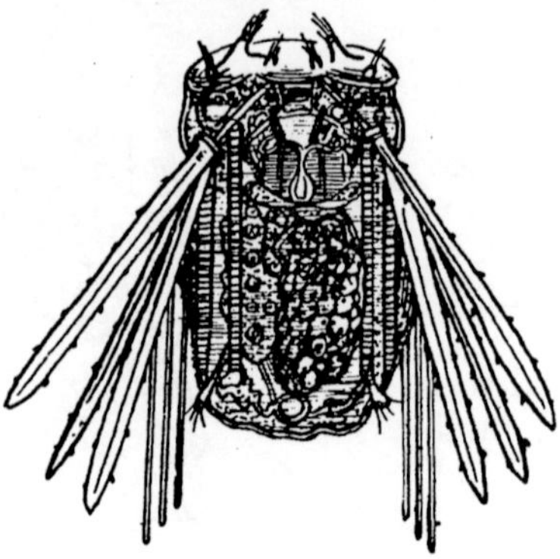

Fig. 222.
POLYARTHRA PLATYPTERA.

The eye is single, placed at the centre of the neck just below the coronal margin. The foot is entirely absent, and the shape of the body in general is sac-like. The stomach portion of the alimentary canal is ciliated and a contractile vesicle is present. The mastax is pear-shaped, with two single-toothed jaws clipping together, masticating the food particles as they arrive.

On either side of the body, running longitudinally, are two coarsely striated muscles, readily seen, and at the lower sides, posteriorly, are two depressions or "fossæ" with several unequal-length setæ extending from them. The body fluid is generally of an orange or yellowish red colour.

There is but one species, and the eggs adhere to the exterior of the body, hatching out in the water. There is only one at a time. The embryo shows several bluish spots upon its surface.

The rotifer is found plentiful at times among Confervæ and aquatic plants. It is often infested with the flagellate protozoan Colacium, one of the family of Euglenidæ, which roughly in appearance is like a number of Euglena adhering to a stalk, mouth first, one above the other, and which is usually surrounded by a jelly-like envelope over all. They attach themselves to the surface of the rotifer and are carried about in its foraging preambles as a parasitic growth, but without any apparent detriment to their host. This parasitic organism is often seen attached to Cyclops and other fresh-water crustaceans also.

Length of body about 1/140th inch.

Polyarthra differs from the Synchætidæ by the absence of

"auricles." These are ear-like flaps on the frontal disc which are lobed and strongly ciliated, giving an additional surface, the extra cilia increasing the power of movement and the collection of food.

The animal has been dubbed the "Swordbearer" on account of its "pinnæ."

Synchæta (Figs. 223 and 225)

These are typical open-water rotifers occurring in the surface skimmings or plankton frequently. They are broad, cone-shaped animals that have well-developed "auricles" or extra lobes on the coronal margins, which give them considerable power for speedy swimming. They are with difficulty followed in their peregrinations as seen living under the microscope.

There is a single cervical eye and a rotary organ with eight or ten lobes armed with two to four stout bristles or styles usually curved and projecting directly forwards.

The body is short and tapers to a pointed foot, which is furcate and rarely absent. The bristles are situated between the clusters of cilia and are used probably as tactile organs.

The eye in *Synchæta pectinata* is a bluish colour, differing from the general red hue in rotifers. Length about 1/120th inch.

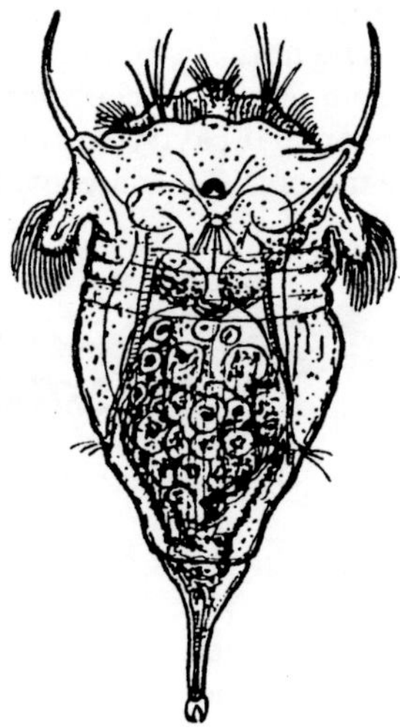

Fig. 223.
SYNCHÆTA STYLATA (ROUSSELET.)
Var.: *S. pectinata.*

Brachionidæ

The rather large family of the Brachionidæ are known chiefly by their lorica having several tooth-like projections or spines upon it. The lorica consists of a convex dorsal plate and a flat ventral one. It is stout and usually armed with the spines both anteriorly and posteriorly, but not in all the species.

It is open in front for the ciliary disc and behind for the long strong foot, which is often covered with rings closely set together. The foot ends in two toes, or is forked at its free extremities.

The rotary organs are double, which are usually divided into five parts, three central and two lateral. The three central are ciliated tufts, chiefly used as feelers and generally stouter than the lateral cilia, which constitute the main rotary disc. There is a single red eye placed cervically, and the body is somewhat compressed. The rotifer is able to withdraw itself completely under its lorica. Frequently Vorticella, Epistylis and similar ciliated infusorians are found parasitic upon the loricas of the larger species.

Brachionus militaris (Fig. 224, facing p. 168)

The lorica in this species is tessellated, its surface divided into twelve regular pentagonal plates or " facettes," the frontal border having ten long spines, the central one, dorsally, being the largest, and is usually bent almost to a right angle.

The posterior border has four stout projections and a deep median excavation above the central curved " horns." The head is expanded in a funnel-shaped manner and surrounded by its circlet of cilia. Its eversion or turning outwards being prevented by the stiff spines of the lorica. The foot is small and short.

Upon each spine forming the outer angle of the lorica posteriorly is a circular depression sharply defined, from which proceed a bundle of unequal-sized bristles, as in " *Polyarthra*".

The muscles, both in the foot and head, are striated transversely. There is a large contractile sac occupying almost two-thirds of the abdominal cavity to the right side of the rotifer, consisting of two chambers, the more oval posterior one being the larger, and the contractions are continued alternately. There is a short passage from the posterior one to the cloaca opening. Length of lorica about 1/80th inch.

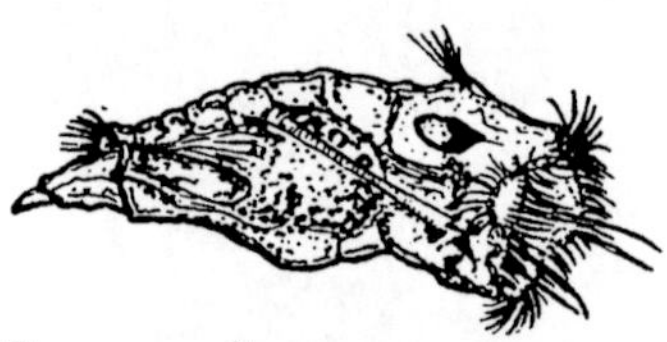

Fig. 225.—Synchæta tremula.

The ova are distinguished as winter ova with thick leathery opaque walls, and summer ova with thin transparent walls, and the male ova, all about 1/35th inch long, attached to the exterior of the rotifer and hatched in the surrounding water.

The eye is large, situated at posterior end of a large " brain." The gastric glands are triangular.

The rotifer is fond of rotating upon its longer axis in one spot, very like " *Synchæta tremula* " (Fig. 225). No trace of any anchoring cable can be seen as with other species of rotifers.

Brachionus pala (Fig. 226)

The lorica is smooth with four spines in front and two obtuse ones posteriorly, near the opening for the foot. The two toes are cleft or bifid.

The rotifer has the habit of swimming in a perpendicular position with the brow directed upward, giving it a distinctive appearance. It occurs prolifically at times, rendering the water quite milky and turbid.

The body is sac-like, and the foot is ringed transversely. The eggs are carried along with the rotifer in a group of five or six usually, near to the base of the foot externally, and are hatched in the water.

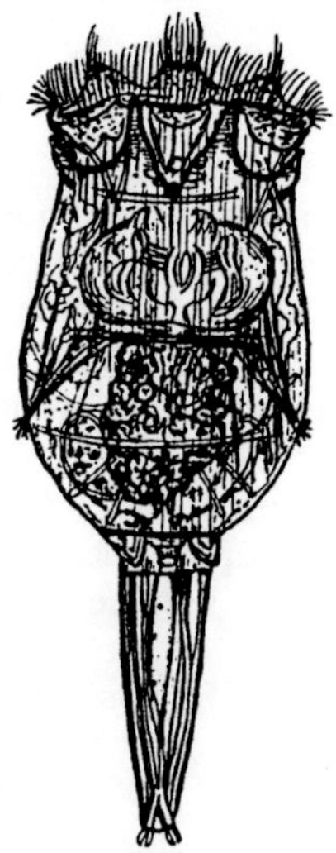

Fig. 226.
BRACHIONUS PALA.

Brachionus urceolaris

The lorica in this rotifer is smooth, the posterior extremity being rounded. The anterior end has six rather short spines, and the whole is slightly granulated. The jaws have each five teeth.

The eggs of the male are smaller than those of the female and are adherent to the posterior end of the lorica in greater numbers generally than those of the female. The male eggs are almost spherical, and reach a length of 1/55th inch. They are distinctive in appearance by their palish yellow hue and are much more transparent than the female ova, which are of a dusky grey.

The red eye-speck may be seen within before hatchment, which takes place exteriorly in the surrounding water.

The males of the species are seldom seen later than the end of May and are much rarer than females, quite a common feature of the Rotifera sexes generally. This scarcity of males seems, as yet, to be unaccounted for, especially as the development of the animals all appears to be either bi-sexual or diœcious.

Towards the end of summer, when the Crustaceans, Cyclops, Daphnia, Cladocera, etc., have become scarcer and to a large extent eliminated, the Rotifera take a fresh lease of life and ofttimes are quite prolific in the month of September, both the free-swimming forms of Brachionus, Polyarthra, Synchæta and others, as well as the fixed forms of Melicerta, Limnias, etc., and if the microscopist will gather the Lemna, Elodéa, and such aquatic plants growing at this time he will surely find many Rotifera to interest and study amongst their roots, leaves, and stems.

Many ponds that have dried up with the heat of summer will begin to fill again with the cooler days and the rains, and dried up organisms and eggs fill out again and become active life once more, a provision of nature which does not appear to have any deleterious effect on their existence whatever, but merely a suspension of energy and effort for a time in their erstwhile busy lives.

Noteus quadricornis (Fig. 227, facing p. 168)

In this rotifer the eyes are absent. It is a similar lorica to the Brachionus and has a two-wheeled ciliary disc, but without any long bristle-like feelers upon it. There is a deep depression in the centre, and three lobes or " auricles " arise from its free margin.

The lorica has several spinous processes both on the anterior as well as the posterior border. The surface is roughened with small knob-like, warty protuberances. It is flattened ventrally and open at either end. At the front border are four stout projections and two upon the rear. Two lateral depressions upon the lorica's dorsal surface have short uneven-lengthed bristles arising from them.

The foot is telescopic-jointed and terminates in two extremely fine points which can be curved and used as hooks. It is not ringed like Brachionus.

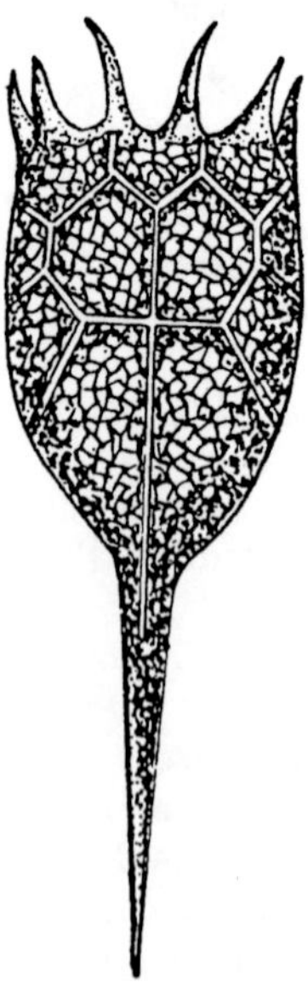

Fig. 228.
ANURÆA COCHLEARIS.
Var.: " macracantha."

The rotifer is common among Oscillatoriæ, and is also frequently found among the debris on the ooze at the bottom of still waters. It is about 1/70th inch in length.

Anuræa cochlearis (Figs. 228 and 229)

This genus may justly be described as "*Brachioni*" without feet, the general organization is so similar to that group. The males, which are rare, only retaining a foot.

Fig. 229.
ANURÆA COCHLEARIS.
Var.: " tecta."

There are several species. The lorica is convex dorsally, and with a flat ventral plate. This is not striated longitudinally, but in most there are rows of " facettes " or plates upon the back. In a few species it is smooth.

A. cochlearis, a species with many variations, in the variety *A. cochlearis* " *macracantha*," there is a long, pointed, fairly stout spine from the posterior margin, and in another variety, *A. cochlearis* "*tecta*" this is entirely absent, while upon its anterior margin there are six spines, the middle pair of which are longest and are curved strongly forward.

Both varieties have the lorica facetted or tessellated. There is a single cervical eye. The eggs are expelled and then carried attached to the lower border, hatching in the water. These rotifers are among the commonest found in surface collections of wide open stretches of water. Length about 1/200th inch.

Anuræa serrulata (Figs. 230 and 231)

The lorica of this rotifer is ovate or sub-cylindrical, with six unequal and prominent spines upon the front end. The two central ones are the longer and curved forwards. The posterior end has two short, rather indistinct ones, at the outer lateral margins. The dorsal portion is tessellated, and with all such loricas these plates have warty verrucose knobs standing out upon them. The plates are not even-sided, some long, some short, or five or six sides in number. There is a single eye placed cervically.

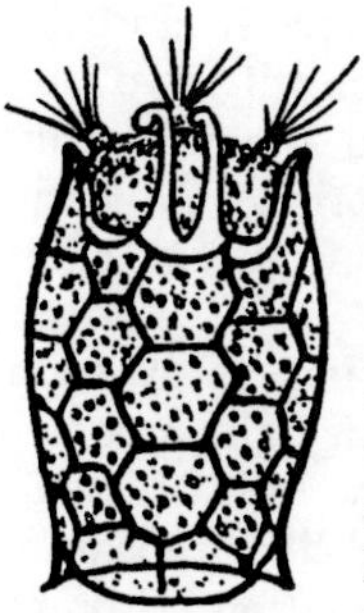

Fig. 230. ANURÆA SERRULATA. (Dorsal view.)

Fig. 231. ANURÆA SERRULATA. (Side view.)

The coronal disc has two ciliary wheels, and independently there are three ciliated cylindrical tufts truncated at their extremities, not unlike shaving brushes, standing out. They are used as antennæ or feelers, and take little or no part in the swimming movements of the rotifer.

The alimentary canal is similar to the Brachionidæ. The length of lorica is about 1/210th inch. The rotifer swims freely but not rapidly, and is fond of the light, being generally found near the surface or upon some aquatic plants close to.

Euchlanidæ (Fig. 232)

This family comprises those rotifers having a lorica composed of two separate plates, one dorsal and the other ventral, not always the same in size or counterpart, the dorsal invariably being the larger.

They are smooth and quite transparent. The plates may be connected by a membrane which folds into a furrow each side as in the genus *Distyla* with its narrower lorica, or they may have no connecting parts, as in the genus *Euchlanis*. Other genera of importance included being *Cathypna* with its broad cup-shaped lorica and *Monostyla* with its one toe, distinguishing its species.

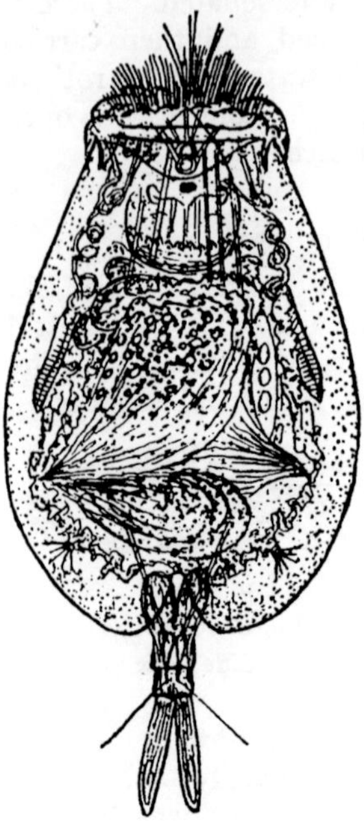

Fig. 232.
EUCHLANIS MACRURA.

Taking *Euchlanis macrura* as representative, the lorica of this is ovate, large and depressed, with two prominent setæ to the foot and situated upon either side its base. The toes are long and fairly broad, tapering to a blunt point, which gives them a distinguishing stronger appearance than others of the genus.

The jaws have five teeth and two maxillary appendages each with two more. The eye is cervical, and the ciliary disc has three little tufts of bristles projecting forward as feelers apart from the ciliary fringe.

It has a rotary motion as it swims along, and the play of light upon its highly refractive, glassy lorica gives it a very beautiful crystal-like appearance under dark ground illumination. It prefers clean, clear water seemingly, where the Confervæ and aquatic plants such as Riccia and others give out plenty of oxygen, where sunlight is freely accessible, and in such places will often be found in quantities.

The length without foot is about 1/100th inch. Occasionally the stomach and alimentary canal will appear of a reddish hue. This is probably and in part at least due to the fine cilia which line these parts, but which is difficult to observe distinctly.

Dinocharis pocillum (Fig. 233)

The lorica of this rotifer is roughened with small knobs and marked out with strong dividing lines. The skin is thick, and the flexibility of the lorica is obtained chiefly from these ridge-like divisions.

There is a single eye near to the frontal margin. The rotary organ consists chiefly in tufts of cilia rather than a regular sequence around the margin. Near the base of the foot two stout spines project, one either side. There are two long toes and a very short one, giving the general appearance of only two. With these its movements are often of a leaping character.

It is a very beautiful rotifer. Its glassy, roughened skin shows well with dark ground illumination, stamping it as one of the finest of the loricated species, its proportions being uniform and well balanced.

The lorica is without spines either end, but is seen to be fairly formidably armed with a sharp lateral margin. In shape it is nearly cylindrical, slightly convex dorsally. The toes vary in length slightly with the species, some are rather short while "*D. pocillum*" are always long.

The male of the species has short toes and a somewhat constricted foot, amounting to ringed, telescopic parts. The head can be withdrawn entirely beneath the lorica. The length of body is about 1/120th inch.

It is found frequently among the Hornwort (Ceratophyllum) and Lemna in ponds and still waters.

Fig. 233. DINOCHARIS POCILLUM.

Furcularia longiseta

(Fig. 234)

This rotifer might be styled "the clown on stilts." Its long pair of toes are used in stalking about at a great distance from the ground apparently. One of the toes is longer than the other, and both are simple unarticulated prongs. The muscles operating them are situated at the base, where a series of two or three folds of the body, narrower than the bulk of this in width, are situated. There is one red eyespeck below the ciliary disc.

The body is short and cylindrical. The rotary organs are wheel-like bundles of cilia either side, with two others near the front. These together form the swimming apparatus as well as for drawing in the food particles. The long toes are generally dragged after the

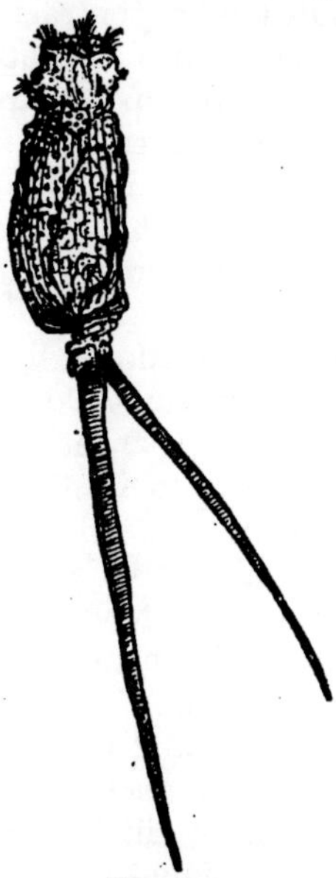

Fig. 234. FURCULARIA LONGISETA.

rotifer, and are quite flexible. Often one will be seen curved like a bow over the other in its twistings and turnings. They are two and a half times longer than the body in some specimens, ungainly, and almost comic to watch, seemingly to get in the way when the other end is nibbling around for food.

When the animal tries to stand upright beneath the cover glass, as observed, he literally knocks his head against the ceiling owing to these outrageously long "pins," and has to bend at the waist to accommodate himself, or else the thoughtful microscopist must ease the lid for him a little. The body length is about 1/140th inch.

There needs little plant life or other substance for Furcularia to roam upon if the observer desires to see any particular markings upon it while beneath the microscope.

Its body is striped longitudinally and the toes transversely. It is found among Riccia and Converfæ generally, in quiet waters.

Floscularia (Figs. 235, 236, 237 and 238)

This lovely rotifer rivals closely, if not equals, the Stephanoceri in beauty of plumage. It may be found attached to the stems and leaves of aquatic plants either in an expanded or a contracted position. It surrounds itself by a transparent gelatinous envelope.

When fully expanded it is of a fluted-funnel or trumpet shape. Contracted it is broadly oval, and in this position is covered by its hyaline sheath. The sheath is not easily seen, unless one is familiar with the object, or small particles adhering upon the outside disclose its outline. Especially difficult is it when the green stems of plants and alga form its background.

It is also of a pellucid substance essentially, and affords little contrast to distinguish it from the water surrounding. Do not therefore be disappointed if it is not observed at first glance.

Every now and again the body will withdraw into its cover, leaving but a thick bundle of its long cilia protruding. The foot may then be seen to have a dumpy, sucker-like disc at its attachment and a telescopic arrangement of interfolding parts for varying its length. At times the animal takes on a sluggish, dormy attitude, and will remain in an encased position for several hours.

The principal parts of the body as seen extended and in action consist of the open cup-shaped "infundibulum," or funnel, around the rim of which are five lobed prominences, profusely ciliated with long straight threads. These cilia do not play rapidly, like most other rotifers, but once extended are kept quite straight and motionless. They perform their duties by tilting inwards together, forming a mesh or net, through which any small creatures once within the "charmed circle" are unable to get out. These small objects constitute the chief food of the Floscule, and consist mostly of small monads, alga spores, etc., Chlamydomonas, simple cells with a pair of little whips, being an especial delicacy.

Should a large and suitable specimen be captured the whole body will contract with it into a bunch within its case, then "All hope abandon ye who enter here" is literally true and largely writ with Floscularia. It likes, however, to play with smaller fry a while, and at a distance down the funnel is a diaphragm, horseshoe-shape, with an opening. Upon this are many short cilia and to one side of it a tuft of longer ones. The action of these together causes all food particles to perform a rotary spin above and to touch it occasionally as they come round. This gives an opportunity to discriminate between the suitable and unsuitable and also to obtain, probably, some gelatinous substance from the outside of them as lubrication in aid of the working parts of the animal. It takes toll of their presence.

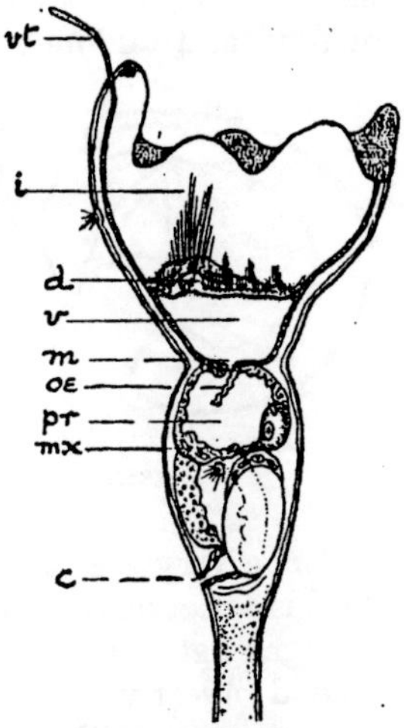

Fig. 235.—FLOSCULE CAMPANULATA.

vt = vascular tag.
i = Infundibulum.
d = Diaphragm.
v = Vestibulum.
m = mouth.
oe = œsophagus.
pr = proventriculus.
mx = mastax.
c = cloaca.

Having been selected and passed this ordeal they are then ushered into the "vestibulum" or entrance hall. Here they come in contact with more viscid digestive juices, any straggling ends of cilia neatly wrapped around, revolved and disported about and suitably prepared before being ready to pass on to the next. This is the mouth or taster, containing a short, generally flattened, flexible pipe, the œsophagus, through which each particle must "run the gauntlet" before emerging in to the "proventriculus" or pro-stomach, otherwise known as the provender store. On the lower side of this is the masticator or "mastax" in which the final preparation is completed. It consists of two arched uncinate or horny jaws, opening and closing in the centre like a double drawbridge, having two pointed ends facing each other. There the food is broken and crushed by the constant forcible pressure as they are brought together and passed into the stomach.

The jaws are rather curiously shaped bodies, and an illustration of their appearance, from the side, has been given (Fig. 236).

The œsophagus or throat passage mentioned is lined with very minute cilia, in regular motion, which urges forward each morsel of food and adding more digestive juices with it. It is quite a severe test to actually see these while in use. Apart from their motion there are several layers of structure to penetrate and their smallness to cope with. They have been likened to a "cascade of diamond points" seen under good illumination apparently flowing through, but not emerging.

Fig. 236.—Jaws of Floscule campanulata.

It is a very pretty sight and worth the effort to accomplish. To do this select as clear a space as possible, not too much food about, a bright light, a dark velvet background, free from white specks, a good objective, and a keen eye and you are there. The writer can see them with a half-inch objective and the long tube when so placed.

Beneath the outer coat of the body many small granules and fatty particles can be seen rushing frantically hither and thither as the body is flexed about, lubricating the inner walls. These form a very distinctive feature of the Floscules generally.

Occasionally the rotifer will forsake its anchorage, releasing its foot and slipping away with its sheath, capable of swimming freely by its frontal cilia and dumping itself once again in some new position.

The construction of the outer walls of the funnel is double, similar to a soft indiarubber ball that is depressed and becomes folded, leaving a narrow space between. As the centre is drawn down by contraction the other follows, bringing in with it the prominences, and their cilia in a tuft together. It then is not unlike an ordinary shaving brush, some of the long cilia being left protruding from its oval body outside in a close bundle.

A part seldom described or figured can be observed situated upon the outside of the expanded funnel, and always on the highest prominence near the margin. It consists of a hollow finger-like tube projecting, containing a few small refractive granules at times, irregularly placed within. It is not unlike the processes on Melicerta, Philodina, and other rotifers, but is a simple and flexible tube, not perfectly straight, which moves itself occasionally and is also pushed aside by the long cilia among which it is surrounded, and is without any brush of cilia at its extremity. At its base it opens into a wider part between the folds of the outer wall, dividing either side and also continuing as a tube or tubes down the rear side of the funnel with other ramifications towards the mastax. It is probably an outer appendage to the pair of fine tubules connected with the excretory organs in the removal of nitrogenous matters. Its projected length is generally about half that of the long straight cilia beside it. It is often hidden among these and may need careful search (Fig. 235 vt.).

About half-way down the funnel may also be seen two depressions with little tufts of cilia, one either side, forming some kind of sensory organs, probably the ends of the "flame cells."

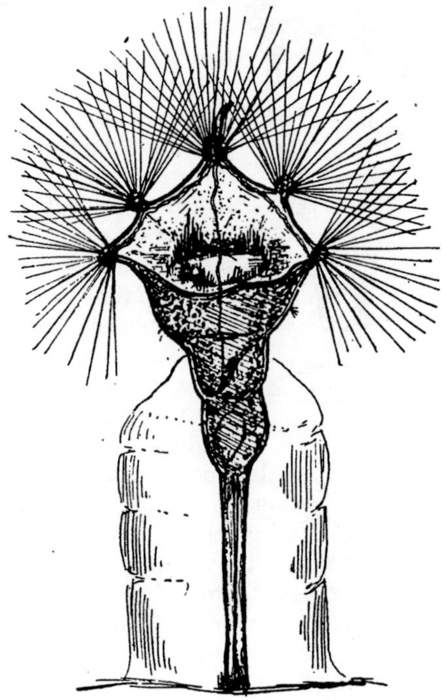

Fig. 237.—FLOSCULE CAMPANULATA.

Within the sheath situated near the foot may be noticed several oval transparent bodies occasionally. These are the eggs, and show the specimen to be a female Floscule. Thanks to the transparency of the sheath and the egg's shell, the outlines of the little embryo Floscules within may be readily observed. The eggs may number anything from three to thirteen, but five or six is a usual quantity.

The short little foot of the embryo with its dumpy centre and the rudiment of a pointed toe either side can be made out situated in the pointed end of the egg. At the broader end is the head with its "two sparkling eyes" of ruby red. These persist after hatchment for a time, but in the stationary adult are absent or unobservable.

Witnessing the hatching out of a young Floscule is an interesting experience. The movements within the egg become gradually more energetic; there is a greater commotion of twisting and turning, the head turns instinctively and forcibly, butting the shell, first to one side and then the other. Finally it bursts through, and once this is effected it has little difficulty, apparently, in penetrating the outer sheath of its parent.

The baby Floscule is now free and in the water, its head surrounded with a complete crown of tiny cilia, its two red eye-spots showing up in contrast with its transparent body, the little foot wriggling about and used as a tail. It seems astonishing that with no previous observation or instruction it knows at once how to swim and ply its cilia in perfect rhythmic movements, use its foot, and steer its direction. As a preliminary effort one was seen to swim out and round back again to its parent, having covered a distance equal to fifty times the length of its body, and then to hover about for some time near by.

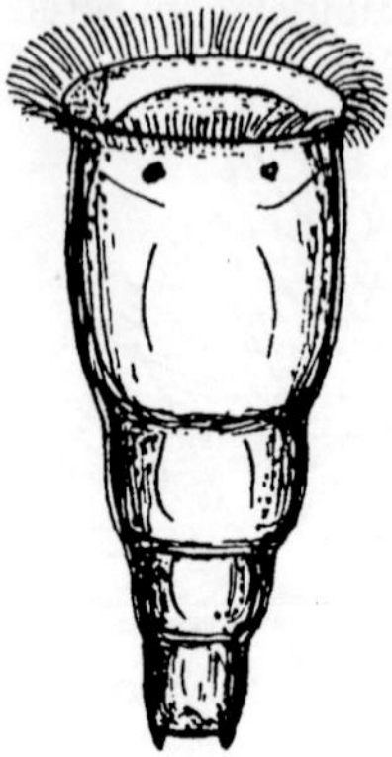

Fig. 238.
BABY FLOSCULE.
(Just hatched.)

An evident sense of direction seemed to be possessed by its returning "home" after its initial trip. It also formed a test of its ciliated organs and their capabilities, and as the sailor might say, "Just see if the ship was seaworthy."

When an adult Floscule becomes dull and retracted it shows no ability or notion to withdraw further or move its position away when buffeted by other rotifers, etc. Seen between a large bulky Stentor and the busy pair of ciliated wheels of a common rotifer, lashing the water into a whirlpool, it placidly kept its ground until finally many of its long cilia were broken off and seen floating around in the maelstrom set up by the pair.

There is little nervous matter at such times in the animal, either in the body of the organism or its cilia, a lash or two missing being of little consequence or concern to it.

The male Floscule is a very insignificant creature in comparison to the female, being little more than a quarter the size, either in length or breadth, and its cilia does not attain the majestic dimensions of "her grace." It is usually seen without a sheath, is a free swimmer, and seldom, if ever, attached to any object.

The diameter of the female is 1/110th inch, sometimes smaller, without its envelope.

Limnias ceratophylli (Fig. 239)

The crown or cilia disc of this rotifer is two-lobed with a broad gap on its dorsal side. It has two small antennæ on its back or dorsal side and two longer ones on the opposite or ventral portion. It lives in a tube formed by its own secretions and not built up as in Melicerta by the adding of matter taken out of its surroundings.

Usually the gelatinous sticky nature of this tube causes many particles of flocculent debris to accumulate upon its sides, giving it a rather dilapidated appearance, but which is stigma upon the indefatigable workmanship of its inmate. He is a shy creature, but quite dainty in the selection of his food.

When young it is a colourless and smooth-looking organism; this, however, changes with age to a brown or darkish colour. Its sheath is fairly rigid and almost cylindrical in shape, widest at its upper margin, and permits of a slight elasticity there. It attaches itself to aquatic plants by a sucker-like foot, and may be found upon Ceratophyllum, hence its specific name, or quite as usually in the writer's vicinity on Myriophyllum as well as Elodéa and Lemna.

Fig. 239.
LIMNIAS CERATOPHYLLI.

Another species is worthy of note in that it has as many as seven projections near its corona and upon the dorsal side set in three rows, the first with two, second three, and the next two. Its tube, moreover, is ringed or in annular ridges closely set together, and forms a distinguishing character to the species illustrated. It is called *Limnias annulatus*.

Limnias may be observed with several eggs within, and indeed most species one finds are females. Two red spots or eyes are present, more distinctly seen in the younger specimens, becoming separated and changed to a browny colour as age advances. When the rotary disc is expanded four large globules are noticeable near the centre.

Melicerta ringens (Figs. 240, 241 and 242)

A tube-dwelling rotifer and one frequently met with. It is especially admired by microscopists, and is found attached to Myriophyllum, Elodéa, and almost all the commoner aquatic plants at one time or another. It is a builder, a most painstaking and efficient one, too. Around its colourless and somewhat delicate body it builds its own house, quite a substantial projection, as fast as it grows and needs more cover. Cosy and compact, it is fashioned with bricks moulded by its own "hands" and in its own mortar. These are part of its anatomy and set apart for the purpose. Frail and nervous in structure, it has found a necessity for a covering to fit it to withstand the rough elements, unforeseen enemies, etc., surrounding it, and has surmounted the difficulty admirably, so that it lives in peace and safety from any ordinary harm. An excellent instance of the survival of the fittest.

Fig. 240.—M. RINGENS. (Front view.)

Never known to strike, this little builder does his duty truly and well. Without plumb-line or set square he fashions his dwelling in the utmost order. By rotating on an axis in his sheath he obtains his distance for the next brick or pellet to be laid, and it is a singular fact that he always makes his bricks upon the one side and lays them on the other. This rotation smooths the lining of his tube, and his little feelers, by touch, tell him where to place the next, and without hesitation he nimbly and unerringly pops the brick alongside the last at the right spot. So the circular piles grow around him. No grumbling about the quality of the mortar or materials, he fashions anything suitably to hand. If you give him particles of carmine he will lay his bricks in red; if indigo, of blue; Indian ink, black; and so on, so that you may induce him to produce a tower of many colours like unto Joseph's coat.

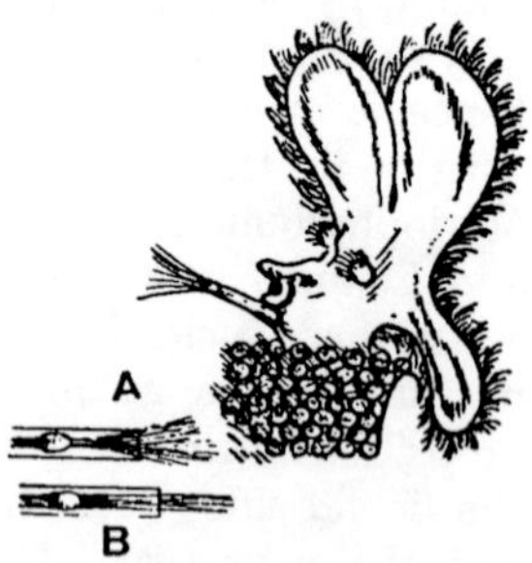

Fig. 241.—M. RINGENS. (Side view.)
A. Brush extended.
B. Brush withdrawn.

The coronal disc of the rotifer has four lobes, around which are a double series of cilia separated by a groove often also ciliated. The inner cilia of the pair are the larger, and along the groove food particles are swept towards the mouth. The mouth is upon the ventral side. There are two red eyes, not readily discernible, and in this particular, Melicerta differs from most of its neighbour rotifers, which usually only have one. There are two powerful jaws opening and closing as food is being dealt with. They are known as of the malleo-ramate type (see Fig. 242 A). Below the mouth is a lobe which forms the so-called "chin," and is used as the mortar in which the pellets are moulded and rounded.

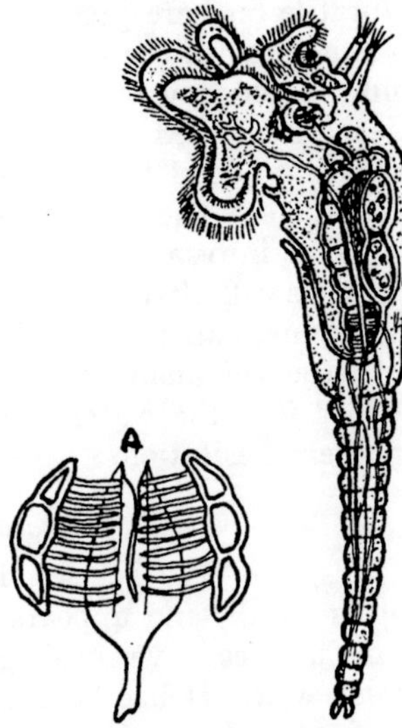

Fig. 242.—M. RINGENS. (Out of its sheath.)
A. Jaws of same.

Mr. Gosse as far back as 1851 described the process very interestingly. He says the rotating organ is of a cup-like figure and seated immediately above the projecting tube. This organ he saw fill and empty itself many times in succession until a goodly array of dark pellets were laid down somewhat irregularly, the animal effecting their distribution by bending its head downwards so as to bring this cup and the margin of its sheath into opposition. After a certain number were deposited in one part the animal would suddenly turn itself round in its case and deposit some in another part. It took from two and a half to three and a half minutes to make and deposit a pellet. Coloured particles in the water are hurled round the margin of the ciliated disc until they pass off in front through the great cleft (sinus) between the lobes and the "atoms," if few, glide along the facial surface following the irregularities of the outline with great precision, dash round the projecting "chin," and lodge themselves one after another in the little cup-like receptacle beneath, in which again they are whirled round with great rapidity and prepared into pellets for the building up of the case of the animal.

There are two hollow projections on each side of the disc, from which a little bundle of straight setæ issues. These may be extended or withdrawn by delicate muscular bands (see Fig. 241 A and B). They are the brush-like antennæ. A narrow œsophagus leads downwards to the jaws, and below this is an oblong-shaped stomach lined with vibratile cilia. A constriction separates this from a spherical portion, also ciliated, which opens in turn into a long cloaca (Fig. 242), which turns sharply upwards to its outlet below the lower lobes. There is a water vascular system, as in Floscularia, to which Melicerta has many points of similarity, consisting of two canals arising below the stomach and which ascend on each side of the alimentary canal towards the head, where they branch. Vibratile tags are present at intervals.

The study of the water currents set up by the cilia is interesting and will occupy one's best mental resources to gather just how they are effected. There are three different currents, and a little colouring added to the water will allow the following of them to advantage. The rotifer on alarm rolls up and contracts its "lobes" into a rounded mass and retires quickly to its sheath. The eggs will often be seen as oval bodies in the lower portion, with the young, half developed, moving within, and probably they may be witnessed finally bursting their case and being liberated fully fledged into the water, where they swim away without tuition or apparent hesitation.

Stephanoceros eichornii (Fig. 243)

This, the only species, is justly considered the most beautiful rotifer of all, and appears under dark ground illumination, as its name implies, a veritable garland in wax. *L. Stephanos*=garland, *ceros*=wax. It has been termed the "Queen of the pond," most specimens observed being females of the species. The males are indifferent, undersized, meagre-looking creatures beside "Her Majesty." As the long pointed arms, five in number, are extended and their exquisitely fine and long cilia are displayed in wave-like movements it is really a wonderful object, eliciting the admiration of all who behold it.

Stephanoceros secretes a thick gelatinous sheath around the lower half of its body and attaches itself by a cleft pad to the leaves of aquatic plants, such as Elodéa, the Yellow Water Lily, or filaments of algæ, as Spirogyra.

The sheath is quite hyaline in texture and apparently solid, forming a substantial protection in several wide folds, and not tubular in any way. Beneath the outer coat of the body proper is a granulated layer containing nuclear cells of an oily appearance and consistence. These are used as lubricants, rushing pell-mell about between the two layers of the coat as the animal flexes and contracts itself this way and that.

The arms are carried normally erect, slightly curving inwards at the tips, which are narrower there than at their base. They are not perfectly round, but approach a five-sided section, having the outside flat, and on the inner side the oblique rows of cilia are placed along the length. All thus face towards the centre, and to give some idea of their number there are nineteen or twenty rows on each arm with an average of sixty in a row, so that on the five arms there are six thousand in all. The length of these long cilia is about 1/250th inch.

Upon the mouth entrance and also upon the arms are many short cilia not included in the above. The tentacles are hollow or of sufficient porosity to allow the passage of small highly refractive bodies to make their way from the large oval cells in the body to the tips, and often a number of these may be seen congregated there, reflecting the light brilliantly. These eventually escape into the water, possessing two cilia which enables them to swim freely near by. They are zoospores developed by the parent cells.

In the gullet there are two entrances for food particles to pass through, both being thickly lined with cilia, kept in constant movement, directing and sorting the various food substances as they arrive in the maelstrom of the "vestibule" or upper open portion of the buccal cavity.

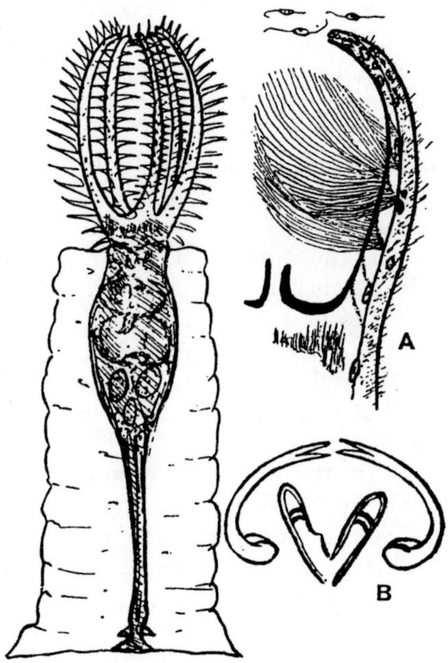

Fig. 243.—STEPHANOCEROS EICHORNII.
A. Single tentacle showing long and short cilia, also zoospores.
B. Jaws (Trophi).

The short cilia particularly, both upon the arms and within the throat, exhibit quite an electric spasmodic method in their action, and the note is hazarded for further observation whether there is not some effect in this in disarming or stunning small fry and assisting in their capture as food. These jerky twitchings, however, appear to persist whether food is obtained or not quite impulsively, and to darken the object or flood with light suddenly gives no variation to them while the animal is in natural and comfortable surroundings. There is an intermittency in their continuance, notwithstanding, which occurs at intervals, and not connected evidently with light.

The large cilia on the arms rises from a layer above the external cuticle and may be detached in tufts by suitable pressure, so that they evidently are but delicately affixed.

Below the vestibule the cavity described under Floscularia, called the "proventriculus," follows and consists of two membranes separated by a narrow passage through which the food is again tasted and mixed with digestive juices before reaching the mastax. Here the jaws or "trophi," which are two chitinous, forked, arching arms, work backwards and forwards crushing and masticating it before passing it farther to the next compartment or stomach proper, where it is now in a more or less comminuted condition and is dealt with by juices that finally reduce it either for assimilation or as waste matter to be ejected (see Figs. 235 and 243 B).

Upon the outside of the body opposite the vestibule are two little patches having short hairs standing out; these are considered to be antennæ. A little distance to one side within, may be seen a circular-lobed spot, this is a nervous ganglion, and close to it is the red "eye-spot." The eye-spots in the embryo are usually two, but as age is advanced they may be lost or only one may be apparent.

Within the body the large oval cells are the ova or eggs of the female from which the young arise in due season. The long extensile foot has four muscles, each enclosed in its separate sheath, which run the whole length to the attached end; these shorten the body and withdraw it into its case. They are branched in pairs from the sides at intervals. The cloaca or external opening from the intestines is situated to the side at the lower portion of the body.

A water vascular system of tubes or canals with vibratile stops or tags supplies a counterpart to the circulatory apparatus in the higher animals. To see the ovary cells to best advantage add a little acetic acid to the water, they are then rendered much more clearly.

Upon the long arms running longitudinally may be seen lines in short lengths end to end with a space between as in Vorticella stems, which has some connection with the contraction and expansion of them; ten or twelve are set in parallel around the tentacle. Within the tubes are short cilia used to pass along the refractive bodies to their tips.

Œcistes melicerta (Fig. 244)

is found attached to Moss fronds, Riccia, Alga, and other plants where the water has not recently been disturbed. It forms a close-fitting sheath, like Melicerta, but without the addition of any little pellets. It has no organ for that purpose. It is a soft-bodied creature, and is very timid, retreating into its sheath at a sharp rap near by.

Ordinarily Œcistes produces an irregular hyaline envelope upon which small particles adhere, showing up its outline and presenting a rugged appearance. Its cilia crown is complete, without any indentations or flutings. It has a longer set of lashes upon its outer rim and a shorter series within these. From its retracted posture it gradually pushes forward a rounded surface having a slit across equivalent to the mouth. This opens out, as if gaping, as it further advances, and the outer row of cilia are extended. Finally, when fully open, the thick lips of the mouth allow the inner set to appear. The mouth edge is ridged circularly, and the cilia are placed upon these elevations.

Fig. 244.
ŒCISTES MELICERTA.

Œcistes attaches itself by a sucker-like foot to a plant, and has a folding-bellows action for lengthening or shortening its extremity. Upon the central portion of the ciliary disc are two whorl-like ciliated apertures through which food selected is passed or otherwise from which rejected.

The "buccal funnel," as it is called, from this entrance to the "mastax" carries the food by a rhythmic action, there to be dealt with, crushed, or in cases where large, curiously passed whole through to the œsophagus and stomach. The œsophagus is capable of a very wide expansion, and the jaws simply pass anything above the normal size through it, to be dealt with solely by the stomach juices. It is fairly long, flexible, and irregular in outline. If the rotifer is suitably situated the action of it can be seen quite clearly. The small granular food is deftly transported and disposed of into the stomach.

A larger unicellular protococcus was found a delectable morsel at the mouth entrance and an extra commotion of the cilia instantly set up ; the single cell was passed in and, touching the jaws, they widely opened immediately, stopping their usual to-and-fro motion, letting it through to the œsophagus, which as rapidly expanded and without delay passed it direct into the stomach. It was an ocular demonstration of the fleeting nature of life.

Upon the neck are two fairly long and transparent curved tubes, having a bunch of cilia at the tips. In the usual descriptions of this species only one is given. This must be an oversight. They are not readily seen at all times, but with favourable conditions, a good specimen, light and objectives, may be made out with certainty. There is an inner tube running through the length, containing a few refractive granules here and there. Their use is a problem requiring solution.

Many of the Rotifera have similar projections, and are usually said to form antennæ, Floscularia, Melicerta, Philodina, etc., and some possess warty knobs in place of them, Megalotrocha, Limnias, etc. They may be sensory organs, but in the writer's opinion in Œcistes are connected with the alimentary apparatus and a suitable food supply.

Food passing the stomach is discharged from an aperture at the back of the neck. The number of bites taken by the jaws has been counted as averaging one hundred and thirty per minute. Length with tail fully extended about 1/50th inch.

Œcistes has a peculiar twist of its head before ensheathing, and the cilia at play along the " buccal tube " to the mastax is something worth seeing with dark ground illumination.

CHAPTER X

BRYOZOA OR MOSS ANIMALCULES

These animals are also known as Polyzoa. The forms of all of them are exceedingly beautiful and have a superficial resemblance to many of the marine Polypes and Anemones with their feathery, waxen plumages. It is fortunate that the greater part of their structures are so transparent and delicate that under ordinary favourable conditions they may be fully observed under the microscope, in all their parts, while living.

Their habitat is upon the decayed twigs, branches, etc., that have sunk to the bottom of ponds, or it may be upon the under side of Water Lily leaves, or around the rootlets of Lemnæ, the branching stems of Water Crowfoot, or the rootlets of trees growing upon the banks which may be near to the surface. In any or all of these situations they have been found thriving equally well. Nevertheless it requires some experience to be able to know their general appearance on sight. As they form colonies of themselves, consisting of many individual animals or zooids, they will grow into large patches sometimes and may be seen by a hand lens alone, and not infrequently the mass is large enough to be visible to the unassisted vision with practice. Thick crusts of their gelatinous bodies will form upon stones or in great profusion twine around the stems of aquatic plants. From out these masses they project their graceful " heads " expanded into a circlet of flowing finger-like tentacles, generally in a crescent or horseshoe shape and called the " lophophore." Should any disturbance take place near, back into their recesses they shoot, so exceedingly timid the creatures are, and their delicate parts are bunched together instantly and retracted under cover of their tube or sheath.

All the clusters of Bryozoa are essentially adherent to some object or other, save a few like Cristatella and Pectinatella, which possess a slight creeping movement of the colony as a whole, but this is never of any great extent.

The young animals after leaving the eggs have also a short free-swimming existence, but this again is only brief before coming to rest in some fixed abode like to their parentage. Owing to the secretion of an outer covering or "cuticula" the actual body wall is very thin.

The alimentary system is comparatively simple, consisting of the tentacular "lophophore" and its U-shaped tube, the mouth and œsophagus leading to the capacious stomach and the rectum cavity. In the U-shaped tube and within the epithelium lining its walls the principal digestive glands are situated.

The mouth is furnished with the funicular lophophore, and in Plumatella at the narrow end is a fleshy tongue, "the epistome," which can close over the opening, preventing the escape of food, and may also be seen to contract as particles pass along to the œsophagus. The œsophagus is fairly long and winds about before reaching the stomach, the external outlet from the rectum being near the base of the tentacles at the side.

Reproduction is effected both by ordinary sexual reproduction and by budding similar to the Hydra group. Occasionally a colony will undergo fission, separating into individuals. This process occurs frequently with "Urnatella." Winter eggs or "statoblasts" are also produced from which the embryo develops.

The animals possess neither heart nor blood, and the only circulation which takes place is the general fluid contents within the body cavity, during its frequent contractions and expansions, and the compressions thereby engendered.

The food consists principally of diatoms, algæ, and infusoria, which is swept into the vortex made from the currents induced by the ciliated tentacles. Once within these they seldom evade the net and are deftly swirled to the opening of the mouth at the lower end. Should a comparatively large and dainty morsel be secured the whole fringe will suddenly collapse. This strong and forceful contraction literally crushes it by the compression brought to bear from the whole muscular action of the "lophophore" in retracting.

The tentacles show great sensitiveness to different kinds of food particles, meticulously accepting some and rejecting others in a manner that is somewhat remarkable to observe. About the lower half of the tentacles is stretched a transparent inter-tentacular membrane between each, forming a strengthening to their flaccid structures and a more secure trap in the retention of their food captures.

Bryozoa do not constitute a natural group of fresh-water animals, but have descended from ancestors whose families were distinctly marine.

Plumatella repens (Figs. 245 and 246, facing p. 231)

This is a very common species and forms a variety of *P. polymorpha*. The colony grows in tubular branches along its substratum, bearing many short side branches in which the little animals dwell.

The "lophophore" has sixty tentacles set side by side in a conspicuous horseshoe shape, their free ends frequently bending over outwards when fully expanded, giving a graceful vase-like lip to the whole. They have many short cilia, set principally upon two sides down their length. The interior forms a deep hollow to the

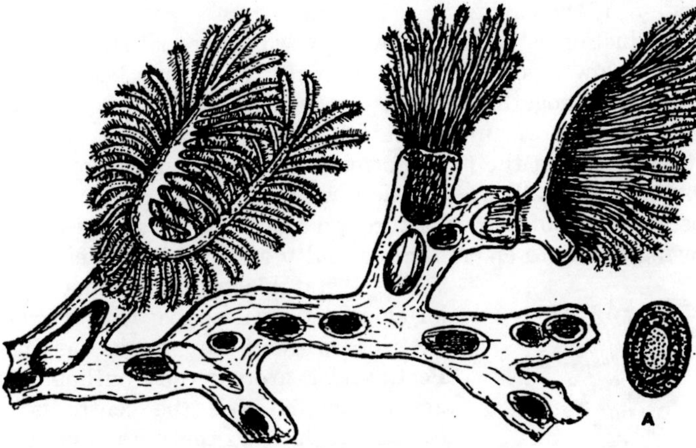

Fig. 245.—Plumatella repens.
Var.: P. polymorpha.
A. Statoblast of same.

mouth opening, which is covered by a kind of tongue, "the epistome."

The inter-tentacular membrane connecting the lower half of the tentacles is plainly visible. The colony is rather straggling in its manner of growth. The animals are very shy in protruding their lovely plumes if the slightest disturbance is near. Apart from the cilia upon the tentacles are many very fine hairs or setæ, which can be observed upon small cushions or papillæ, standing off rather stiffly and straight, but which can be moved or swayed from side to side. Their duty appears to be as feelers of some kind or as special organs of touch to disarm the living food particles as they come through the mesh after the manner of the tentacular protozoans. They are about four times longer than the motile cilia.

The species generally have a curious habit of slowly turning and touching with the tentacles first one object and then another that comes within their net before deciding whether to hold or reject it. The mouth or gullet is ciliated around its upper margin, and is of pear shape, with the narrow end downwards.

The food as it passes along the œsophagus may be seen alternately driven up and down with a continual surging movement, elaborating and extracting all possible nutriment from it after admixing the glandular juices of the epithelium to it. As this is accomplished the mass becomes of a browny yellow and finally quite dark coloured. The excretory orifice is situated near the base of the tentacles at the side near the narrow neck portion.

Reproduction is produced by eggs which are situated within a sac at the lower portion of the animal. They may be seen singly or four or five together, of a rather dark colour, floating and sometimes revolving within the cavity. The winter eggs or "statoblasts" remain in the tubes forming the colony until such time as necessity compels their exudation, which generally means the death or breaking up of the cells through age. These lay by until the following spring to emerge into activity and continue their species.

The "statoblasts" in Plumatella are an elongated oval shape, with a smooth exterior, without any hooks, as in the Pectinatella, around their outline. They are reticulated and the centre is of a lighter brown than the outer and is also oval shaped. They appear strongly fashioned and able to withstand considerable buffeting in currents, etc., as they may occasionally find themselves during their long wait through the rigours of winter until the spring arrives.

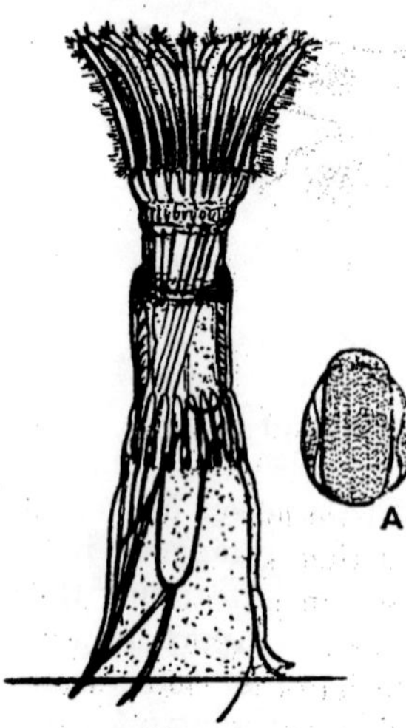

Fig. 247.
FREDERICELLA SULTANA.
A. Statoblast of same.

Fredericella sultana (Fig. 247)

These colonies are not unlike antlers in their manner of branching and are often found along with "Plumatella" near the sides of shady pools and shallow backwaters. They are massed together with some of their tubes recumbent and others upright, and generally all more or less covered with algal growths, sand, etc.; rarely are they clear and hyaline. The colony as a whole will sometimes be affixed at its base only, having its branches freely floating in the water. It is seldom of large extent, covering but a small surface.

The structure is tubular, with terminal apertures at their widened ends where the " lophophore " protrudes. Sometimes these ends are bifid and a pair of them will be surmounted. The tentacles are fewer in number than " Plumatella," seldom exceeding twenty-four. They are arranged nearly circular, and this forms a good feature to distinguish them. The whole polypide is of a long, slender, trumpet shape, and only one inhabits a cavity at a time. The inter-tentacular membrane is prominent around the bottom half of the tentacles, and a muscular connection to the base of the lower portion, when in contraction, gives a decided twist to the whole as it retreats within its sheath on alarm.

The winter eggs or " statoblasts " are a dark brown, more or less elliptical or reniform and with a smooth upper surface without hooks or spines. The usual oval and lighter centre called the " float " is absent in Fredericella. Their habitat may also be upon the stones or wood lying in still or slowly moving waters. A swift stream would be too rough for their small and delicate structures.

They are, generally speaking, below the normal vision and must be searched for with a low-power hand-glass in likely situations, and microscopists most frequently obtain their specimens when foraging for other objects. They are not common probably from this cause, and their colour may be very similar to the object they are attached to, as well as encrusted with a fair amount of sand or other debris.

Paludicella ehrenbergii (Fig. 248)

The animals or zooids of this colony are separated by partitions from one another, and have a jointed appearance, not unlike the algæ Batrochospermum. Each joint is one-celled and club shaped, and upon the larger end is situated at one side a quadrangular aperture through which the little polypides protrude. The cells are placed in a row end to end, irregularly branching from the broader ends, so that many side extensions follow.

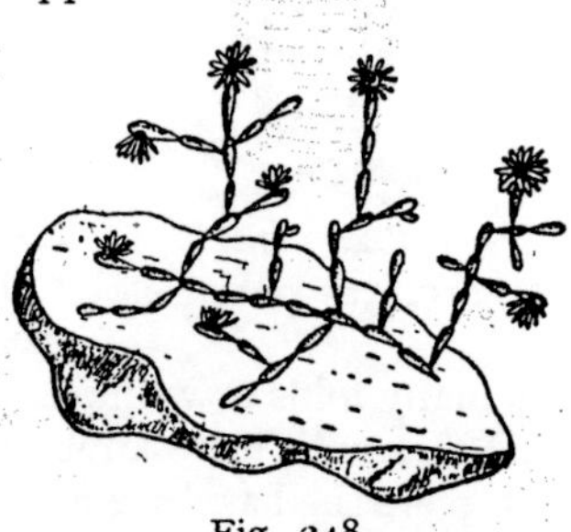

Fig. 248.
PALUDICELLA EHRENBERGII.

The colony may start as a row recumbent upon a stone or other surface, and from these will arise lateral branches each with the cells end to end and their tentacles displayed, one zooid upon each. The zooids themselves are about 1/50th inch in length and possess but a small number of tentacles, about sixteen in all. Their habitat has often been within water-pipes with water constantly circulating, so that their substance is fairly strong and adherent.

Urnatella gracilis (Fig. 249)

The stem-like growth of this polypide consists of a basal plate from which arise two segmented stems terminating in the tentacled animal. Each of the segments is made up of urn-shaped cells united end to end, which hang suspended from its support. A frequent habitat is the under side of a stone beneath which the water is constantly flowing, with the two colonies of urns side by side; pendant from one base of attachment. There are alternate dark and lighter browny bands around each segment, and the whole has a beady appearance. The central portion of each urn is bulged and of a lighter colour to the narrow ends, which are almost black. About ten or twelve of the urns form a single stem or colony. The colonies may be straight or hang curved from the side of another branch, their length rarely being more than a quarter inch, and must therefore be searched for by a hand-glass when in situ, otherwise they will in all probability evade the normal vision.

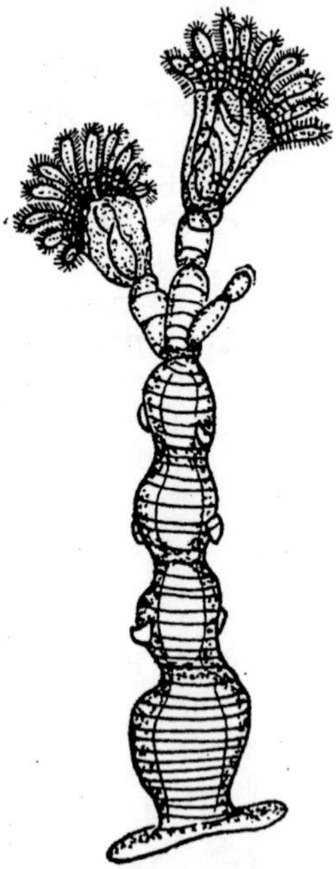

Fig. 249.
URNATELLA GRACILIS.

Upon each side of every segment in mature stems is a cup-shaped, small projection appearing like handles to the urns and are representative of those segments that have decayed or have been shed to form new colonies elsewhere. At some time or other each urn has had two formed from it, one at each side, and is traversed by many folds or wrinkles which are frequently spotted with brown and also with little knobs of a similar colour. When the stems are complete the basal segment is always the largest, each urn gradually lessening in size to the tentacles' base. Through the centre of all runs a cylindrical cord, adding strength and flexibility to the structure. The two urns near the apex are generally more transparent than the others and are evidently immature in their growth, but these alone are those which will be found bearing the tentacles, as when branches are formed, otherwise the flowing plumes are solitary upon the end.

The tentacles and their base are shaped much like a bell with a wide expanded oval or nearly circular lip of these finger-like projections. Upon the narrower end of the bell is a band, fairly broad, constituting the inter-tentacular membrane, forming a kind of stiffening collar to them. From twelve to sixteen tentacles are the usual number, and are ciliated with active and comparatively long cilia. The anal opening is inside the corona, and the tentacles are incapable of complete retraction.

Urnatella is a timid creature, like most of the Bryozoa, and retreats quickly on the slightest agitation near by. The "statoblasts" have not yet been observed, neither the eggs, and a budding process of reproduction appears to be their general mode of increase. The urns appear capable of tiding over the winter season without any separation, and remain securely anchored until the favourable conditions of spring arrives to bud forth and branch anew and found fresh colonies of their species.

Pectinatella magnifica (Fig. 250)

This forms in a thick, gelatinous, opaline mass of several feet in diameter sometimes, and often as much as six to eight inches in thickness, and is really an aggregation of colonies which have secreted a common basal stratum. It is first commenced by a single polypide budding and becoming two. These again continue the process, and as the number increases, so the colony also increases in its rapidity of production and extent, finally bearing many hundreds of the tiny animals in rosette-like colonies. Summer-time is the period of greatest growth. In autumn the mass begins to decay, and the separate groups of rosettes are set free and may be found in floating clumps, which may again attach themselves to some object until these in turn degenerate and die. The colour of the polypides is generally of a pinkish tint.

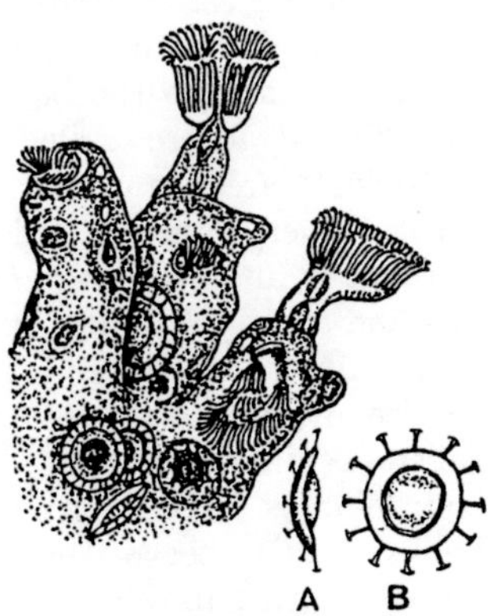

Fig. 250.
PECTINATELLA MAGNIFICA.
A. Statoblast (side view).
B. Statoblast (front view).

They are formed in a double row around each lobe, or may be scattered irregularly about the mass.

The gelatinous substance as it ages becomes browner owing to many small rounded bodies appearing in it, which as the animal dies are released and float out or adhere to the surface upon which they are affixed. These are the winter eggs or "statoblasts," the hibernating cells formed at the end of their active life to tide over the climatic conditions at a period unsuited for them to continue in, to one of more congenial surroundings in spring, then to develop out once again into active existence anew. These "statoblasts" or "hibernacula," as they are sometimes called, serve a further purpose for dissemination, being carried by the Water Fowl within the muds adhering to their feet and legs, or driven by winds among the dust from dried up localities on the air and so to fresh pastures, and in this manner they are spread abroad, becoming almost cosmopolitan.

"Statoblasts" are capable of withstanding the freezing in water, and are known to resist the action of the digestive fluids of many aquatic birds. Their shape is sub-rectangular to almost circular, and they are large. The central lighter portion or "float" follows the shape of the outer "annulus." In size they measure about 1/25th inch in longest diameter. Upon the outer margin is a single row of double or anchor-like hooks, set upon the ends of short stalks. Their colour is chiefly dark brown, and the annulus is always much reticulated with innumerable hexagonal cells.

Running down the centre of all the Bryozoa from the lower side of the stomach to the base or posterior end of each polypide is a rope-like structure called the "funiculus," and from this the statoblasts are formed by a budding process. At the lowest end they are very minute, and as they slowly push their way towards the higher and stomach end, gradually enlarge themselves, where in some cases they affix themselves to its exterior; all, however, eventually reach the outer water either at maturity or upon the demise of the polypide, as it softens its surrounding gelatinous substance in decay and so frees them. Upon the margin of the "statoblasts" there may be from ten to twenty of the anchor-like hooks, which serve the purpose of holding them to aquatic objects until suitable season arrives for their activity.

Pectinatella finds its most congenial habitat in shaded positions, upon submerged logs, walls of reservoirs, boulders, lock gates, etc., and, like most of the Bryozoa, prefers still waters. Exposure to heat, air, or too much sunlight is fatal to them. It is a noticeable fact that in England they have seldom if ever been found. Probably

Photo by] [*F. J. W. P.*

Fig. 246.—PLUMATELLA.

Tip of tentacle showing the delicate cilia (highly magnified).

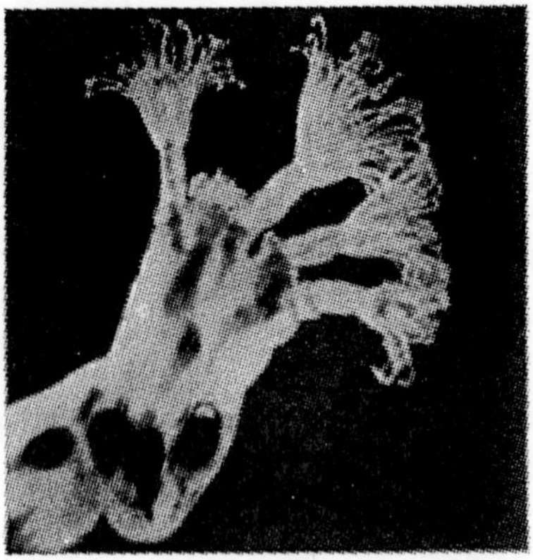

Photo by] [*F. J. W. P.*

Fig. 251.—LOPHOPUS CRYSTALLINUS.

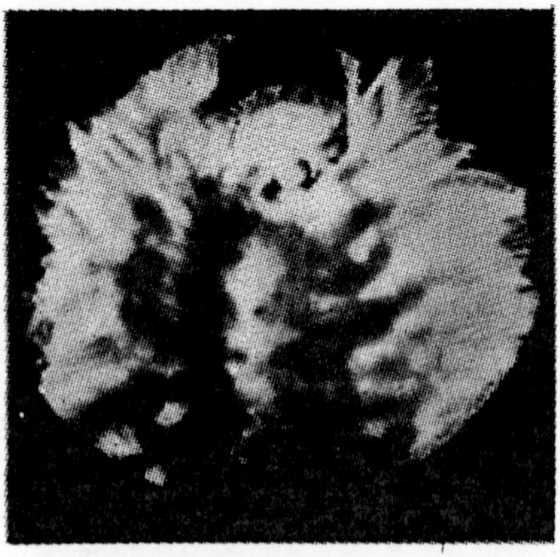

Fig. 252.—CRISTATELLA MUCEDO.

Portion of colony "*en masse.*"

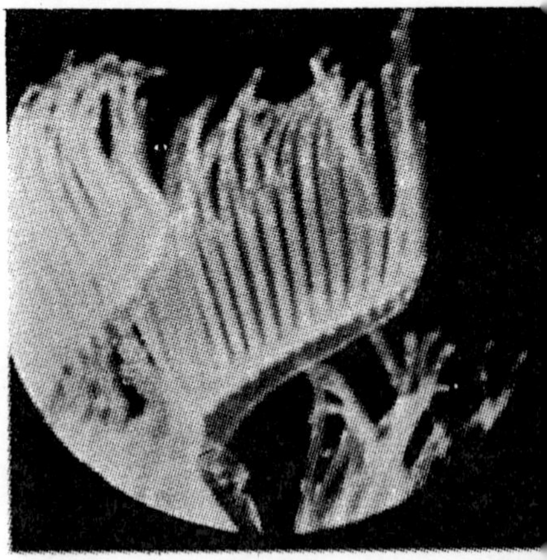

Photo by] [*F. J. W.*

Fig. 253.—"LOPHOPHORE" OF CRISTATELLA MUCEDO.

[*To face page 23*

the haunts the average micro-fisher frequents have not been the right ones, as there seems no reason why Pectinatella should not exist congenially here in our summer climate. In America, Hamburg, and Japan it is frequent.

Lophopus crystallinus (Fig. 251)

The colony of polypides is sac-like and often with finger-like indentations and constrictions along its margin, which give it a somewhat lobed appearance. The mass is opaline or sometimes very transparent. The outer covering or "cuticula" is hyaline and delicate near its surface, becoming more incrustated towards the base. The polypides generally arise in a group from each lobe. With age the colonies become cloudier and more or less opaque. At times the mass becomes much branched and the polypides irregularly scattered about. The size of the colonies is from 1/10th inch to ½ inch long. Each "lophophore" has about sixty tentacles forming its coronet.

Reproduction is effected by splitting apart along the constrictions between the lobes. This has been observed taking place to completion in about three days. Their habitat is preferably quiet waters, as with Plumatella; very rarely are they found in running water, unless but slowly; they frequently are attached to the rootlets of Lemna or plant stems generally. The "statoblasts" are large and elliptical and at each end are drawn out into a sharp acute point which at times is prolonged. Two forms have been observed, however, one bluntly pointed and the other obliquely and sharply so, as in the illustration. There is a central broad "float," and they are without hooks, as in the Pectinatellæ and Cristatellæ.

Lophopus and also Cristatella are notably distinct from other forms in the rapid way they always expand their tentacles after being contracted into their sheath. The withdrawal of the lophophores is readily accounted for by the action of the retractor muscles, but the protrusion of them is a more intricate problem. It is assumed at present this is due to the flexibility of the "cuticula," which is a double fold, and consequent upon a fluid pressure and extension of the substance from within, somewhat like the turning inside out of a stocking or a rubber tube.

Cristatella mucedo (Figs. 252 and 253)

This is an unbranched, greenish, gelatinous substance, forming an elongated colony of polypides about 8 inches in length or more and sometimes ¾ inch wide, especially about autumn. The plumes are all upon the upper side of a flat "sole," as the lower mass is called. Around the margin is a zone of tissue, which gives rise to the new zooids by the process of budding; hence it is known as the "budding tissue." External "cuticula" is absent save for a thin layer at the under side of the "sole."

The polypides all contract into one common cavity at their bases, forming a continuous space, interrupted only between the animals by a vertical septa. In this way each lives to help all the others belonging to the colony as well as himself. Truly a communistic policy of government. Each for all and all for each is the principle with Cristatella.

The tentacles number from eighty to ninety in each "lophophore," set in a double row around the U-shaped margin. Their base is united by a delicate web known as the "intertentacular membrane," similar to other species of the Bryozoa. Each tentacle is tubular and lined with an epithelial substance in which are embedded, cells of various kinds and sizes. These cells constitute the chief digestive glands, secreting the specific juices used in the admixture and preparation of the animal's food. The body wall also consists of two epithelial layers known as the ectoderm (outer layer) and the mesoderm (inner layer). Between these are the muscular fibres. Along the œsophagus and parts of the stomach lining are many cilia, very difficult to make out in life, which assist in the trituration and circulation of the fluid pabulum. Upon the tips of the tentacles are several buds, similar to "taste buds," which are used for the selection or rejection of unsuitable foods. The whole substance is much granulated throughout.

The colony as a whole, and unlike almost all the other Bryozoa, is capable of gradually moving along, usually in the direction of its longer axis, or may spiral along its support. Either end may go first. It amounts to a very tardy gait, about an inch in twenty-four hours. No necessity to carry a red flag in front of this locomotive. The tentacles and zooids generally, however, are of a pinkish tint, in life.

The "statoblasts" of Cristatella are the most ornamental in appearance of all, and of a circular crenated outline, and with a paler "annulus" and large central "float." They possess from ten to thirty-four dorsal hooks and twenty to fifty ventral ones, splayed at the end like anchors, and used for the same purpose in a bifid manner. The hook stalks are usually waved, and the central "float" is much knobbed with round, raised protuberances upon it. The habitat is in still or slowly moving waters, upon the branches of submerged dead trees, or upon the under side of Lily leaves and other aquatic plants.

Cristatella seems to have no objection to the direct rays of the sun, although it is usually found upon the shady side of objects, but during its peregrinations the writer has found it on the surface of a branch and also upon the upper side of a Yellow Water Lily leaf with the sun shining upon it. The "statoblasts" will measure generally about 1/25th inch in diameter.

"Lophophore" means "wearing or bearing a crest." Reproduction occurs by budding, and the colonies will also occasionally undergo fission and move apart and so create new masses.

CHAPTER XI

ENTOMOSTRACA

THIS word literally denotes an "insect or animal within a shell," and the larger group of crustaceans, including Crabs and Lobsters, are such and the Fairy Shrimps or Phyllopoda also, so that its translation has little characteristic significance, since both these extremes might be included. For the present purpose we are dealing with the microscopic life found among our fresh-water collections, and shall confine ourselves mainly with the Copepods, Daphnia, etc., and their nearer minute relatives.

Writing generally of the Entomostraca, these are more abundant in our fresh waters than the Phyllopoda. The latter inhabit the smaller pools, often formed during the rains of spring, and which dry up during the summer. They are peculiarly adapted to this varied kind of life, where the mineral salts alter considerably with the quantity of water, and the eggs are capable of prolonged desiccation. Indeed it seems almost necessary they should be first dried and afterwards immersed before hatching. In this manner the dry mud often contains large amounts of the parched eggs, which are carried long distances by the winds and birds to fresh pastures, there to continue and open out into life anew.

Phyllopods usually swim on their backs, and may be distinguished among our microscopic Entomostracans by this peculiarity. Daphnia and the small crustaceans generally, are found universally in every kind of fresh waters. When the sun is bright they may be seen in great quantities near the surface, and if a cloud comes over they will soon depart to the lower regions, to rise once more as the sunlight returns. Daphnia belong to the Cladocera, a group of Entomostracan crustaceans, usually considered a sub-order of the Phyllopoda. Their inter-relationships are therefore very much bound up together.

The Cladocera is divided into two sections, the larger of which is called the Calyptomera, and includes the Daphnia, or those having a large "bivalve" shell which covers the body and legs, and the Gymnomera, which includes those having the body and legs free, retaining the shell only as a "brood sac" for the cover of their eggs. This latter section includes but two species at present, Polyphemus and Leptodora.

In the Calyptomera there is a distinct head and a "bivalve" shell, which is simply a fold of the skin extending over the body from the back downwards. In the head is a large compound eye which is capable of being rotated. It is generally black or of a dark pigment and in constant movement during life. Near to it frequently are one or more pigment spots or "ocelli." The head also carries the brain and numerous nerves to the antennæ, eyes, and digestive tract. It has appendages also, generally in two pairs. The first are the small antennules with their tiny olfactory sense rods placed usually at the extremities, and with one or more side-placed sensory hairs upon them. Second are the prominent large antennæ, which form the main instruments of locomotion with their stout joints, forked branches, carrying the long plumose setæ attachments for swimming. These are moved by powerful muscles usually occupying the major portion of the head. There are various methods of locomotion, and upon the size of the antennæ and the number and length of their setæ these mainly depend. The smaller Daphnias hop rather than leap, with quick actions, but in the heavy forms there is a rotary and unsteady motion induced by vigorous thrusts of the antennæ.

The mouth parts consist of the mandibles, which are strong chitinous bodies of one piece and without palps. They are toothed and ridged, set in opposition, and grind the food very completely. The maxillæ, rather small organs whose office principally is to push the food between the mandibles, work like a pair of hands, and the labrum or lip, which closes the mouth from below. Usually the head is depressed downward to form a beak or "rostrum," as it is called. The shell takes many forms, and may be oval, round, or very nearly square, and though called "bivalve" is really in one piece, curved dorsally, but never showing any join there. It is reticulated and lined superficially with various cross markings. On some are spines or hairs or both, and along the lower margin especially they are frequent.

Daphnia has a single spine posteriorly extending the point of junction of the valves. Scapholeberis has two, upon the lower and hinder portion of the shell, projecting backwards to long points and to a considerable length sometimes. The shell case is a double layer, between which circulates the blood; the inner and more delicate surface serves as a respiratory layer, the interchange of the gases, carbon and oxygen, amounting to breathing in the crustacea. In the legs, too, breathing, curiously enough, is also carried on, in addition to the valve surfaces, and they are used as gills.

Just behind the head is an elongated oval sac. This is the heart and in the living animal can be seen rapidly pulsating. In many of the transparent Cladocera species the movement of the colourless or yellowish corpuscles may be seen entering openings each side the heart and being expelled in front. There is no system of tubes for the circulation to pass through as in the higher animals, but a complex series of definite courses can be observed all over the body as the blood flows rapidly along, under and over. Below the base of the antennæ is a flattened glandular tube with several loops; this is the "shell gland," supposed to function as a kidney.

The intestines run in a simple manner through the body, which lies free within the valves, and the main portion of the latter bears the feet, arranged in pairs, usually five, sometimes six. The feet are leaf-like structures, bearing numerous complex hairs and setæ, whose chief use is literally to create a current and "flick" or "kick" the food into the mouth. This action at the same time brings fresh oxygen for respiration and flows it over the inner surfaces of the valves.

The food particles are chiefly algæ, and as they are collected and fed towards the mouth by the feet, the maxillæ push them between the jaws as the labrum or lip opens, and the mandibles do the rest, grinding them up and passing them on to the œsophagus. There is little break in this procedure, and the jaws of the Cladocera are continually chewing all the time, whether anything comes up to munch or not: their sole business is to keep moving, seemingly. In the matter of diet there is definite taste shown, and generally speaking the animals prefer a diatom or a portion of Spirogyra to an Oscillatoria, or in fact any of the blue-green algæ. These evidently are too rank for them. They are known further to reject particles after partial mastication, showing that the taste sensation can arise in the mouth. The eye, too, is very sensitive to light and shade, and also its intensity; many instances have been observed showing considerable peculiarities, one species with another and of varying temperature effects. Yellow rays are said to be the most attractive, while blue repels and red is inoperative.

The large cavity on the posterior dorsal portion of the rotifer is the "brood case," where the eggs are deposited and hatched. Unlike the general rule of the Rotifera, the young embryos are well grown before being set free into the surrounding water.

Entomostraca in general forms an important article of diet for the large fresh-water fish, and there is a one-celled microscopic infusorian that is a natural enemy to them, the Stentor. This will devour them whole, and as these are sometimes very plentiful, can account for many of the Entomostracans apart from the larger fishes.

Entomostracans are useful and act as fresh-water scavengers, eating up decaying vegetable matters which would ultimately pollute the waters if allowed to remain and so confer in their life a benefit both to themselves, their neighbours, and, indirectly, to ourselves. Innumerable flies and the dismembered parts of such, as the wings, legs, etc., of the ephemeridæ, gnats, midges, mosquitos, litter the surfaces of lakes, ponds, and streams at all times, and these broken fragments are alike acceptable to the Entomostracans, and so animal matters are also readily assimilated and removed from polluting the waters.

In examining the crustaceans a deep cell should be used, as their bodies are delicate and easily injured. With care, in a screw-top live-box, the animal can be just touched with the cover glass sufficiently to hold it in one place, and this is the right time to see the active movements of the legs and antennæ and the structure generally. The water, if coloured with pigments, carmine, etc., often tends to become cloudy owing to the rapid lashings of the appendages, but otherwise if fairly still it will frequently give a pleasant contrast to the usually hyaline body under observation.

In collecting specimens the net is about the best appliance, and in using it it is better to take frequent hauls and emptyings rather than a prolonged sweep. In this way many will be taken which otherwise would escape. Empty into a white receptacle, a cup or basin, and clear away as much of the algæ and plant life which is allowed to settle at the bottom before transferring to your tubes and bottles. A separate bottle holding some of the weeds may be carried, to add upon arrival home.

Some species can only be found near the water's surface at night, notably the "Gymnomera," others only upon the mud or in muddy pools, as "*Moina*," experience being the best guide to such matters in collecting.

The abdomen posteriorly terminates with the intestinal canal in a hinge-like portion free to move forward and backward and known as the "post-abdomen." It is a flat piece, roughly in the shape of the letter "L," having the intestinal opening at the distal anal end and guarded by a long sharp claw or pair of claws near to. The claw itself may have several small spinules upon it and often a longer one at its base. Near to the opening is a tooth-like margin or comb which usually is denticulate, or may be pectinate. This is called the "pecten." At the upper angle is a pair of sensory bristles which spring from a common base called the "abdominal setæ," sometimes very long, and the margin posteriorly may be furnished with various dentures and points, or even two "pectens" along its length. Collectively this portion of the body is a very specialized organ, therefore, and is quite distinct from the feet, which usually are under cover of the valves.

The post-abdomen works principally outside, easily seen, with its sharp curved claws and combs, and in its general regular movements and compressions assists the removal of the unused foods from the intestines. The margins on the ventral sides of the valves are usually set with a row of several bristles, opposite to each other, and it is these that the comb "fiddles across," keeping them clean and free from adherent debris in its backward and forward movements.

In writing of the antennæ, feet, valve, and claw it must be remembered these are in pairs on the living animal, and to describe one pair of antennæ is usually understood to include the two pairs to save repetition, and very often the illustrations will only give one side. Similarly the claw is generally two, unless stated differently, upon the post-abdomen. The beginner might be inclined to think a new species was before him if these details are not borne in mind.

Fig. 254.—Daphnia pulex.

Daphnia pulex (Fig. 254)

There are a large number of varieties included with this species, but it will be right to assign to "*Pulex*" all the heavy-bodied opaque forms that have comb-like or pectinate claws upon the feet. Another feature of distinction is the presence of a sharp spine on the posterior border. This varies in length with the different species and curves upward slightly. Apparently it is shorter with age and usually longer in the young. It is always present in mature life. The body is oval, stout, heavy-looking, and rarely transparent. The antennules are very small, but the antennæ are prominent, usually divided into two branching parts, each having several feathery smaller branchings spreading from them. The feet are kept normally in energetic movement, bringing in fresh oxygen to the respiratory layers and throwing food towards the mouth. The heart can be seen near the head rapidly pulsating. The eye is large and compound with small lenses around its margin. The several oval bodies usually present near the back and posterior portion are the eggs deposited within the brood " case."

The body surface is reticulated. The feet aid in clinging to plants as well as producing currents of water for breathing; the hooks and spines of the first foot especially so, probably also in detaching the soft algæ used in food. The post-abdomen has a " pecten " of twelve to fifteen spines, and the anal opening is at the end of it. Upon the upper rear angle are two stout radiating spines or bristles.

The colour of the body varies from a yellowy brown to a red almost. The length of the female is about 1/10th inch. The young differs in appearance considerably from the parent. An " ocellus " or pigment spot is often present near to the eye. It is small and immovable.

The " pecten " in the Cladocera is principally to keep the legs clean and free from extraneous matters and from Infusorians, Vorticella, etc., and other parasites attaching themselves and finding lodge upon a soft and delicate portion of the body.

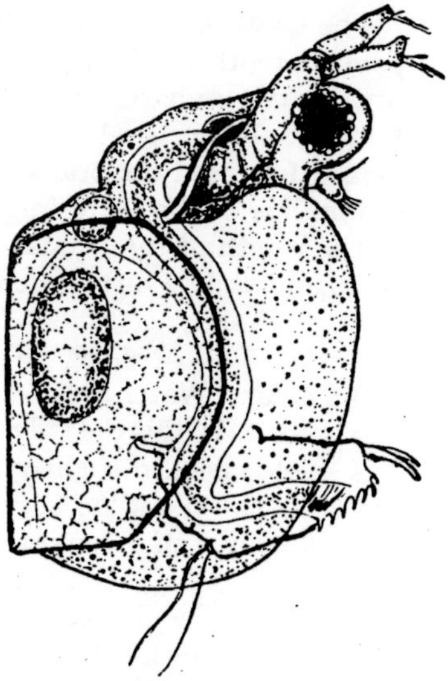

Fig. 255.—CERIODAPHNIA RETICULATA.
(Antennæ foreshortened.)

Ceriodaphnia reticulata
(Fig. 255)

This little animal may be easily recognized in the aquarium by its peculiar " flap and rest " movements. It will give a few strokes upwards, carrying it a short distance, and then immediately cease, only to allow its body to sink nearly to the same level again before it commences once more. Its progress therefore is not at all rapid, reckoned in distance, whilst in its normal habits. It is a singular procedure, and the significance and reason of it may well be left to the microscopist to study.

The body is rather more square than rounded, chiefly owing to a saddle-like plate upon the dorsal side being developed called the "ephippium." The ephippium is a semi-elliptical portion of each valve upon the back, connected with the covering of the future egg and embryo. When the moult of the whole shell takes place this separates with it and encloses the eggs within, which then are capable of laying by to the next favourable season before hatching.

The head is but slightly angulated in front, not amounting to a horn as in some species. The valves are reticulated, the lines of network being readily apparent, almost raised in some, and though occasionally they may be rather faint in a species, they are always present. The antennules are small, and a short curved sensory hair is situated just above the apex. The eye is large and prominent, as in most of the Daphnia family, the two antennæ also resembling theirs very much. The antennæ are fairly long, and divided into two "rami," each branch having five long bristles. There are three equal joints on each branch in "*Ceriodaphnia*."

The claws have from eight to ten denticulate teeth, the sharp-pointed tooth at the tip being stout and conspicuous. There are usually two radiating bristles from the upper posterior angle of the post-abdomen, which are not readily perceived unless specially looked for, used doubtless as organs of touch, extending beyond the outline of the body frequently.

The length of the animal is about 1/40th inch. It is common in pools and widely distributed.

Scapholeberis mucronata (Fig. 256)

A similar "Water Flea" to *Ceriodaphnia* belonging to the family Daphnidæ. In this the valves are striated and obscurely reticulated. The shell is beaked slightly, and the ventral and posterior margins are quite flat in outline, seen squarely placed The body is not compressed, and with a convex dorsal border. The head is bulged, and carries in it a large compound eye, black, with several small lenses upon its margin, almost filling the part. At the posterior end of the shell's lower border this extends into two pointed spines, generally short, but in the "cornuta" form of Mucronata they are long and well developed. On the ventral margin are many short fine setæ. In the variation "*Armata*" (Herrick) the spines are as long as the ventral margin of the valve. The claws are denticulated or toothed rather than pectinate and comb-like. There is one short projection on the anterior portion of the abdomen. The antennules are very small, with little movement, set just within the beak protuberance. There may be several eggs in the brood cavity upon the posterior portion of the back, but usually only one is seen in some varieties.

The animal has a habit of making use of the surface film of the

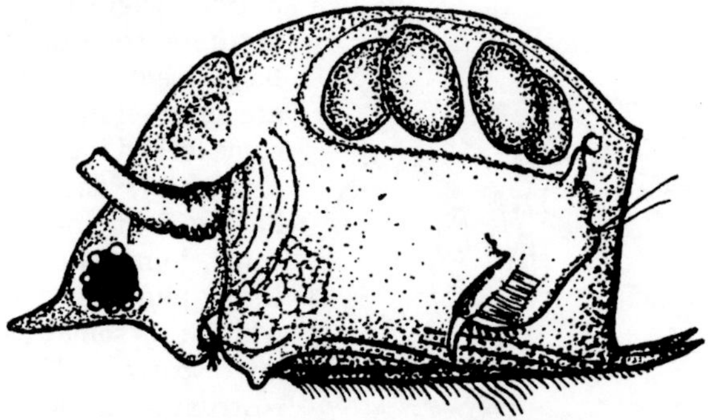

Fig. 256.—SCAPHOLEBERIS MUCRONATA.
(Cornuta form without antennæ.)

water to suspend itself by, and in this position will stay a considerable time with its feet busily on the move. The writer collected from a fresh-water lake several of the cornuta form of Mucronata having a short pointed beak on the front of the head. The two black eyes seemingly coalesced, so that they occupied the major portion of the head. The body was striated longitudinally, and the rear spines were very long.

Scapholeberis can be seen readily with the unaided eye in clear water, and measures about 1/10th inch in length, without its spines. At the right seasons, spring and autumn, it is fairly common in some localities.

Bosmina longirostra (Elephant Water Flea) (Fig. 257)

The family Bosminidæ have several genera and may be recognized by their six pairs of feet, short rounded or oval body, not compressed, quite a simple tract for an intestine, without any convolutions or cæca, and by the antennules of the female being large and immovably fixed to the head. The back is much arched, and the valves cover the body and abdomen entirely. The anterior and ventral portion is curved and smooth, without any projection. The lower angle of the posterior margin of the body, however, bears a single short spur or spine called the "mucro." The shell is normally quite transparent, and upon its surface delicate longitudinal lines and reticulations are traceable. It is very thin structurally. The antennules lie parallel to one another approximately, curving slightly upward, and at the extremities appear to be segmented. The antennæ are short, stout, and soon become forked into two "rami" (branches), one, the dorsal, having four and the ventral three, joints, before reaching the plumose branchings, which are somewhat slender and sparse. The beak or rostrum as it continues down from the head becomes forked into a double trunk-like pair of probosces, and it is these which give it the resemblance to the trunk of an elephant, from which its sobriquet is taken. The post-abdomen claws are set on a cylindrical process, often with two series of spinules on them.

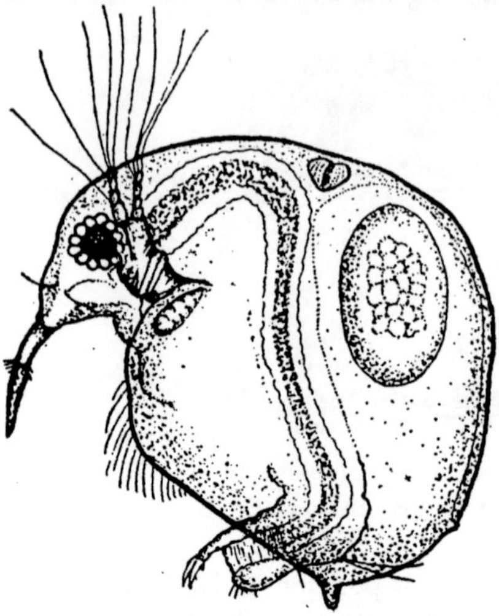

Fig. 257.—Bosmina longirostra.

Just above the junction with the antennules is a solitary bristle or sensory hair which stands out prominently and forms quite a distinguishing point on all Bosmina. It is not readily seen at all times and may need careful search. The heart is situated dorsally in front of the "brood case," and may be seen pulsating actively. The eggs in the "cavity" are oval, or as more mature, circular in shape, and seldom number more than three or four at a time. The eye with its black pigment has a circlet of lenses around its circumference. The rostrum as it narrows to the tip has several transverse joints upon it.

The animal is variable in its outline. It is not uncommon in Britain, and measures about 1/25th inch in length. Found in lakes and ponds frequently.

Chydorus sphæricus (Fig. 258)

This is probably the most abundant of all the Cladocera, its yellowy brown body being present in almost all fresh waters from the smallest pool to the largest lakes and in every quarter of the globe. Its shape is broadly elliptical, the posterior portion cut off straight, as seen from the side. At the upper angle, posteriorly, of the junction of the two valves is a short but definite point or spine. The shell is partially reticulated, the ventral half completely, and the upper punctated with small spots. This latter portion lies above the "brood sac" and shows a large oval cavity there.

There are several varieties of shell markings in "Sphæricus." In var. *Nitidus* it is smooth, in var. *Cœlatus* there are elevations, and in var. *Punctatus* small punctæ or knobs. The feet are covered beneath the valves for the most part, the stout, broader post-abdomen with claws alone working outside. This latter has a dentate row of sharp points, ten to twelve on its rear margin.

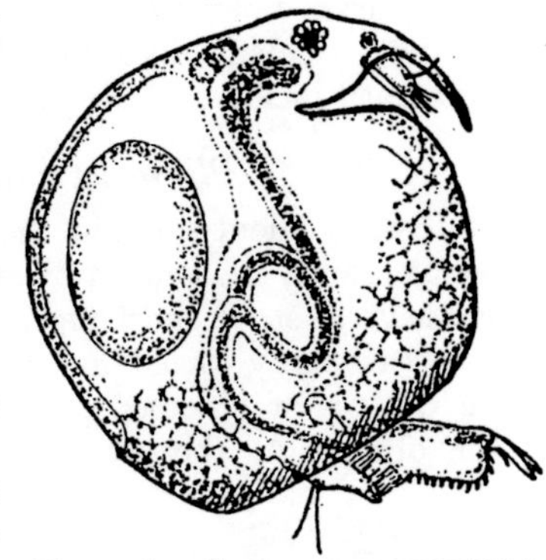

Fig. 258.—CHYDORUS SPHÆRICUS.

The lower edges of the valves are set with many straight flexible spines, opposed to each other, and the post-abdomen member every now and again sweeps these with its comb-like "pecten," similar to the action of a bent arm striking backward, depressing them in its course, cleaning away any debris or other extraneous matter, and springing them into parallel position again. At the upper angle of this hinged member are two radiating sensory bristles extending beyond the valves seen from the side. The setæ of the antennules are all set upon its extreme tip. The "rostrum" is semicircular, following the body's outline and curving downwards parallel to its circumference at a short distance apart. The eye is black and prominent, with several small transparent lenses around its margin. A little below this at the base of the "rostrum" is another black pigment spot, smaller, which gives one at first sight the impression the animal has two eyes. It is the "ocellus," and if watched will not be seen to move, while the true eye does. It has no circlet of lenses around it, and is often of an irregular shape and sometimes called a pigment "fleck." The heart in front of the "brood case" will be seen busily pulsating rhythmically, and the mandibles situated close by will also be munching away, whether there is anything to masticate or not. These evidently are two organs with reflex actions, although the mandibles can be hastened at times as if a certain amount of will enters into it. The intestinal canal is a fairly wide and convoluted tube with one or more large loops in its length, ending in the post-abdomen near the claw, which protects it outwardly. The food may be seen within the canal passing backwards and forwards extracting the last particles of nutriment possible from it.

The diet is chiefly decaying algæ. The length of the animal is about 1/60th inch.

Pleuroxus truncata (Fig. 259)

A fairly prevalent form in the autumn is this little Cladoceran. It will be readily recognized by its valves on the posterior margins, having several stout short teeth, about fourteen each side, though this is occasionally varied one side from the other. In the specimen before the writer nineteen are upon one valve and only fourteen on the other. As the spines follow upon many bristles or setæ situated all round the remaining portions age may make a difference in the development of them. On the anterior margin are another group of five each side. The head develops into a long pointed rostrum, continued in the curve of the back, and rarely bent forward. The body is of an oval shape with the posterior portion cut off straight, almost, and upon which the teeth are placed. The back is arched and the shell longitudinally striped about the middle and obliquely from the ventral margins. The antennules are broad based and cone shape with several short bristles at the extremities, and at least two are rather longer than the rest. The post-abdomen is long and stout, tapering slightly towards the apex. The denticles consist of a "pecten" of from twelve to fifteen, which become larger nearer the claws. The opposite side in all species of the genus is smooth, the denticles occurring upon the margin only. Each of the claws has two basal spines with a few shorter and smaller ones. The pair of abdominal setæ at the elbow are short and do not project far to the rear of the valves. The feet are in five pairs.

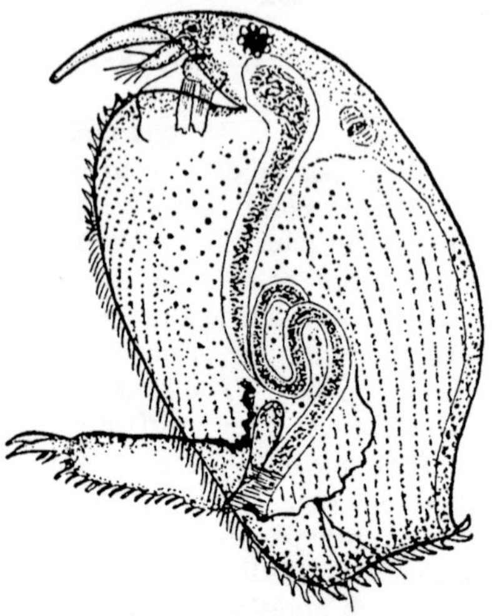

Fig. 259.—Pleuroxus truncata.

The shell, the colour of which is a yellowy brown, frequently has small globules of an oily nature about the middle of the valves beneath. The valves seen from the ventral side show an outer curve at the centre, and when both are near to one another give a circular opening in their outlines.

The length of the body is about 1/30th inch. The animal is found near the margins of quiet ponds. All the species are "littoral," i.e. belonging to the shore, as distinct from the "limnetic," or those preferring the larger open waters.

Leptodora kindtii (Fig. 260)

This is the only species of the genus and at present the sole representative of the family. It belongs to the section "Gymnomera." Neither the body nor the feet are enclosed in a shell.

Fig. 260.—LEPTODORA KINDTII.

It is large and can be readily seen by the unaided eye, the female measuring as much as ¾ inch from tip to toe. It is a rather strange-looking creature at first sight, but its very transparent body makes it a beautiful object under dark ground illumination.

The body is long, cylindrical, and fairly straight, ending in a fine tail-like appendage. The head end is narrow, the eye large, nearly filling the front portion, and the body has four joints along its length. The first portion carries the six free spinous feet on the ventral side, and on the dorsal what has been called the "packman's bag" or "brood sac," which is only loosely attached at its mouth end, the larger part bulging out in the water and containing the eggs. This sac is of a stouter substance on its upper side and apparently of more than one coat. The antennules are freely movable, the female's being the longer.

At the base of the antennæ is a large strong joint and an arm produced into two separate branches or "rami" having four joints on each, which bear numerous plumose setæ along their length. The whole form the antennæ and the animal's chief swimming apparatus. The mandibles are long and slender, pointed, and with three spines near to the apex. The stomach is situated nearly at the end of the long body, upon the last abdominal segment, the œsophagus reaching to it in a fairly straight, narrow passage along the centre of the body. In the winter eggs the young hatch out first into a larval state or "nauplius," similarly to Cyclops, Artemia, and others.

The animal is a rapacious eater, living principally upon the smaller crustaceans, rotifers, protozoans, etc. Seen with transmitted light, however, it is strange that little more than the dark eye and the yellowy coloured stomach is visible, so delicate are its tissues. Its habitat is in the larger open tracts of water, such as lakes, broads, etc., coming to the surface only at night, a fact which must be remembered when collecting specimens.

In the Middle Lake, Killarney, the writer was fortunate to obtain his only specimens, but they have frequently been obtained in our large English lakes at Coniston, Windermere, and at Bala, and in Devon and other places. They are not uncommon if sought for at the right time, and as they are most remarkable and beautiful objects the trouble, if any, is well spent in their capture.

The legs are very uneven in length, the anterior one seemingly quite out of proportion to the others and much the longest. It has two sensory hairs arising from one point upon the second joint, very transparent, but of a good length. The other legs become shorter as they near the posterior, the final two being very short, little more than a quarter the length of the frontal one.

Camptocercus rectirostris (Fig. 261)

Camptocercus is a genus belonging to the sub-family Chydorinæ. The shell is oval, long, and transparent, the posterior portion traversed by longitudinal lines, becoming more obscure towards the head end. The body is greatly compressed and the head strongly arched. The valves are rounded at the ends, smooth, and without spines of any kind. There are five pairs of feet, covered completely by the valves. Three or four small teeth are situated at the lower curve of the shell posteriorly and several short hairs ventrally. The eye is small, and a little below an "ocellus" (about half the size of the eye) is present. The rostrum portion is thick and blunt, which may be carried slightly extended from the body or depressed near to. The post-abdomen is very long and slender, with numerous small marginal denticles along its length. The claw is long and straight with one basal spine and a few very small spinules. At the upper angle of the arm are two radiating abdominal setæ, usual to the Cladocera generally, furnishing sensory hairs to that region. The heart is particularly well situated in a transparent position for easy observation, and seen pulsating rapidly. It is an oval body with a transverse membrane dividing it into two nearly equal-sized chambers. The alimentary tract shows one or two coils along its course, the anal orifice opening behind the marginal denticles on the claw arm. The eggs when present are carried near the dorsal posterior portion within the "brood sac." The antennules form a brush-like projection behind the short rostrum, the setæ being longest in the male, and all of them situated at the extremity. A small notch or spine projects about half-way along its length.

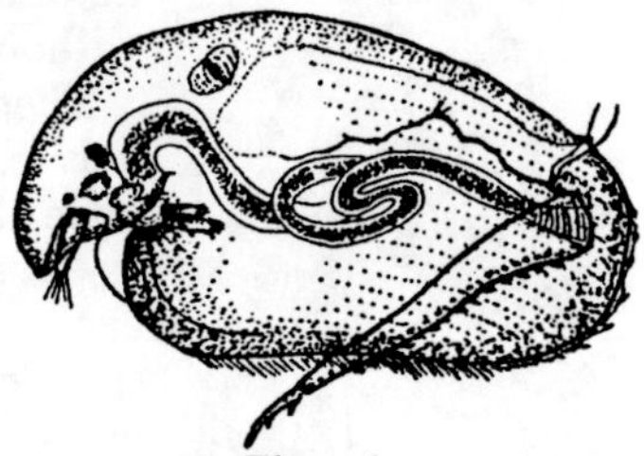

Fig. 261.
CAMPTOCERCUS RECTIROSTRIS.
(Minus antennæ.)

The animal is common in all pools and lakes amongst the weeds and confervæ along the margin. It measures about 1/12th inch in length.

Sida crystallina (Fig. 262)

The shell of this is an elongated oval, the base of which is cut off almost straight. The head is large, with a conspicuous gland situated on the upper side. The rostrum is short and pointed, and the antennules, in the female, are attached to one side of it. Beside the little bundle of setæ on the end of the antennules is a short whip or flagellum in frequent movement. The eye is large, with a number of small lenses around its dark pigment, and a small "ocellus" is generally present near to the base of the rostrum, sometimes rather obscure. The intestinal tract lies almost straight, and has a broader portion at its anterior end. The heart is an elongated sac lying in front of the "egg cavity" on the dorsal side, and may be seen divided into two chambers by a transverse partition, beating regularly. The antennæ (remembering there are always a pair of these) are large and divided into two flattened "rami," the dorsal branch having three joints, and is much longer than the ventral one with its two only. The branches are furnished with numerous delicate setæ, and about the joints are several spines, without the feathery hairs.

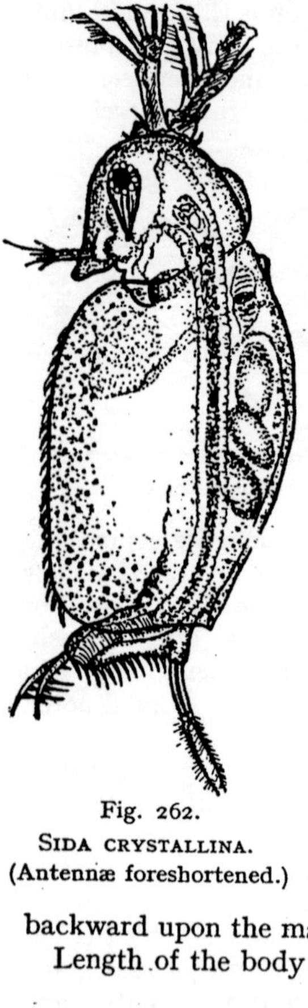

Fig. 262.
SIDA CRYSTALLINA.
(Antennæ foreshortened.)

There are six pairs of feet covered by the valves, bearing many long hairs, and in the male the first foot has a special crook, used as a grasping organ. The post-abdomen has two fairly straight long claws, with four spinules on each. Upon the margin are many sharp denticulate teeth (or spines) which form a rather long comb or "pecten." The anal canal by the side can be seen strongly convoluted in its course, and upon the upper posterior elbow, or angle, are two abdominal setæ, stout for half their length and ending in fine plumose hairs. The valves are flecked with many dusky pigment spots upon the lower sides and sparsely so in other parts. The colour of the shell is a light yellow, transparent, but can hardly be called crystalline. Its surface is covered with exceedingly fine hairs; these can also be well seen along the margins. To the posterior end on the ventral side many sharp-pointed teeth are situated, directed backward upon the margins.

Length of the body is about 1/20th inch.

Cyclops (Figs. 263 and 264)

This Entomostracan is one of the most constant objects among the amateur collector's gathering. It belongs to the Copepoda, an attractive and arresting group of animals, the structure of which, however, is rather complex when their classification is studied. The body is divided into two principal regions, the cephalothorax (head and chest) and the abdomen. In the first is included the anterior five or six segments and in the latter the remaining three to five, varying in the species. The feet are in five pairs. Each foot, besides having two basal segments, branches into two "rami," each with three joints, one of which, the outer, termed the "exopodite," is much longer and stouter than the inner or "endopodite." The fifth pair differ in the female from the male, and the males themselves are dissimilar right side from left. While the animal is alive it is difficult to get a good view of the feet under the microscope, and the better plan is to see them in a properly mounted specimen.

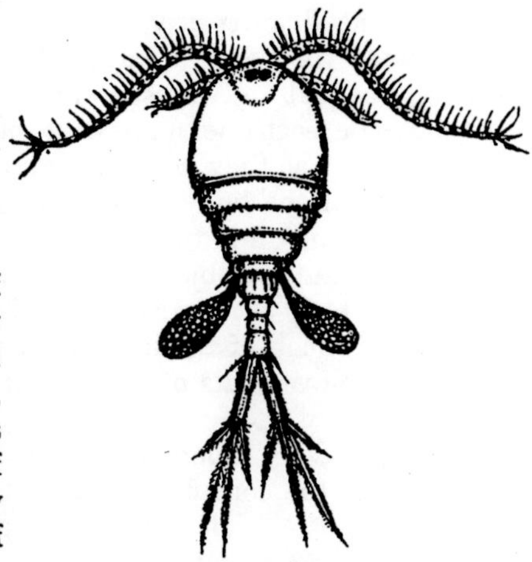

Fig. 263.—Cyclops.

There are a great many varieties of Cyclops, and consequently a good deal of detail that cannot be fully described here, both of form and in general appearance, and this complexity has given rise to a host of specific names which really turn out to be only phases of one and the same organism. Intermediate forms showing all stages between the extreme varieties have been found and accounted for. Where the animal is found in comparatively shallow waters the divided tail end or "furcal rami" are comparatively short and stout, while in deep water specimens they are long and slender. *Cyclops viridis* is taken as the typical form, and to this most of the species can be allied.

There are two pairs of antennæ. The first are quite characteristic and always the larger, with a graceful curve or sweep, and are flexible, active, and powerful. They are made up of a considerable number of segments from as few as six to as many as twenty-five. Where the species is definite they are usually invariable. The second pair are considerably shorter, much less conspicuous, and with fewer divisions. Each segment is armed with hairs or setæ, sharp and as "transparent as spun glass," definite in their number and location. There are also definite sensory structures situated along their length, and the segments are finely crossed by transverse rings, requiring careful observation to see. The male antennæ are rather less showy than the female, being the shorter and often thicker, swelling out sometimes towards the tips and ending in a hinge joint peculiar to the sex and quite distinctive. On some species crowns or circlets of spines are seen on certain of the antennal segments, giving quite a feathery, ornate appearance.

Beside the antennæ are included under the first and large shield-shaped segment the mandibles, the mouth, and the first pair of feet. The other segments cover the remaining feet, abdomen, and viscera, and hold the egg sac's attachments.

In general outline the form of Cyclops is pear-shape, or with a club-shaped head anteriorly, tapering to the two long lobes or "furcal rami" of the tail end. These "rami" are embellished with long, unequal and transparent setæ, and are usually more or less plumose, i.e. each one can be made out to have short sharp spines upon them, finer as they near the point, and always set with their ends away from the body at an angle, like the feathers upon an arrow head. Sometimes they may be long, comparatively, and are then indeed beautiful objects with a bright light and dark background. The body being also ordinarily colourless, the beating of the heart and much of the internal structure can be readily seen, including the movements of the alimentary tract and other points of interest.

The young is very different in appearance in its first stage from its parent. In the conspicuous egg sacs carried either side by the female, where perhaps thirty to forty eggs are contained, the embryo passes through a complicated series of forms. On issuing from the egg the first appearance in the water shows the creature as a flat oval body without any separation into head and abdomen parts, and with only three pairs of appendages, viz. the two antennæ and the mandibles, all of which are used at this stage as swimming organs. This is the "nauplius" stage. Then a series of moults follow, three generally, up to the "meta-nauplius" stage, and still further until the mature form is reached, each stage adding more and more appendages towards the final approximation to its parent.

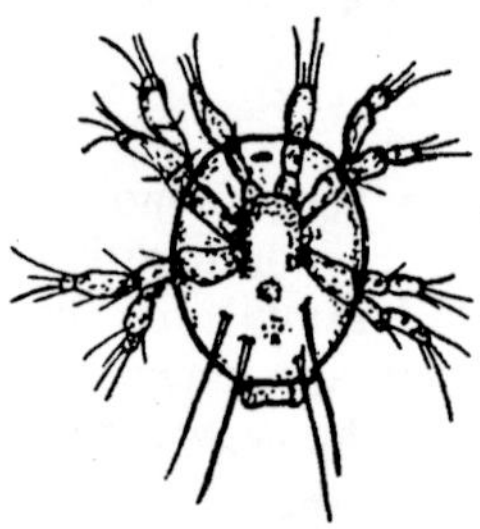

Fig. 264.
YOUNG OF "CYCLOPS."

The life history of a Cyclops and of the Copepods generally is therefore a most interesting and intricate study in itself, which needs a good stock of patience and care to bring it through successfully. In getting rid of its cast-off skins in moulting it often falls a victim to the process. These exuviæ will often be seen in the water as perfect casts of the animal's outlines, with all its articulations and delicate plumose setæ intact. It is a marvel how the creature slips out of it so perfectly; even the beautiful reticulations of the valves are there just as in life.

Cyclops multiplies very rapidly and abundantly. It has been calculated that the number of eggs one female could become the progenitor of would reach several hundred millions in the course of twelve months, but like all other creatures they have their enemies, and the larvæ of land and aquatic insects, such as Corethra, Dragonfly, etc., and Hydra and the Stentors, all play their part both in the eggs' and the parents' destruction.

Trout are even said to owe their delicate flavour in a great measure to the plentiful supply of these minute living crustaceans.

The eye is single, of a ruby-red colour, placed centrally in the forehead, and from this feature the name Cyclops was given in recollection of those one-eyed monsters of heathen mythology whom Vulcan employed to forge thunderbolts for Jove.

Diaptomus (Fig. 265)

This Copepod will be recognized at first sight by its extremely long and almost straight front antennæ, which frequently exceed the length of the whole body, so that when turned back in line they will overlap it completely. Its movements are principally twofold, the shorter pair of secondary antennæ acting for normal crawling in distinction to the long jerking leaps which the body makes by the use of the frontal ones. These frontal pair have about twenty-six articulations and are usually carried widely extended at right angles to the body. Each of the articulations has either a long or short setæ attached to it at the joint and generally in the sequence of one long and then the next short, up to the tip, where the end ones are rather specialized with plumose setæ in addition and number five or six in all.

The secondary antennæ are held in a bent-arm fashion, seldom straight, and play rapidly, with the terminal fingers or long seta, similar to the open hand, striking either side of the head. This rapid beating produces a gradual crawl forward when seen upon a flat surface, a shimmering kind of movement, very interesting to watch. Suddenly the long primary antennæ give a backward lash and the animal jumps out of the field of the microscope. The motion comes so suddenly as to startle one at first, until the creature is understood.

The eye appears like a pair that have coalesced and is irregular in its outline, placed centrally at the anterior end of the body above the joints of the front antennæ. The body itself has several divisions upon its surface, about six, which give it an insect appearance, and at the junction of thorax to the abdomen portion are several stout projections to this segment. In the female there is but one egg sac, round or slightly oval in shape generally, which when filled with its eggs covers up the posterior portion, leaving several of the hinder spines projecting, with their finely plumosed feathery ends. There are five upon each side, a central pair, these latter being finer and slightly longer.

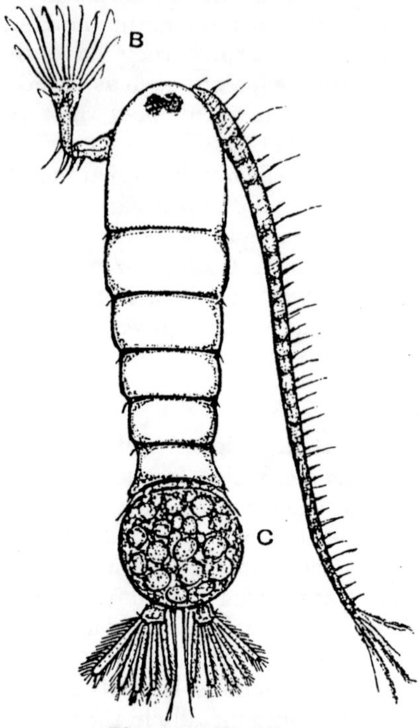

Fig. 265.—DIAPTOMUS.
A. Large antennæ.
B. Small secondary antennæ.
C. Egg sac covering the abdominal segments.

Diaptomus varies considerably, each individual needs a special observation, and very little of the complete life histories of these animals indeed is known. So far it is recognized that the Diaptomi are not only peculiar or specific for a continent, but also for a particular region, and are considerably controlled in their distribution by temperature conditions. The most important means for specific identification are the modifications of the fifth feet, which seem constant in a given species, furnishing the genus. The fifth feet are the last pair, and in the female are symmetrical, though shorter and not so well developed as the others, but in the male the right foot differs from the left and is always modified into a long sickle-shaped, hooked extremity which is used as a grasping organ. The feet are branched into two "rami," as in the Cyclopidæ, the outer being the longer. The inner or "endopodites" of the first swimming feet are composed of only two segments, while the third and fourth pair have three. On some species the last but one segment of the right frontal antennæ will have a special appendage, in the males, or again it may be absent. Such variations as these are taken into account in the classification of species. There are about fifty different species tabulated at present, and doubtless many more will be added.

Some of the animals are distinctively coloured, and deep red, brilliant purple, and bluish with purple tipped antennæ to white and colourless specimens are found, which add greatly to the interest and attractiveness of their complex forms.

The body is composed of six joints or segments, the front one forming quite a shield above the head, the remainder narrowing over the abdomen into five more, though two of these in the female are often united, making ten in all. The body is large, measuring 1/10th inch and often less, but in a few cases even more, so there is little difficulty to see it with the unaided sight when separated in clear water. The large antennæ are ringed transversely with very fine lines.

As with the Cyclopidæ, they form a considerable portion of the diet of the larger fish generally, and in turn reduce in their own diet the decaying portions of algæ and other aquatic vegetation which would tend to accummulate and foul the waters in which they live, thus usefully aiding sanitation and the cleansing of our great lakes and pools wherein they are so plentiful.

Canthocamptus (Fig. 266)

In some localities this is quite as plentiful and prolific as Cyclops. It belongs to the family Harpacticidæ, a division of the Copepods. It is a small animal, in which the two usual portions of the body

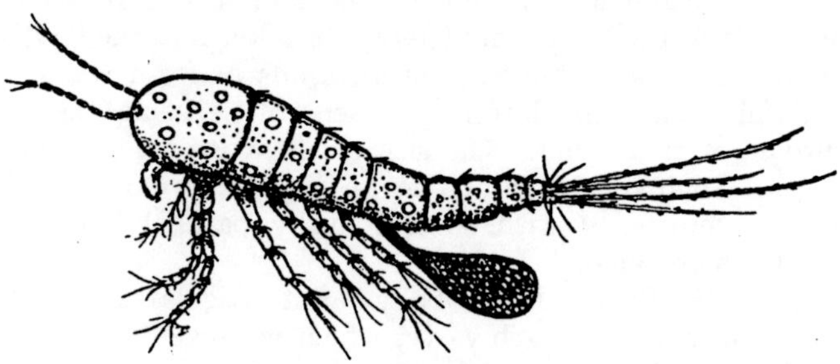

Fig. 266.—CANTHOCAMPTUS.

divided into cephalothorax and abdomen are not distinctly separated, giving it a more or less continuous segmentary worm-like appearance.

The antennæ are always short, and never composed of more than eight articulations in all. In one genus, "*Marshia*," there are only six. In Canthocamptus their number is eight, and the swimming feet which are branched into two have either two or three segments upon the outer ones, and in the third foot this is usually much longer than the inner branch or "endopodite." The fifth foot of the male is always composed of three segments.

The eye is single, and placed similarly in the middle of the head to Cyclops. The body is somewhat cylindrical, widest at the head end, tapering narrower to the tail. At the tail are two end projections, the "furcæ," which are furnished with fairly long plumose spines. The last abdominal segment also has several short setæ. The furcæ of the male and female differ slightly from one another in all the species. The ovary or egg sac is carried exterior to the body; it is single, and is a flask-shaped bag-like appendage, attached by a very delicate filament beneath the abdomen at the sixth division of the body. Although so thinly fastened it is very strongly so, taking much force to separate it, and carries thirty to forty eggs at a time. These hatch out by stages, much as the Copepods generally. The young therefore differ at first very much from their parent. The illustration of the "nauplius" stage for Cyclops (Fig. 264) gives a general idea. They prefer shallow, still water, but are capable of accommodating themselves within variable latitudes.

Length about 1/60th inch.

Cypris (Fig. 267)

This little animal is not unlike a bean in outline, its shell covering the whole of the body completely. It belongs to the "Ostracoda" group. They are abundant in all kinds of fresh water, and also plentiful in salt, the latter specimens often being the larger and sturdier of their kind. The size of them averages 1/20th inch in length. Their colour may be brown or green or of a limy-white, and marked commonly with several dusky parallel bands or canals called "pore canals."

Unlike the Copepods, the body is not articulated. It is enclosed in a shell, similar to a bivalve, which is hinged at the upper side and is opened and closed by muscles attached near the centre of each valve. There is a single eye at the upper forward end, and below this are the two pairs of antennæ with their swimming setæ. The stomach lies along the dorsal margin, and beneath this are the maxillæ and feet, both elaborately feathered for their particular purposes. Behind this on the lower posterior portion is the furca with its several terminal setæ and claws. The female of this genus is usually the larger. The shell may be outwardly covered with short hairs or smooth, and is sometimes verrucosed with small tubercles. It is seldom opened wide, the plumosed antennæ and feet working through a very narrow slit during their creeping or jerky progressions.

Their chief food is algæ and the different decaying aquatic vegetation, and they will frequently be taken in the gatherings of such material, upon which they delight to ramble. It seems rather strange that Cypridæ are rarely found in spring or well water, but can adapt themselves to waters which become more or less polluted. Evidently food is scarcer in the former than the latter for their needs, and among the ooze and mud and debris of the bottom they will in the colder winter months often be found in plenty.

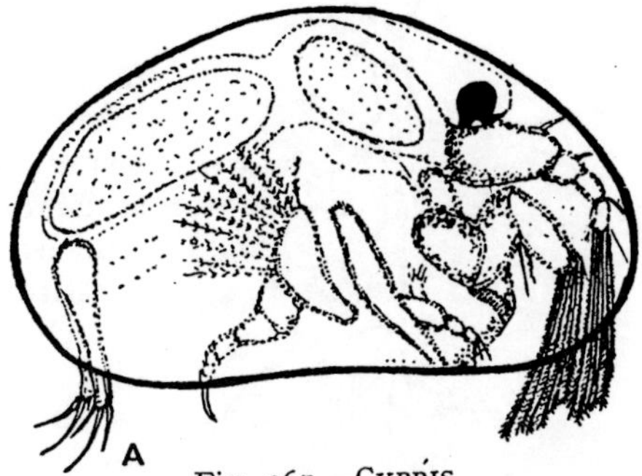

Fig. 267.—CYPRIS.
A. Furca.

The eggs have a small limy shell covering them, and are usually in a cluster of about twenty, attached to the aquatic plants by a gelatinous secretion, which operation has been observed to take about twelve hours. They hatch in from six to fourteen days, according to temperature and the climatic surroundings. Cypris eggs are usually of a reddy colour, and Lemna fronds often harbour little packets of them. The eggs have a remarkable power to resist drying, and will survive long afterwards if placed again in suitable water. This will help to account for their prolific nature. The eggs hatch out into "nauplii" and resemble the parent, but the moults may take place several times before the adult stage is reached. The cast-off shells being a perfect replica even to the most minute hairs, form exquisite clean and clear specimens for preservation.

When the Cypris is threatened with danger it quickly snaps its shell tightly together and sinks to the bottom. In many cases the valves overlap or there is a flange present, so that once they are closed it becomes securely protected against any ordinary enemy.

Artemia (Fig. 268)

This animal is a most curious and remarkable specimen, belonging to the Fairy Shrimps (Phyllopoda). It is said to be only found in brine or salt lakes, but the writer obtained several specimens in the Killarney Lakes, Ireland, which are not in any way salt.

The first appearance of it gives one to suppose it has no body, but consists of feathered appendages attached to a more or less straight rod, continued into a jointed tail with a plumose end. However, these beautifully fringed attachments are the feet, and there are eleven pairs of them, each with many long setæ or hairs along their length. The first portion of them, nearest the body, is stouter than the remainder, and upon the former are semicircular flattened plates, which constitute the animal's lungs, the branchial or "breathing plates." Like all the Phyllopods, it swims usually upon its back with the ventral side up. Occasionally it will keep an upright position when a female is seen, and its large egg sac is extended behind. The plumose feet are in constant, rapid motion, beating the water in rhythmic waves over and about it, and in this way the breathing plates are filled with renewed oxygen and kept aerated.

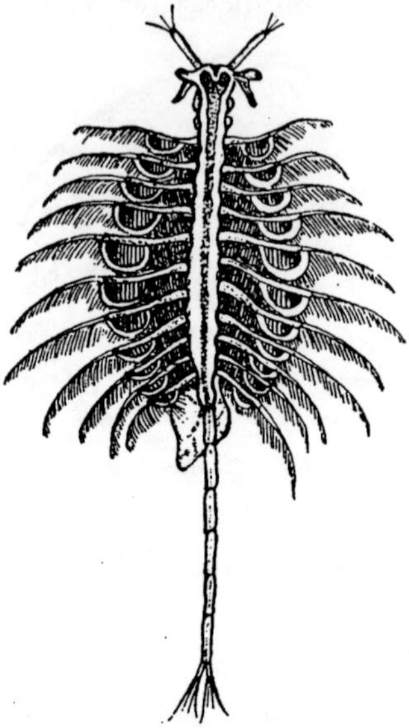

Fig. 268.—ARTEMIA.

It is fairly large, about half an inch nearly in length, but its most graceful movements once witnessed are a delight, and the microscopist having the good fortune to "net" one is favoured indeed. They are scarce. The eyes are non-sessile, black, and placed upon the extremities of two stalks. The antennæ are placed between these and are carried in a forward manner, each having three or more short setæ at the ends. They are not long antennæ comparatively. The egg sac attains quite a balloon size at times, generally after freed from its eggs, sufficient to buoy up the creature in an upright posture. It was in this way the writer first discovered it, fanning the water, without seeming to have any desire to eat or any apparatus to convey it to if it had. There is a central alimentary canal, but the anatomy of the animal has not been fully worked out. The body portion is of a pale flesh colour and very slender.

The young differ a great deal from their parent when first hatched. They are active little creatures, busy in getting about, but with no apparent notion of requiring solid food or of diet at all other than may be dissolved in the water around them. They have but a single eye, which is of a bluish colour, while their bodies are deep ruby-red in contrast.

The form of Artemia varies somewhat, and several varieties have been described.

Eubranchipus vernalis (Fig. 269)

This is another of the Fairy Shrimps (Phyllopoda), formerly known as Branchipus, the "Eu" having been added recently. The body is flesh coloured, fairly stout and large, often measuring 1 inch in length. It is composed of a number of segments or "trunk somites," and to these are attached the feet in pairs. There are eleven pairs in all. The head is distinct from the trunk. In the male the frontal appendages are broad and distinct, hanging down each side and ending in a stiff tooth forming the main

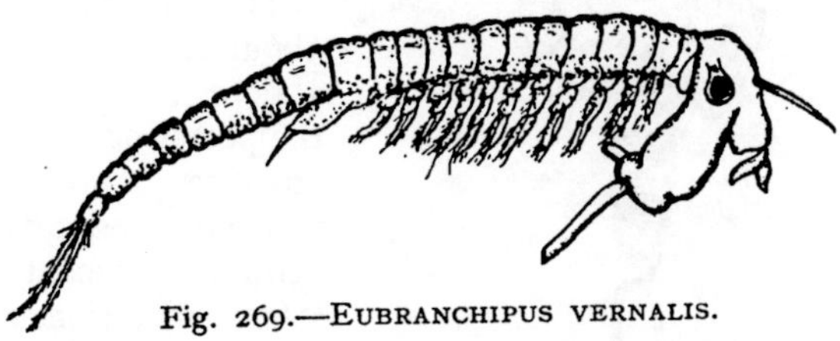

Fig. 269.—EUBRANCHIPUS VERNALIS.

secondary antennæ. The frontal antennæ are small and amount to a pair of short bristles projected forwards. There are two eyes, black, either side the head, at the apex of short stalks, which are movable. All the trunk limbs are leaf-like in form, from which the name Phyllopoda is given. The tail has several joints, the last segment, called the "telson," ending in a pair of plumosed setæ. The heart is an elongated organ occupying nearly all the trunk somites, and having a pair of openings to each of these.

The animal swims upon its back, like the Phyllopods generally, and is of a pale red colour about the body. The young hatch out direct into the meta-nauplii stage, with an oval body having the rudimentary forms of several trunk somites and their appendages apparent.

Eubranchipus is found upon the bottom of quiet waters, and lives upon the debris and decaying organic matters suspended near. It is a cool water winter animal, and is unable to withstand the heat of summer or even late spring. This fact has been attested to by several observers, the eggs alone being left to carry on the life of the species.

Apus æqualis (Fig. 270)

The shape of this little Phyllopod is oval, its carapace being of a greeny colour, as seen by transmitted light. It is occasionally found creeping with its ventral side over aquatic plants or the bottom of ponds and quiet waters upon the muds and ooze. This is rather exceptional for the Phyllopods or Fairy Shrimps in general. Usually they swim upon their backs.

Apus is a very graceful little animal and a good swimmer. It is often seen browsing upon animal bodies, and will " nibble " around insect larvæ, tadpoles, and others that have a gelatinous outer coat, from which it evidently can obtain some matters of food value.

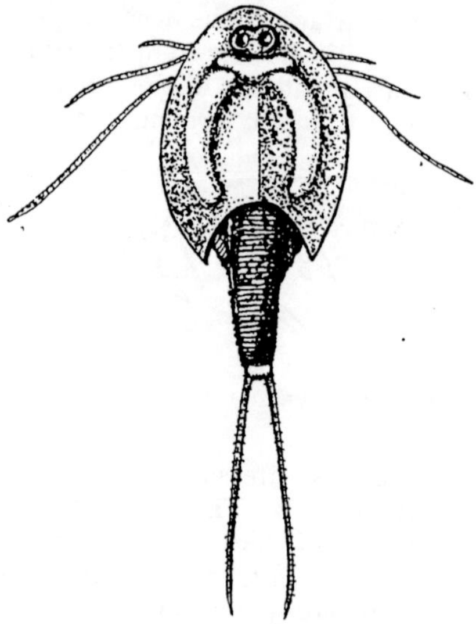

Fig. 270.—APUS ÆQUALIS.

The carapace forms a shield about as long as the tail-like abdomen, which projects from the posterior end. The telson possesses two long spines which may be carried straight behind or flexed in various lateral directions; these are the cercopods or " furcal rami." Beside the two spines other shorter ones are affixed near their bases, and the last segment is hollowed on its posterior side with the carapace forming side extensions, giving a freer movement to the tail-like portion. The eye is single, set in the centre near the front border, a slight separation into two sometimes showing, and the alimentary tracts can be seen in a horseshoe shape about the centre of the back. The " furcal rami " are much jointed into short segments, each having short hairs or bristles near the divisions.

There are many different species, but the majority do not exceed $\frac{3}{4}$ inch in length. One, however, has been measured as great as $2\frac{1}{2}$ inches. This is exceptional.

Apus belongs to the sub-order Notostraca. The number of segments exposed behind the carapace may be as many as thirty.

CHAPTER XII

BRITISH HYDRACARINA (WATER MITES)

(Figs. 271, facing p. 268; and 272 to 277)

ALL micro-fishers will now and again find amongst their captures small spider-like creatures known as mites. They have eight legs, the same as a spider. But the body is formed of one piece only, not in two portions as we find in the spiders. They occur in rivers and ditches, lakes and ponds, in fact in almost any fresh water that does not dry up in the summer. Not only are they found in water on the lowlands, but members of several genera are found in swift-running streams and small lakes in the mountains at a high altitude. Some species are very small, but the most usual sizes are from 1 to 2 mm. in length. They are very interesting little creatures in more ways than one. They far outdo all the other families of mites in colour, for some of these Hydrachnids are most brilliant, and according to the species exhibit every colour in a variety of tones and shades. Thus we have reds, blues, greens, and yellows, many of them looking quite metallic. It is this brilliant colouring which renders them such favourite objects when alive for the microscope. The lower forms are usually red, but the higher forms favour the other colours. Another thing which makes them interesting is their peculiar shape. The lower forms are mostly ovate, but the higher forms vary very much, not only in specific differences but in sexual differences. (See Figs. 276 and 277, which is the male and female of *Arrhenurus bruzelii*, Koen.) Then there is the life history, which to all students of nature is the most interesting. It is also the part we usually know least about. It is so with the Hydracarina. This part we will speak about directly.

There are about twenty-eight families of mites, the Hydracarina is one of them. There are mites found in the sea, but they belong to a different family, known as Halacaridæ. The Hydracarina alone is the family we are at present considering. They are recognized by the body being in one piece. They have eight legs of six sections each, which in most cases terminate in claws. There are a few in which the fourth leg terminates in spines. The legs are usually normal and well supplied with short hairs or spines at the joints and at various places on the sections. The swimmers have long swimming hairs near the joints. These mites are very active. Those without the swimming hairs are known as crawlers. These crawl about on the mud at the bottom of the water and on the stems and leaves of water plants. The movements of some of these are very sluggish. We find the third and fourth pair of legs of the male amongst the higher forms usually modified for sexual purposes. Notice the fourth pair of legs on the male of *Arrhenurus bruzelii* (Fig. 276). The legs are attached to plates on the ventral surface of the body, known as epimeral plates (see Figs. 275 and 273). These plates are sometimes fused into one group, as there may be two, three or four groups according to the species. Those shown on the plate are in four groups of two each. The arrangement of the epimera is an important point in the identification of these little creatures. Placed between the first pair of epimeral plates is the "capitulum," in which is placed the mouth organs and to which is attached the palps. The palps are made up of five segments, in many cases the last two, forming nippers. They vary very much in structure according to the species. The body skin in some is soft, in others hard and brittle. It is found smooth, lined, granulated, covered with papilla as made up in plates. Some, as in Figs. 272 and 273, have a soft skin with numerous chitinized plates distributed about the body. The genital area, placed usually between the epimera as towards the posterior margin, varies very much. Some have two flaps, more or less covering the genital fissure (see Fig. 273), others have plates with a number of discs known as acetabula (Fig. 275). The eyes in most genera are placed near the margin of the body. It is so in the mites shown on the plate. Some have a well-pronounced capsule carrying the lenses, which are often in pairs. There are two well-known genera, *Eylais* and *Limnochares*, which have the eyes in the centre of the dorsal surface near the anterior end. These are on special chitinous plates, and in pairs. Some of those with the eyes on the margin of the body have a median eye as well, with or without a plate according to the species. The anus, which in some species is well defined (see Figs. 275 and 273), is placed well behind the genital area.

We will now say a few words about the life history of water mites. They all deposit eggs; none bring forth their young alive. These eggs can be found on the stems and under the leaves of water plants. No doubt they have to adapt themselves to their environment, but in many cases we find they have quite a conservative taste for what is to be the cradle of their offspring. For instance, the eggs of *Hygrobates* are found on Lemna, *Piona* on

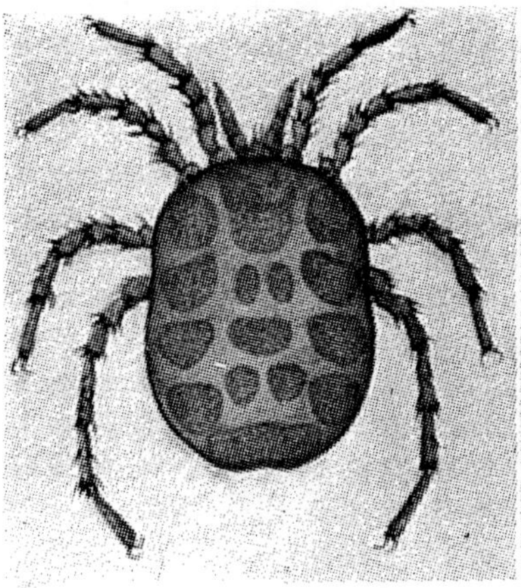

Fig. 272.

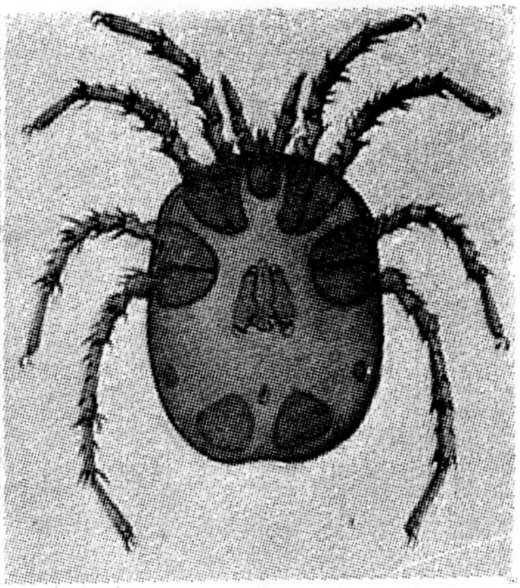

Fig. 273.

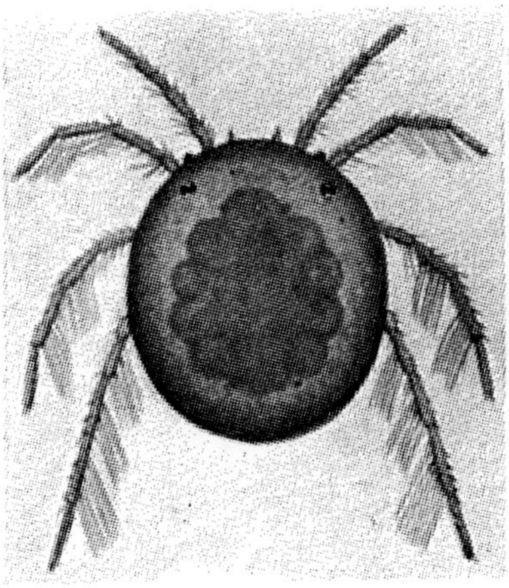

Fig. 274.

Fig. 275.

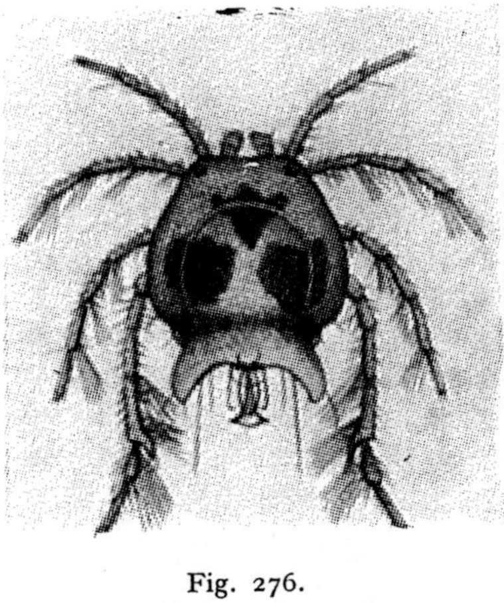

Fig. 276.

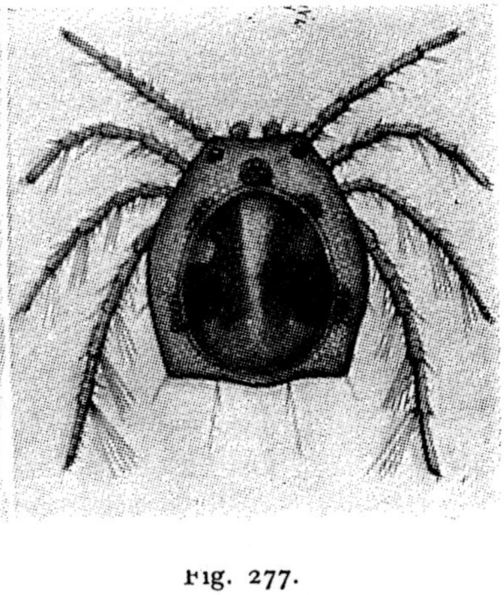

Fig. 277.

BRITISH FRESH-WATER MITES—HYDRACARINA.

By Chas. D. Soar, F.L.S., F.R.M.S.

Anacharis, *Sperchon* on the green slime and moss on stones, *Hydrachna* in holes bored in the stems of reeds, *Unionicola* in the fresh-water mussel, and so on according to the species. These eggs are usually fastened and protected with a gelatinous deposit, which covers the eggs all over. This film gets quite hard and remains intact until the larvæ are ready to escape. It is quite a transparent film. The eggs, usually of a bright red or orange colour, can be plainly seen inside, and all the internal growth can be studied as the eggs develop, if kept in a small tank under the microscope. These eggs, usually deposited in masses, when first ejected are quite round. They gradually grow larger and more oval, until at the time the larva is ready to leave it must be two or three times the bulk it was originally. The time of incubation varies; some take over a month, others only a few days, and in the case of one genus at least (*Eylais*) the hatching has been known to hang over until the following spring. This no doubt occurred when the eggs were deposited late in the year. There is one record of some ova being deposited in a glass tube by an Eylais. The water was thrown away and forgotten for two years, but on again filling the tube with water the larvæ very soon came out and were found swimming about quite as lively and perfect as others that had not been subjected to such drastic treatment. This shows how valuable the protective film in which they are deposited is to the eggs, not only preventing other organisms from having eggs for breakfast but in holding sufficient moisture to keep the vitality of the egg intact.

When the time has come for the larvæ to escape, the envelope of the egg splits into two portions and the larva wriggles its way out and through the protective film into the water as a free individual. The larvæ are quite different to the adult. The skin is very soft, sometimes with plenty of colour, but they only have six legs, not eight as the adult. Some species at once make for the top of the water and run about quickly in all directions on the surface film. Others remain near the bottom; but as far as we know they all have the same object in life, and that is to find a host on which they can become parasitic until the next stage. We say as far as we know, because those we do know are parasitic, but that does not prove that they all are. This is one of the questions we require more light on. It may be we shall find that there are some species which grow from the newly hatched larvæ to the adult without becoming parasitic at all. However that may be, those we do know are, and they are real parasites: they draw their nourishment direct from the host. Let us mention one in particular as an example, *Hydrachna globosa* (De Geer). The larvæ in this case become parasitic on several water insects. *Nepa cinerea* is a common host. They attach themselves to this insect with their mouth organs to various parts of the body and legs. After a short time the six legs, not being required, fall away and the parasite looks like a small pear-shaped appendage hanging on by the stalk to the Nepa. Here they remain from the summer of one year to the spring of the next, growing larger and larger, and the nymph stage forming inside. This can be plainly seen if one is removed and examined under the microscope. The early naturalists, finding these appendages hanging on to insects and knowing when the envelope burst a water mite came out, thought it was the eggs they were finding and not the parasitic larvæ. So quite different to other mites, it was recorded these mites had no larvæ stage but went direct from the egg to the nymph. Nepa is not the only insect we find the parasitic forms on. Dytiscus, Ranatra, Carixa, Notonecta, Dragon Flies, Gnats, and several others have to play the host to these little creatures, who make up their mind to patronize them for their board and lodging.

We have found great difficulty in keeping the hosts alive long enough to be able to follow the parasitic stage of any particular mite all through. But we think we have met with fairly good results in periodically examining hosts from the same water and noting the extra amount of growth in the parasitic larvæ. When the time comes for the nymph to escape, the envelope breaks at the posterior margin and the nymph escapes, no more to be parasitic but a free swimmer foraging for itself, the same as an adult. It now has eight legs, as the adult, and in many cases is very like what it will be in the adult stage. The genital area is quite different, being of course in the nymph only in a provisional stage.

The above short account of the parasitic stage of *Hydrachna globosa* must not be taken to represent what they all go through. That must vary with the species, because the life of some of the hosts is much too short to allow the larvæ to remain attached all the winter. In the nymph stage they grow a great deal, and in the meantime the adult is forming inside. When ripe the skin of the nymph bursts near the posterior margin and the adult escapes, not fully grown or in full colour, but we are now able to distinguish the males from the females. They soon attain their full size and lovely colouring, and it is now that they are so interesting from that point of view. What part they play in Nature's economy is not settled. They feed on entomostraca, infusoria, and no doubt bacteria, as they can be always found browsing amongst decayed vegetable matter; thus they must help in a way to keep water sweet and wholesome, and that's something.

In the Britannic area we have recorded about forty-four genera, containing about two hundred and fifty species, which is a very fair record for this small part of the world.

A word or two about the figures.

Fig. 276 is the male of *Arrhenurus bruzelii* (Koen), of a beautiful blue-green colour. We have about fifty species of this genus exhibiting a great number of beautiful forms. Its length is about 1·2 mm., or 1/16 inch.

Fig. 277. The female of the same, also 1·2 mm.

Figs. 274 and 275. *Diplodontus despiciens* (Müll). A soft-skinned red mite. We have only one species of this genus. Very common length, about 2 mm. Male and female in this species very much alike.

Figs. 272 and 273. *Panisus torrenticola* (Pier). This is another red mite, a crawler; it has no swimming hairs. It is known by the number of chitinous plates on the dorsal surface, fifteen in all. Its length is about 1·3 mm. Male and female very much alike.

CHAS. D. SOAR, F.L.S., F.R.M.S.

CHAPTER XIII

MISCELLANEOUS

WATER BEAR (*Macrobiotus*)

(Fig. 278)

THE body is an elongated oval, or may be shortened by the animal's contractions and become wide and almost spherical. It is very flexible and of an opal white or almost transparent texture, according as it is illuminated. It has four pairs of short, fat telescopic legs with two pairs of hooks at the extremities of each, with which it clambers over and round aquatic vegetation or other submerged objects. The mouth is siphon-like and tubular, and connected with two curved, lateral, movable horny pieces or arms to another chitinous cross-piece, completing a triangular figure. The suctorial mouth tube runs in the centre of these to the gizzard, an oval patch at the upper side of the head. It is fixed and striated, without motion, relatively, of its own. The head is of a conical shape. The outer coat is double, of a flexible but firm horny nature, and said to withstand the action of caustic alkalies. The suctorial pipe continues past the " gizzard " as a tortuous elongated tube forming the stomach and terminating at the posterior anal orifice. From this tube radiate several lateral branches. In the dorsal portion of it may often be seen the liquids absorbed as food, giving a noticeable coloured patch there, mostly of a brownish tint, varying according to what has been chiefly imbibed. Sometimes its colour is a golden hue, or again a greener tint is observed. Chief of the substances within the body are fat globules. These are in all parts, including its wide stumpy legs, and are constantly pressed about in any direction as the animal moves and bends in search of food. They form its principal circulatory material. A chain of nervous ganglia radiating from a central nucleus is said to be known. Upon the head are two dark brown or black spots, set widely apart, the so-called eyes.

The animal is an hermaphrodite, producing its own ova, which are usually few and of a comparatively large size, and are situated beneath the dorsal surface near the posterior end. Macrobiotus casts its outer skin from time to time, and the empty cases may sometimes be seen with the eggs attached within, but no Tardigrade in sight. They hatch themselves without any parental attention.

The animal itself has been resuscitated after being dried. Water is not absolutely necessary for its environment, it being quite capable of adapting itself similar to some rotifers among the dust and rubbish of gutters upon house roofs (a locality in which it is said they were first encountered) and upon lichens and mosses that spring up in such situations. Moss plants, as we know, are great absorbers and retainers of moisture, and no doubt the Water Bears can obtain sufficient for their needs of existence from them by the aid of their suctorial mouths.

The animals have been placed experimentally in hot water at a temperature of 115° F., and after such unusual treatment have

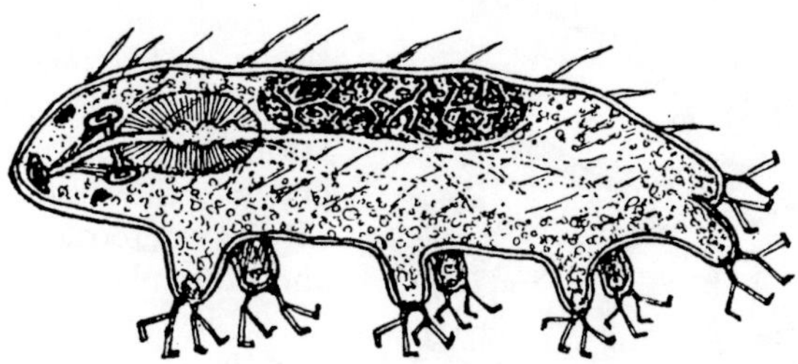

Fig. 278.—Macrobiotus. (Tardigrade.)

revived, proving the wide latitude of adaptability they offer to surroundings and under abnormal conditions. They have affinities to the Arachnida and Acari, and are placed in that class or type of animals, forming probably the lowest section of it. Upon the back and sides are several regularly disposed bristles pointing rearwards, and on some have been noticed soft flexible processes or palpi, with short hairs terminating near the mouth or proboscis.

There is possibly more than one species, and more work upon the organisms is waiting to be done before it can be said to be complete. Their movements are slow and methodical, and consequently the name "Tardigrade" has been applied to them. Altogether they are very interesting creatures to watch, and even comical at times as they persist in trying to catch hold of objects that are not there. They appear quite unable to profit by experience, and will continue clawing downwards and forwards upon a slippery surface or in the open water, seemingly to obtain a foothold, until, rather agitated, there is a final rally and quickened pace, without result, followed by a literal shrug of the head and an expressive turn of the body, evincing their evident disgust of things and probably of a Water Bear's life in general.

Although so well equipped with large hooks and claws to the number of thirty-two, the creature is so palpably weak it is most difficult to get his adipose body about, though his persistence and perseverance never flags or forsakes him.

Hydra (Figs. 279 and 280)

forms the representative species of the class Hydrozoa inhabiting fresh waters. There are but few varieties, and this is the most abundant and widely distributed. It is never found in the sea. Its body is an elongated cylindrical sac, yellowy brown or green in colour, highly elastic and extensile, visible to the eye when isolated. It attaches itself to aquatic objects, the under side of Lily leaves, rootlets of Lemna, and other plants, by its lower extremity, and at the other is situated its mouth. This can expand to comparatively wide extent when large objects are captured in its food supply. Around the mouth are several tentacular arms, swaying to and fro, varying from five to nine according to the Hydra's age, etc. They are roughened with warty protuberances, which consist of special cells, with which it stuns or kills its food. It is a carnivorous animal. Like the body, the tentacles are hollow, the whole organism being in fact a flexible tubular arrangement of both arms and pedicel. There is no other opening than at the mouth.

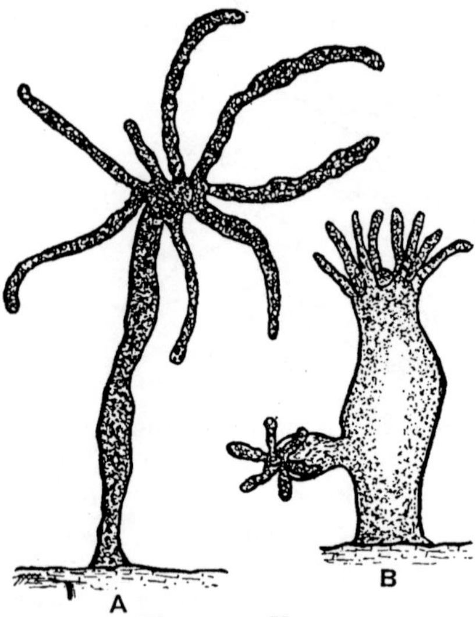

Fig. 279.—HYDRA.
A. Expanded.
B. Contracted. Showing young bud

The body wall and tentacles are composed of two layers of cell structures, separated by a thin membrane, and the whole interior may appropriately be termed the stomach. When the stomach is distended with heavy meals the Hydra shortens and widens its body, and in the lean times will expand to almost thread-like proportions to a great distance. The interior is lined with ciliated simple amœboid cells, all of which are capable of digesting food, and in injuries, where a portion of the body is torn or detached, each part is capable of healing itself and of continuing existence as before. Hydras have been severed experimentally at all different points, including the tentacles, and, left to themselves, have recovered within the period of a day, continuing to grow and eat as if nothing untoward had happened.

Certain cells, plentiful about the mouth orifice and especially upon the tentacles, called "cnidoblasts," produce the "stinging" or "nettling" cells. These are the nematocysts, and are of varied shapes and sizes (Fig. 280). Usually pear-shaped, they consist of a cell within a cell, containing in the interior a poisonous fluid secretion. Within this is coiled a long thread, having four or more minute barbs or spines pointing backwards near the base. Upon pressure these coils are shot out in numbers at passing objects like so many harpoons, lodging in any small organisms encountering them. This results usually in a loss of activity or their death, and so permits of an easy capture and the utilizing of them as food by the Hydra.

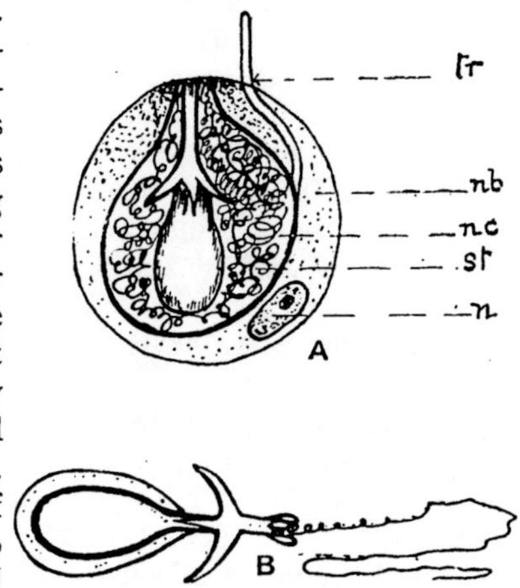

Fig. 280.

A. Stinging cell (*nematocyst*) of Hydra.
tr = trigger or spike.
nb = nematoblast.
nc = nematocyst.
st = stinging thread coiled within the nematocyst.
n = nucleus with nucleoli in the nematoblast.
B. Stinging thread when shot out.

The threads are flexible, thin, transparent pipes, which actually uncoil from within, unfold as it were, similar to pulling a jacket sleeve by the cuff from inside to the shoulder opening. Near the small end of the nematocyst is a short style or "trigger," said to be to give warning when the sting is discharged. Pressing upon the cover glass over an Hydra will often irritate it to discharge several, so that they may be seen in the water near by.

Another method of capturing their food is the direct one, simply coiling a tentacle round the object and if refractory bringing another arm or more to its assistance, and so conveying it willy-nilly to the mouth. The tentacles are apparently sticky, and once the Hydra has fairly closed with its prey little hope of release is offered. Larvæ, Water Fleas, etc., are the chief items of their food. In the aquarium they have been fed with Worms, small portions of lean meat, etc., which they eat readily enough. The stinging cells do not possess the power to kill large fry, which may occasionally be seen escaping the embraces of the tentacles.

Hydra multiplies rapidly, chiefly by a budding process or asexually. Fission into two has been observed. Development of sperm and ovary cells in the outer layer, producing a chitinous ovum or separate spherical and slightly spinous body, is also a frequent method in winter and during inclement periods. It is noticeable that cells producing this latter condition are rarely found upon the stalk or body of the Hydra. When a young Hydra is formed by a bud at the side of its parent, the tube, though short, becomes part of the general tube and any food it ingests mingles with that of its parent in communal interests.

The fable of Hercules and his fight against the Lernean Hydra of mythology with its several heads that ravaged the city of Lerna near Argos and dwelt in a swamp near the Well of Amȳmōnē gives the origin of the name Hydra. "As the story goes" it had nine heads, the middle one of which was immortal. Hercules struck off the heads (usually represented as snake-like) with his club, but in the place of each two new ones grew forth each time. This became a tough proposition for his renowned strength, but finally with the assistance of his servant Iolans he seared or burned the ends of each and the immortal one he placed under a huge rock. He then poisoned his arrows with its bile, the wounds from which were incurable.

Cordylophora (Fig. 281)

Is another similar species to Hydra belonging to the class Hydrozoa. It prefers brackish water, and attains its maximum development there, but it will live in fresh water, and the tanks of the Botanic Gardens, London, had many specimens growing there at one time, although less stalwart and the branches only half as long. It occurs frequently attached to aquatic stems, submerged walls, pier piles, stones, sticks, etc.

Chætonotus larus (Fig. 282)

Has an elongated cylindrical-shaped body usually divided into three portions, head, neck, and abdominal. The under, ventral surface is flattened, and the back, or dorsal, arched; the latter

Photo by] [*F.J.W.P.*

Fig. 271.—ARRHENURUS CRASSICAUDATUS.

Fig. 281.—CORDYLOPHORA.

[*To face page* 268

will be best seen from a side view. The under side has two rows of active cilia near the middle running nearly the whole length of the body. These are its principal means of progression. The dorsal side is beset with strong bristles, sometimes bifid or forked at their free extremities, which in most cases arise from a horny plate or flattened chitinous cells. Some species are almost devoid of these, others have short ones, around the neck portion, and very large and fewer ones upon the abdominal region. Occasionally these will be used to make sudden leaps forwards or sideways or instantly to reverse their course in an opposite direction, and again two strong ones at the rear, called "toes," usually dragged inactively behind, will be "dug in" upon any objects encountered and will aid in propelling. On account of these numerous bristles they have been commonly styled "Bristle Backs" and placed in a group to themselves named "*Gastrotricha*." They swim rapidly at times, their bodies are very flexible, and all movements, in earnest, are done in a smart decisive manner.

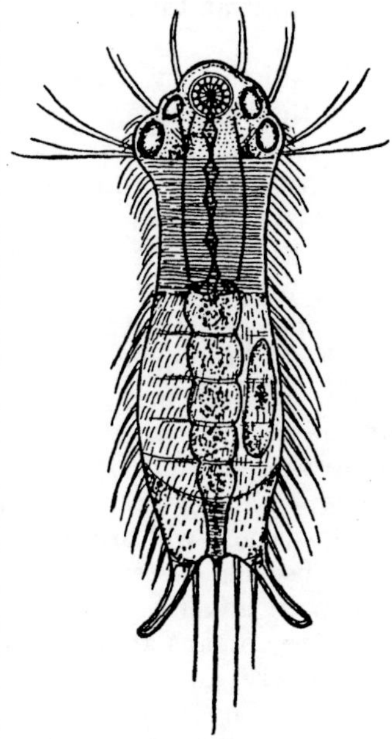

Fig. 282.
CHÆTONOTUS LARUS.

Upon the anterior portion, or head, are several sensory bristles communicating to a saddle-shaped mass of nerve cells there, projecting like cat whiskers at right angles to the body, touching objects with them, but rarely carrying them far from their normal positions. From this centre of nerve cells run others along the throat side and by the alimentary canal to its outlet. The caudal end is bluntly rounded, having either pointed or forked "toes" with a central and longer bristle carried from the back in some, which probably gives warning of objects in the rear.

Chætonotus is very sensitive to touch and will immediately hasten forward, at a bound sometimes, if anything moving brushes past him carelessly. Within the tubular "toes," from cement glands near their base, is carried an adhesive substance which may be exuded to attach them to varying objects and provide anchorage if required. At the anterior extremity, the mouth is situated, and consists of a circular plate having several slits in it, like a stage star-trap, which can be opened forwards and outwards and extended to take in food substances much larger than its normal width. A unicellular alga will be encompassed in this way, and the literal sharp snap back of them into position is usually sufficient to crunch and pulp it and make it ready at once to be passed down into the stomach.

The alimentary canal is a fairly straight, simple tube passing along the body interior to the anal orifice just above the posterior end. The intestinal part is lined with large digestive cells, said to be rich in protoplasm. There are no salivary glands, indeed they would be of little use, so rapidly does he bolt his food past the mouth parts that precious little time would be available to salivate anything.

Chætonotus is not a vegetarian, he is particularly partial to young soft-bodied Chilodons. They form an especial tit-bit to him. The writer has seen him bite with a vicious snap a large mouthful out of one of these and leave the remaining fragment to wriggle off with whatever cilia it had left as best it could. It likes its game flying, so to speak. A dead Chilodon, or even a motionless one, rarely seen, it will brush up to and pass by quite unconcernedly. Its muscular system consists of six pairs of longitudinal fibres, one pair traversing nearly the whole length of the body. The excretory organs are a much contorted coil of tubes near the body centre, with an opening on the ventral side.

A male Chætonotus is not known, which points to the probability that the species is hermaphroditic. Beneath the dorsal surface at the posterior end is carried the eggs of the young. They are few in number and large, increasing the width of the animal sometimes to half as wide again as normally at such times. The head otherwise has the greatest diameter. The eggs are often deposited upon algæ stems or even the cases cast off by entomostracans. They are covered with warty points and hooks to assist their attachment, the young Chætonotus forming inside, and on the egg bursting emerges a full-formed animal. In those which show the scales clearly upon the dorsal side may be noticed the unusual way they overlap, not as in ordinary fish scales like the tiles of a roof, but with their free ends uppermost towards the head of the animal.

Chætonotus acanthodes (Fig. 283)

In this specimen the body is curiously covered with overlapping scales, each of which bears a smaller and supplementary scale or thickening at its posterior end, the anterior portion of this having a short curved spine which is unequally forked at the tip. To the rear of the middle portion of the animal are several transverse rows of larger and stouter spines, which point backward in a curved manner. They do not continue to the extreme end of the body but a space is left vacant or almost denuded, and only the scales are in evidence for protection in that region. To the side of the caudal processes, however, are two large curved spines on each side, while the whole of the body margin is armed with short spines. Upon the ventral surface and between the longitudinal bands of cilia which effect the movements of the animal are many short recurved bristles, while at the posterior end some five or six longer ones project beyond the body margin at the bifurcated extremity, and in all cases the spines are pointed backwards to this portion.

Fig. 283. CHÆTONOTUS ACANTHODES. (Stokes.)

The body is about 1/200th inch long, and its head is lobed into five, which distinguishes it, with the particular arrangement of spines, from its neighbours. Mossy swamps provide its habitat, and among Riccia fronds it has been taken by the writer.

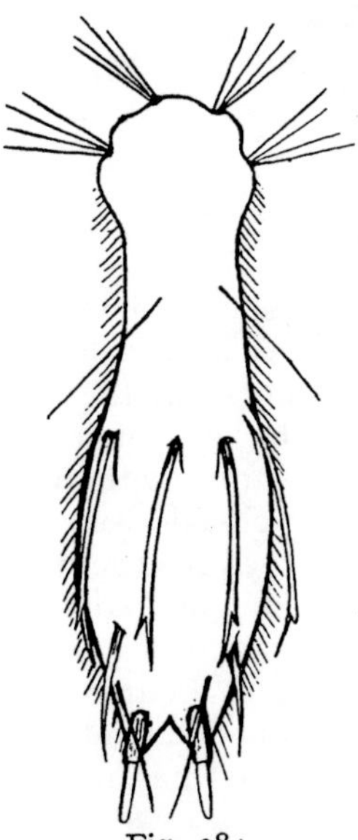

Fig. 284. CHÆTONOTUS SPINULOSUS. (Stokes.)

Chætonotus spinulosus (Fig. 284)

In this the spines are very few and sparsely arranged upon the dorsal side. Only seven or so in all, they are placed below the middle of the body in two widely separated rows across, usually four in the front row and three at the posterior. Each are recurved and bifurcated at the extremity and point to the rear, the front row being decidedly the longest and somewhat stouter than the others. The sides of the body are bordered at the margins with short setæ all along. Apart from these several appendages the upper surface is quite bare, excepting the frontal bristles in bunches of four which are present as tactile organs in all the species. The egg of this animal is rough and covered upon one side with short hairs, and in about thirty hours discharges its embryo.

Mr. W. C. Stokes observes that in another thirty hours the young individual produced an ovarian egg, in which six hours later the nucleus became conspicuous.

Nais (Aquatic Worm) (Fig. 285)

Hardly a microscopic object, measuring 1 inch in length sometimes, yet is frequently met with among the decaying algæ and water weeds near to the sides of the ponds, and well repays a more minute inspection. Principal to notice are its means of assisting progression in the shape of spines, " podal spines," which I would

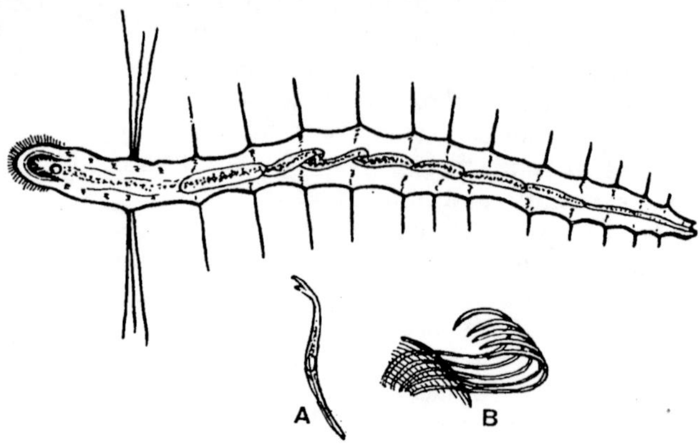

Fig. 285.—Nais.
A. Single spine. B. Hooks extended.

rather term " manal spines," arranged at even distances along the upper sides of the body. These consist of wiry-like chitinous fingers generally of five in a group armed with curved hooks at the tips, which are cleft for a short way down and by which they clutch hold of anything near (Fig. 285B). They are pushed forward or withdrawn as required. At one time it was thought they simply laid back flat to the surface when not in use, but it is now observed they actually slide backwards and forwards through the outer skin or epidermis, strangely enough leaving only the hooked portions outside. When the creature elongates its body the spines slant backwards, and when it ceases this forward move they are stood out full stretch at right angles ready to lay hold of or keep up to anything near to preserve its stability and position.

The animal also has pairs of stiff bristles, one long and one short, set together at intervals along its length, which stand out from the body, and is made aware by these of its vicinity to objects and possibly enabled by them to distinguish by touch differences, say between soft and hard surfaces. Its whole time seems to be taken up gliding about, and it loves to be bending itself in all manner of contortions among the gelatinous algæ and diatomaceous pabulum. It is rare to see it actually eating, but this in a great measure is due to watching the wrong spot, the tip or proboscis, which it is constantly protruding and elongating forward, for its mouth, whereas this organ is situated some distance further back on the under side.

The tube running through the centre of the interior is the alimentary canal and is a much twisted and tortuous passage, lined on the inner sides with numerous cilia. These can be seen in constant movement, especially at the head and anal extremities where the tube is straighter. By these its food is passed along and digestion assisted. Attached to this track are the two circulatory tubes lying one either side, carrying the ruddy tinted blood to and fro. Owing to the transparency of the whole organism everything going on within can readily be seen. There are two granular black spots at the head, one each side, doing duty doubtfully as eyes. The whole body is in sections, or segmented, as it is called. Some of these sections are hermaphroditic, that is, are capable of producing their offspring within themselves. They are the somites, segments specially fitted by nature to effect reproduction by a budding process from the side of the animal. This takes about two to three days.

All the aquatic worms or Oligochæta, including Nais, are, like their neighbours the land worms, instrumental in reducing aquatic vegetation to a finely divided comminuted condition, passing this through their bodies in a similar manner to the land worm with the soil. They are capable of swimming in the open if necessary, but are naturally found where food is obtainable in the greatest quantity. They can be obtained by carefully screening the mud from the bottom and about the roots of plants through a fine-meshed sieve. Where decaying algæ, rushes, or other water weeds are plentiful, simply dividing this apart will produce numbers that are feeding upon them.

Anguillula (Fig. 286)

This worm is often found among decaying vegetation and where the water has become filled with decomposed substances, and is occasionally taken in the microscopist's catch. It belongs to the Nematoda, which are generally of a cylindrical shape and often called " Round Worms " in consequence.

Anguillula and its various members, however, show a somewhat spindle-shaped outline tapering towards the tail end to quite a point, while the head portion is almost as wide as the middle and with a rounded end. Its body is quite transparent and thread-like, without any superficial hooks or appliances as seen in Nais. At the broadest end the mouth is situated; it is circular and leads into an oblong part called the throat or pharynx. It is not difficult to recognize this worm. Its movements are peculiar to its type,

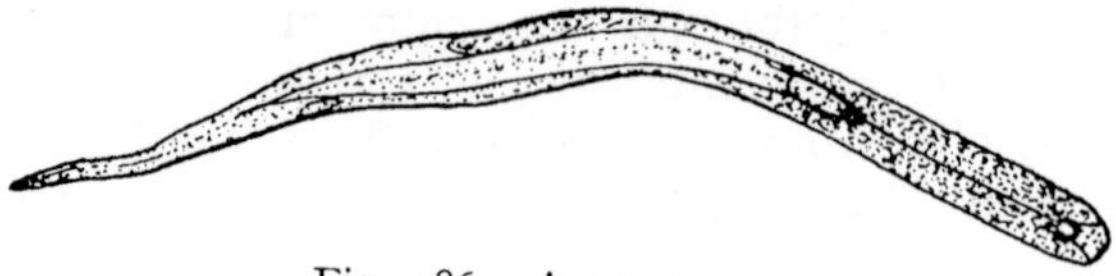

Fig. 286.—ANGUILLULA.

and consist in rather violent coilings and twistings right and left, making little progress thereby, always appearing as if annoyed with its situation or with life in general. It is, moreover, a nuisance when it occurs amongst other fresh-water organisms under the microscope, and is best removed and viewed separately. The outer cuticle is thick, transparent, and non-cellular, and seems to add great strength to its lashing movements. It is not chitinous, being soluble in alkalies, and contains a high percentage of nitrogen. Its substance has been named "cornein," a proteid compound similar to the organic basis of corals.

The worm is produced by eggs, which may sometimes be seen near the pointed end of the body in situ. Some may be outside in the water and have the young embryo plainly discernible within. The worm in paste and the vinegar eel are both Anguillula; the former is *A. glutinis* and the latter *A. aceti*.

INDEX